Shifting Interfaces

Shifting Interfaces

An Anthology of Presence, Empathy, and Agency in 21st Century Media Arts

Edited by
Hava Aldouby

LEUVEN UNIVERSITY PRESS

This research was supported by The Open University of Israel's Research Fund (grant no. 37234)

ISBN 978 94 6270 225 7
D / 2020/ 1869 / 15
NUR: 655, 652

Design: Dogma
Cover illustration: *BRALL* (Breathing Wall), silicone on polycarbonate panel, 145cm × 145cm, 2015.
(© Ece Budak and Ozge Akbulut. Photo © Baris Dervent, Murat Ugurlu).

GPRC
Guaranteed
Peer Reviewed
Content
www.gprc.be

Table of Contents

III
Bacteria and Smart Objects: Presences Beyond the Human

Introduction
Shifting Interfaces

Hava Aldouby

Shifting Interfaces addresses exchanges between art, science, and emerging technologies in the early twenty-first century. The essays in this anthology tap a range of concerns that currently prevail in the field of media arts. In ensemble, the collection relates to cutting-edge projects of research-based art, some involving hands-on lab work and interdisciplinary collaborations, at the same time addressing the theoretical premises and debates undergirding contemporary media arts. This is a rich and varied anthology, in which readers will find academically rigorous theoretical chapters alongside artists discussing and theorizing their own projects. The rich picture of contemporary media arts drawn in this anthology benefits from the diversity of contributors, spanning leading figures in media art history beside active artists involved in hands-on experimental research. Art-based research and art projects based on collaborative lab work are gaining salience in media art studies, and this turn is represented in *Shifting Interfaces* through some intriguing projects-in-progress. The visual design of the anthology makes a clear distinction between artists' essays and purely academic contributions.

The art projects addressed in this collection delve into the anxieties of an age immersed in enormous quantities of data, haunted by ubiquitous surveillance, and encountering overwhelming changes in the lived experience of self and body; of intersubjective relations; and of environmental conditions. This volume focuses on art's mediation of the profound shifts in the sociopolitical and technocultural spheres, affecting, among other aspects, the ways of interaction between human and nonhuman agents. The bodily and mental perimeters of human subjectivity are currently stretched and reformatted. They are extended into the virtual realm and the online world, and expanded to inhabit avatar bodies. New forms of connectivity come into effect under the sway of radical technological changes.

Immersive VR enables border crossings between alternative possible worlds. It thus changes the very notions of border and interface, rendering them porous and endlessly flexible, sometimes to an unsettling degree. This book deals with the age-old desire for immersion in the image. The promise of VR and related technologies to perform as 'empathy machines' is addressed in view of the ongoing debate over the plausibility of mind-changing experiences based on VR's affordances. When the way is open to inhabiting an Other body, be it of the opposite gender or of different ethnicity or age, how does our sense of self shift in the process? On the flipside of this critical position, other essays in the collection point out the challenge and urgent need to retain critical distance vis-à-vis the ubiquity of immersion technologies facilitating emotional and ideological manipulation. Given the omnipresence of immersive technologies, presence and empathy have become concepts of profound concern, as can be gleaned from the chapters of this anthology. Section I focuses on these questions in a dialectical manner, presenting an array of experimental and theoretical approaches.

Further addressed is the engulfing universe of Big Data, affecting the way we connect with networks of knowledge and consumerism, and with apparatuses of intelligence and surveillance. Interfacing with Big Data entails anxiety over the threats to individual sovereignty and agency, posed by looming clouds of unfathomable information, biometrics, and Artificial Intelligence-based profiling. The essays in section II question the idea that agency would become more democratically distributed through online participation in the creation of contents, and via the utopian formation of collective bodies of knowledge. This implicit promise is approached with caution by the contributors to this anthology. These authors attend to systems of dominance that deploy troubling modes of unwitting or uninformed participation. Other contributors foreground activist projects employing bottom-up participatory initiatives, which challenge institutionalized top-down practices. The art projects presented in section II critique the loss of agency, and expose the apparatuses behind the distribution and curatorship of knowledge. By undermining institutionalized interfaces and opening up alternative routes of engagement and knowledge mining, these projects opt to subversively restore presence and agency to subjects interfacing with data giants like Google and Facebook.

Nonhuman agents are gaining increasing presence, both IRL (In Real Life) and online. The realization that nonhuman agents—such as bacteria, smart objects, and artificial intelligences—must be reckoned with is the focus of section III. This section proposes a perspective on the shift of hierarchies that is putting human hegemony in question, even as conventional patterns of engagement with the world are destabilized. The essays address the rapidly spreading Internet of Things (IoT), endowing inanimate artefacts with agency and intelligence, as well as technologies of unmanned imaging, such as satellite and drone photography. The latter technologies make it possible to not only communicate with inanimate objects, but perceive the world through their 'eyes' from previously unimaginable angles, distances, and proximities. The concluding essay of *Shifting Interfaces* discusses the profound effect of these technologies on our scopic interface with the world.

I. Immersion, Empathy, and Presence: Manipulating the Boundaries of Self and Body

Section I highlights the growing interest in mediated presence and empathy, and the rekindled attention to the phenomenology of embodied experience in VR. In the chapter 'From Immersion to Empathy: The Legacy of *Einfühlung* in Virtual Reality and Digital Art,' Grant Bollmer addresses the turn to VR as facilitator of empathy, returning to the nineteenth-century aesthetic concept of *Einfühlung*, or 'feeling into,' as a key to the linkage between empathy and immersion. Bollmer does not concur with the notion that we exist in a crisis of empathy. He therefore offers a critical perspective on the emerging politico-ethical project harnessing VR to the cause of conflict resolution and bridging between social groups. Historically contextualizing the current view of VR as the quintessential 'empathy machine,' Bollmer highlights the biases inherent in this position, claiming that the object of empathy is still the hegemonic, rather than the marginal, body.

With the chapter 'Avatars: Shifting Identities in a Genealogical Perspective,' Andrea Pinotti offers a different angle on the immersive drive. This chapter traces the primal anthropological roots of the desire to cross the threshold between the iconic and the real. The chapter further contextualizes this drive in a historical and cultural perspective. Bearing in mind the imminent social and cultural dominance of 'immersive natives,' Pinotti regards avatars as 'shifting-identity operators,' signaling the multiplication of the self. Avatars, in Pinotti's analysis, are bi-directional vectors of change. They not only influence the nature of spectatorship, but also affect our reality, as manifested by the use of full-body ownership illusions to mitigate gender and ethnic biases. Here, Pinotti addresses immersive empathy in an apparently more affirmative note than Bollmer's doubtful appraisal of such assumptions of social benefit or change.

Discussing the avatar as facilitator of self empathy, Pinotti draws a historical genealogy of apparatuses that enabled viewers to 'meet themselves,' from self-portraits to selfies, through the Rückenfigur of Romantic painting. Daniel Landau, in 'Meeting Yourself in Virtual Reality: A Performative Experiment in Self-Compassion,' speaks to this very issue. Landau, a media artist involved in art-based research, presents a performance cum experiment, enabling participants to 'look themselves in the eye' and have a physical sensation of holding their own hands, while hearing (previously recorded) compassionate words spoken by themselves. Employing a 360° 3D Camera, with a live performer providing tactile stimulation, Landau's Self-Study investigates the potential of VR to affect both the concept and the experience of the self.

In 'Art and Presence: Investigating Embodiment in a Virtual Art Gallery,' I introduce a pilot behavioral experiment, putting VR to use as a versatile experimental platform through which to investigate the impact of born-digital art on participants' sense of embodiment. This ongoing collaborative project, involving a media artist, a VR researcher, and an art historian, explores the potential of experiencing artworks in VR to re-anchor participants in their physical body. This project seeks indications of somatosensory arousal driven by haptically effective images, assumed to suggestively activate the senses of touch. The underlying hypothesis is that images which emphasize texture, albeit 'merely' digitally suggested, are effective in eliciting the somatosensory state tentatively defined as *presence*. Avoiding actual physical stimulation of the skin, this project deviates from the race of industry to provide 'real' haptic-feedback apparatuses.

The next chapter in section I proceeds from presence to telepresence through the experimental oeuvre of Paul Sermon, a leading media artist who pioneered telepresence art already in the early 1990s. In 'Shared Objective Empathy in Telematic Space,' Sermon presents three art projects in which physical and virtual presences were merged through real-time video, projection mapping, chroma-keying, and videoconferencing technology. Sermon's *Telematic Dreaming* (1992), *Peace Talks* (2003), and *Touched* (2017) foster inter-subjective engagements between remote participants who are brought together in merged telepresent environments. This essay outlines the history of the artist's 25-year engagement with telematic presence.

Wendy Coones wraps up this section with an original take on empathy-enabled emotional manipulation in museum exhibits. In her chapter, 'Elaborating Mediation through Media Arts,' Coones emphasizes the important role of media arts in enabling the maneuvering between immersion and critical distance. Coones traces a historical trajectory—from the touchable Wunderkammer of the Renaissance to the 'look but no touch' of the Early Museum age, and from walk-in Dioramas to immersive science exhibitions—examining the extent of critical distance that these cultural establishments afforded their visitors. For Coones, retaining a self-reflexive distance while being *in* the experience is a paradigm to be adopted from media arts.

II. Subjects of Surveillance: Reclaiming Human Agency in the Big Data Universe

Section II addresses the nexus between contemporary media arts and globalized anxieties, pertaining to surveillance apparatuses, big data clouds, and endangered ecosystems. Derek Curry and Jennifer Gradecki present a creative intervention in Open Source Intelligence (OSINT) systems that are employed to monitor online contents. 'Qualculative Poetics: An Artistic Critique of Rational Judgment' presents Gradecki and Currie's project: The Crowd-Sourced Intelligence Agency (CSIA), a partial replication of an OSINT surveillance system created by using information from leaked documents, technical manuals, ethnographic reports, and Freedom of Information Act (FOIA) files. The CSIA is designed to expose to critical scrutiny the prevailing claims to technological rationality and the underlying and unacknowledged dangers of attributing infallible judgment to algorithm-based apparatuses. To say it with the artists, 'when the CSIA is placed into an art context, users experience OSINT surveillance with the aesthetic contemplation reserved for art objects.' Shifting the threshold between the calculable and the 'qualculable,' Gradecki and Currie stress and reaffirm the complexity and vulnerability of human agency within the rapidly changing sociopolitical sphere.

Ziv Ne'eman and Tsila Hassine present yet another artistic intervention in the Big Data systems shaping our knowledge about the world. *Shmoogle* is an alternative search engine that serves all of Google's results, but in random order, overturning Google's PageRank™ hierarchy while at the same time disclosing the original positions of the results on Google's pages. In 'Google, *Shmoogle*: A Critical Twist on Search Engine Ideology,' Ne'eman and Hassine address the overwhelming dominance held by Google over the gateways between the online world and IRL (In Real Life). The authors maintain that Google's secretive policy regarding its search algorithm and the practical monopoly that it holds over information mediation have become a threat to subjective and social freedom. *Shmoogle* emancipates user experience, promoting critical distance and reclaiming control over the interface with Real Life and at the same time with the online world.

The third chapter in this section, 'Sculpting Time: The Art of Collective Memory' by the prominent media art scholar Ryszard Kluszczyński, addresses the mediation of collective memory in the age of the digital archive. Presenting archival art projects that reflect on this aspect of information culture, Kluszczyński focuses on bottom-up participatory apparatuses of memory creation and preservation, in the genre of archival art. The works discussed in this essay reclaim memory from the grip of dominant ideologies, social structures, and systems of political power. Recorded sounds and testimonies and living re-enactments of photographically preserved memory form collective and participatory 'memory rhizomes.'

Unlike Kluszczyński, Manuela Naveau critiques the very assumption that participation equals democratization. In 'The Logic of Participation in Digital Art and Network Culture,' Naveau emphasizes the actual loss of agency that occurs when participation is 'hijacked,' as it were, leaving its subjects trapped in an enormous network that they cannot decipher, let alone partake in the formation thereof. Naveau's essay focuses on activist interventions, in the form of art projects exposing the largely overlooked mechanisms of unwitting and/or involuntary participation in online ventures. Participation is here questioned under the assumption that voluntary and knowledgeable partaking is only an increment of the entire picture, and that the potential for uninformed or involuntary participation begs urgent attention. Asking 'are we merely taking part, or do we have a share?' Naveau critiques overly celebratory approaches to participation, arguing that the institutional powers shaping and using online participation demand skeptical examination. In accord with Ne'eman and Hassine's sabotage of Google's PageRank™ algorithm, Naveau argues for decentralized connectedness and a self-organized online society.

Concluding this section, dedicated to cultural/political critique, Yvonne Volkart focuses on the material impact of digital technology, discussing research-based art projects that investigate the dark and uncanny side of what is conceived, ironically, as immaterial information technology. Nondegradable electronic waste, CO_2 emissions, the loss of biodiversity, and the exhaustion of raw materials are some of the topics of 'Aesthetic Strategies in the Wasteocene.' In resonance with the title of the present anthology, Volkart approaches art as a cultural interface. As such, it shifts normalized views, introducing skeptical reflection and opening up transversal potentials. Waste is conceived in this chapter as a disruptive force, attention to which might destabilize relations between 'living' and 'dead' matter, the domesticated and the alien, culture and the abject. Volkart highlights the endless cycle of exploitation and waste, compellingly questioning the idea that matter could just dissipate, technology notwithstanding. Waste, Volkart demonstrates, has primordial roots. This is strikingly pointed out by one of her interviewees, remarking that 'it is quite likely that a modern smartphone comprises at least a few atoms of copper originally mined during the Bronze Age.'

III. Bacteria and Smart Objects: Presences Beyond the Human

Section III reflects on nonhuman agents, which seem to be assuming increasingly autonomic agency. With his chapter 'Rehabilitating Bacteria: An Epistemological Art/Science Interface,' Jens Hauser opens this section and discusses the contemporary trend in hands-on biomedia art, rediscovering bacteria as the oldest, smallest, structurally simplest organisms vital for all other life forms, and increasingly staging them as agents. While in earlier biomedia art bacteria remained ontological blind spots—cells, tissues, and genetic sequences were considered more suitable biological entities to microscopically 'embody' or 'encode' individual organisms for which they stood in *pars pro toto*—Hauser demonstrates how bacteria and their related metaphors can currently be seen as epistemological indicators in both the arts and the natural sciences, becoming omnipresent in art-based research and research-based art. Through the lens of his theory of biomediality, Hauser analyses how the perceived status of bacteria has shifted: from mere breeding containers to actors, from programmable 'workhorses' to complex functioning ecologies, and from 'neutral' prokaryotic cells to 'cellular cities and societies.' In this sense, bacteria cover the whole range of what Hauser defines as the modes of biomediality—as milieus, means of transformation, or tools for measuring—and become at the same time subjects, objects, and tools in artistic displays. These range from intestinal flora and ecologies in Winogradsky columns to photo-tactic bacteria as photographic media, extremophile bacteria as cleaning agents, or programmed bacteria in synthetic biology to become analytic tools.

In 'Exceptional Matters,' the artist Adam Brown presents three Bio Art projects that utilize inorganic matter and bacteria in order to question the boundaries between the human and the nonhuman, the organic and the inorganic. Through these projects, the author ponders the role that 'exceptional (i.e. nonhuman) matters' may have had, unacknowledged, in shaping our culture, myths, and even history, and thus our very humanity. Brown's ongoing engagement with in-vitro life processes as new media for art has yielded collaborative art-science projects that probe the primal origins of life, ancient myths, and deep-rooted religious beliefs. These projects assert the inherent agency of matter, thus stripping the human subject of a millennially sanctioned supremacy.

Also testing the seam lines between living and nonliving matter is the aesthetics of 'softness,' or bio-morphism. Jonas Jørgensen's chapter, 'From Soft Sculpture to Soft Robotics: Retracing an Entropic Aesthetics of the Life-like,' links biomorphic or 'soft' robotics with the art-historical precedent of soft sculpture from the late 1960s and early 1970s. Highlighting the intuitive association of softness with the 'life-like,' Jørgensen investigates media art installations that make recourse to what he paradoxically calls 'an entropic aesthetics of life.' Soft matter implies life while bordering on the entropic, just as the living organism itself is immersed in exchanges of entropic matter with the environment. In the soft-robotics art installations discussed by Jørgensen, the computed environment acquires qualities that in the past had been reserved exclusively to living beings.

In the foregoing chapter, Jørgensen addresses the 'interlacing of human existence with a set of intelligent and agential technologies.' In the chapter that follows, 'Natures Mortes, Live Data: The Art Object in the IoT Era,' Tsila Hassine and Olga Kisseleva ask how the presence of intelligent objects—more 'alive' than 'still,' empowered by the Internet of Things (IoT)—may reshape our concept and experience of art. In a project interconnecting gallery visitors and art objects, Hassine and Kisseleva devised a complexly relational experience in which the 'still life' images respond to the viewers' very act of looking. The project *Natures Mortes; Live Data*, speaks to the history of still-life painting, interrogating the representation of inanimate objects from the Roman *Xenia* to the seventeenth-century Dutch School, in an attempt to understand how humans had related to their objects throughout cultural history. The authors inform us that 50 to 100 billion 'things' are expected to be connected to the Internet by 2020, a technological revolution that harbors a profound experiential shift—on the move towards increasing interconnectedness beyond the human. With an interdisciplinary team of students recruited from art, design, and engineering departments, Hassine and Kisseleva attributed 'personalities' to selected photographs from

Kisseleva's earlier series *Divers Faits* (2010). The personality traits attributed to the photographic objects were based on predefined values, inspired by visual and semantic aspects of the image. The accumulated length of time spent by visitors in front of the work, defined as 'popularity,' affected the intensity of illumination cast upon it by the gallery's lighting system. At the same time, the 'personality traits' of each photograph modified the rhythm and dispersion of the lighting, for example reaching full extent and then dying out, or sharing the limelight with predefined 'friends.' In the era of IoT, Hassine and Kisseleva conclude, art objects will be 'looking at us looking at them.'

In yet another take on perception through a nonhuman viewpoint, Aaron Burton concludes this collection with a consideration of unmanned photography. With 'Unmanned Imaging of the Anthropocene,' Burton demonstrates that removing human presence from behind the camera shifts the historically dominant human-centric perspective, significantly altering our paradigms of representation. Burton's chapter discusses satellite images of planet Earth, as well as drone photography and action (GoPro) cameras, which offer an unprecedented degree of interpersonal and ecological proxemics. While Volkart (see section II) suggested that dedicated art interventions might effectively mediate the Wasteocene, Burton posits satellite and drone photography as extremely effective in mediating the global ecological crisis of the Anthropocene. At the same time, drones and action cameras enable a more participatory scopic experience, offering extreme proximity with the object. Closing with these shifts in our scopic paradigms, this anthology concludes the story it opted to tell—of the cultural and political role played by contemporary media arts, partaking in the attempt to make sense of the foundational shifts in our lived experience.

I

Immersion, Empathy, and Presence

....

Manipulating the Boundaries of Self and Body

From Immersion to Empathy
The Legacy of *Einfühlung* in Virtual Reality and Digital Art

Grant Bollmer

Virtual reality carries with it a number of questionable associations. These include, for instance, informational disembodiment and the emergence of a mediated, synthetic collective; cybernetic subjects, liberated from biology, eternally alive upon an idealist plane of digitality; an infinite spatial logic both 'inside' computers and ethereally immaterial. These beliefs presume technologically augmented minds that transcend the body, space, and time; they locate 'thought' not corporeally, but in a realm of representational imagination that overflows embodiment (cf. Hillis 1999). This list may seem to be a relic of the 1990s. It's difficult to see how understandings of an 'immaterial' or 'disembodied' technology can persist, as they require an adherence to both a Cartesian worldview and a particular kind of hylomorphism. They neglect the geological and energetic requirements to run computers and other digital technologies (cf. Parikka 2015).

Yet these beliefs have returned with a vengeance nearly two decades after VR's first demise as a consumer technology, coinciding with a renewed round of interest and speculation about virtual reality and its potentialities for perception, experience, and space, albeit in a subtly different form. The purpose of this chapter is to theorize assumptions of presence implied by digital simulation, not *only* as a recurrence of the digital idealism sketched above, but to demonstrate how this idealism has been associated with claims of *empathy*, about 'feeling into' the subjective experience of another as an aspect of mediation with political, ethical, and social implications. My interest, I should note, is less about VR, in and of itself, than it is about epistemological changes that descend from a genealogy that intertwines immersion with empathy. I find VR—both in popular culture and in contemporary art—to be a particularly salient example to begin charting this genealogy, though what I describe here has implications beyond VR. VR demonstrates how 'presence' has been reframed in light of the supposed 'crisis' of empathy ongoing today, a crisis that suggests many of our political, social, and moral problems arise from a technologically-fostered inability to understand and relate to the experiences of others (e.g. Turkle 2015).

I do not think that we exist in a 'crisis of empathy.' Rather, I suggest that the metaphysics of immersive transcendence have been reframed through a normative aesthetic directive, one implicit in representations of VR from the 1990s but with an added social demand: the task of simulation should represent for a spectator the immersive sensation of *being* another and of *feeling* what it means to live as another. While we can see ideas such as these when we return to, say, Kathryn Bigelow's *Strange Days* (1995), in which VR's potential use as a means to inhabit the body of another is employed for prurient pleasures, today this potential is thought to have certain social benefits: of 'walking a mile in another's shoes,' to use a popular definition of what 'empathy' might mean. The crisis of empathy is a rearticulation of a digital metaphysics that suggests the power of simulation enables us to transcend our bodies, and this crisis emerges from the failure of simulation to achieve the metaphysical goal of inhabiting another's life. Or, the crisis of empathy is another name for the failure of immersive media to fully immerse the body in another world, escaping corporeality.

This normative demand has a history, one that, perhaps oddly, turns us to debates in aesthetic theory in Germany from the late 1800s and early 1900s and its theorization of *Einfühlung*, or 'in-feeling.' These Germanic debates have been foundational for any psychological theory of empathy, and the turn to empathy today obscures the more radical beliefs about informational disembodiment by displacing them with a particular kind of intersubjective relation that permits one body to enter into and inhabit another.[1] While we think of empathy as a psychological concept, its significance today—and its implications for VR, digital art, and aesthetics—presumes a political, aesthetic norm in which 'knowing' comes not from listening and deliberation, from affirming the otherness of others and the ethics of responsibility to the other, but from direct and supposedly 'unmediated' simulation, in which the other effectively becomes part of the experience of the observer, vampirically or cannibalistically absorbed into what Emmanuel Levinas called 'the same' (1969; cf. Han 2018; hooks 1992, 21–39). My goal here is to critique the limits and potentials of empathy and digital media by contextualizing and historicizing the recent resurgence of VR. To be clear, this is not a critique of empathy as such, but a critique of particular political goals assumed by the move from immersion to empathy—which, again, is more about the epistemological context implied by the crisis of empathy today rather than something specific to VR.

Virtual Reality's Second Death?

While, throughout the past two decades, media theorists have turned to materiality to analyze the agency of technology, rejecting the idealism of 1990s cybercultuture (see Bollmer 2019), the metaphysics of '90s VR returned, partial and refracted, in a cycle of hype that began when Palmer Luckey's Oculus Rift raised over $US 2.5 million on the crowdfunding platform Kickstarter in 2012. The Oculus was thought to represent a renewed version of VR that avoided many of the more technical problems it had had throughout the 1990s—problems such as motion sickness and disorientation, combined with clumsy and cumbersome technological interfaces. The Oculus, investors imagined, could provide virtual immersion at a technical scale previously impossible, at a price accessible to most everyday gamers and tech enthusiasts. Less than two years after its successful Kickstarter, on March 25, 2014, Luckey's company was acquired by Facebook for a staggering $US 2 billion, cementing the renewed excitement around VR as a medium (see Kumparak 2014).

Yet, as I write these words, the VR bubble, much as was the case in the '90s, seems again to be deflating, with VR dying a second death. The potentials of VR for consumer entertainment have been marred by the high costs of VR equipment and the extravagant technical requirements for computers to run VR software; the belief that the Oculus represented a technology affordable to many has simply not been realized. The lack of consumer interest in VR has led tech writers to claim that 'This VR Cycle is Dead' (Lomas 2017) in the face of flagging sales (Bradshaw 2017), with industry hype turning towards Augmented Reality devices, like the Magic Leap headset, for hopeful fantasies of a material world remade in accordance with human imagination. The hype for VR, in and of itself, has not completely waned, however. A variety of media companies—be they in videogames or in film—still see VR as the future of entertainment, investing in projects like the Oculus-based VR film *Spheres* (2018), written and directed by Eliza McNitt, produced by director Darren Aronofsky and Ari Handel. *Spheres*, an abstract VR documentary about cosmology narrated by actors and performers Millie Bobby Brown, Jessica Chastain, and Patti Smith, was acquired at the 2018 Sundance film festival for a sum exceeding seven figures (Watercutter 2018) and won the Grand Prize in VR at the Venice Film Festival—one of several major awards that highlight a popular investment in VR as a medium for (often experimental) forms of narrative filmmaking and documentary.

In the world of contemporary art the attention to VR is also obvious. The art world, while certainly subject to its own forms of market-based economic speculation and hype, is at least slightly isolated from the boom and bust cycles of tech industry speculation, and VR has become a notable medium for media art, with works reliant on the Oculus or other, similar VR technologies, like the HTC Vive and the Samsung Gear, appearing in countless exhibitions over the past few years. While, throughout the 1990s, notable works like Char Davis' *Osmose* (1995) gained an almost excessive amount of theoretical commentary in spite of the relatively small number of people who actually experienced them (see Grau 2003, 193), VR has become a popular medium for art in a staggering number of venues. Be they 'blockbuster' shows intended to attract large audiences, such as the touring *Björk Digital* show which debuted at Carriageworks in Sydney, Australia, as part of the 2016 Vivid Sydney festival; be they challenging artworks intended to confront and even enact a kind of violence on audiences, such as Jordan Wolfson's 2017 *Real Violence*, which was included as part of that year's Whitney Biennial (see Bollmer and Guinness, 2020); be they works that embrace VR's telepresent potential of transporting the viewer to a physically inaccessible location, such as the collective Chim↑Pom's *Don't Follow the Wind* from 2015 (Guinness 2017); be they critically acclaimed works, such as the director Alejandro González Iñárritu's *Carne y Arena* (2017), that use VR to dramatize narratives that have explicitly humanitarian goals, VR is clearly a prominent medium for art today, deployed for a range of different purposes, within a range of different styles.

In many ways, the aesthetic criteria used to describe these works perpetuate the lengthy history of sensory 'immersion' through media, albeit with a synesthetic specificity that accounts for the failures of 90s-era VR—and current consumer-grade VR—to achieve the utopian metaphysical goal of virtual disembodiment into a technological space. Miriam Ross (2018) suggests today's VR perpetuates a more measured drive for immersion than in the past, seeming to acknowledge the failures of the '90s. But this drive for immersion—restrained or not—continues not only the logic that legitimated VR previously, but a logic that grounds a vast amount of Western art history. As Oliver Grau demonstrates in his overview of artistic and technological immersion, the debate over sensory immersion and critical distance has long characterized debate over art, illusion, and simulation. Digitally simulated virtual realities perpetuate this history, combined with a tendency to 'almost wholly visually seal off the observer hermetically from external visual impressions, appeal to him or her with plastic objects, expand perspective of real space into illusion space, observe scale and color correspondence, and (…) use indirect light effects to make the image appear as the source of the real. The intention is to install an artificial world that renders the image space a totality or at least fills the observer's entire field of vision' (2003, 13).

This is still the case with the Oculus, certainly. Grau suggest this tendency descends from techniques associated with French and Italian frescoes and landscape rooms, the religious use of ceiling paintings, panoramas as mass entertainment in the nineteenth century, and the 'total' multimedial logic of the Wagnerian *Gesamtkunstwerk* as it moved from opera to cinema and beyond (cf. Smith 2007; Ross 2018). But, while immersion often carries with it the desire of losing the body and physical space in the face of simulated reality (see Hillis 1999), it also relies on particular, embodied potentials for sensory motion. Immersion, as Alison Griffiths defines it, is 'the sensation of entering a space that immediately identifies itself as somehow separate from the world and that eschews conventional modes of spectatorship in favor of a more bodily participation in the experience' (2008, 2). This would suggest that immersion *requires* the body rather than transcending it, though what 'body' or 'embodiment' may mean is clouded by particular neurological understandings of cognition and sensation, not all of which require the existence of a physical body to ‘sense’ or ‘move’ (cf. Bollmer 2019). The dreams and fears of immersion, dreams of a world unbounded by anything but mental imagination contrasted with fears of a loss of critical distance and the ‘real,’ are probably as old as representation itself. But while the tendency to seek immersion is exceptionally old, the particular conditions for medial immersion—as Ross implies—are reinvented every time a new technology remakes the potential for visual and haptic representation. Thus, we should look at these particularities—not at immersion in general, but in the ways that immersion is framed within a specific context.

Immersion as Empathy, Empathy as Immersion

Today, the desire for immersion is explicitly linked with the psychological state of *empathy* which, I suggest, remakes past fantasies of bodily transcendence as an ethical demand in addition to a metaphysical desire. Works like the aforementioned *Carne y Arena* are framed through their ability to foster empathy by simulating the personal experiences of a particular individual for the observer. Following the claims of music video and VR director Chris Milk in a popular TED talk, VR is the ‘ultimate empathy machine’ (Milk 2015; cf. Bollmer 2017), expanding on a term film critic Roger Ebert once used in the documentary *Life Itself* (2014) to describe cinema and its ability to realistically represent another’s experience for the viewer. If cinema is, as Ebert suggests, a ‘machine’ for generating empathy by allowing the viewer to see—and ‘inhabit’—the life of another, then the totalizing, immersive potentials of VR supplant cinema by making the moving image interactive, enclosed, and responsive to the embodied motion of the viewer.[2] These arguments have some support from the scientists and literary critics working in the emergent field of neuroaesthetics, who argue that emotional investment in art emerges from a particular mirroring capacity of the human brain, strongly linked with the simulation of embodied motion, operationalized as ‘empathy’ in neuroscientific studies (i.e. Freedberg and Gallese 2007). The turn to empathy signifies not just a concern with immersion, but a concern with how medial representations relate to the very possibility of intersubjective, self-other relations, and how the possibility of self-other relations grounds the ethical, political, and social effects of technology.

Empathy—given what appears to be a Greek origin as *empathea*—would seem a concept as old as fears and desires for immersion. But this is not the case. Empathy was a neologism employed by psychologist Edward Titchener in his *Lectures on the Experimental Psychology of the Thought-Process* (1909) to translate the German term *Einfühlung*—literally translated, ‘in-feeling.’[3] Titchener was using empathy to refer to a particular sensation of simulated motion, of being moved. He states: ‘“feelings”—feelings of how and why, and nevertheless, and therefore—normally take the form, in my experience, of motor empathy.

I act the feeling out, though as a rule in imaginal and not in sensational terms. It may be fleeting, or it may be relatively stable; whatever it is, I have not the slighted doubt of its kinaesthetic character' (185). This is a similar definition of empathy to the one more recently given by art historian David Freedberg and neuroscientist Vittorio Gallese: 'viewers often experience a sense of bodily involvement with the movements that are implied by the physical traces—in brushmarks or paint drippings—of the creative actions of the producer of the work' (2007, 197). Freedberg and Gallese argue that the pleasure a viewer gets from a painting by Jackson Pollock or Lucio Fontana emerges from the indexicality of motion captured in a work, be it the splattering of paint for the former or the slashing of the canvas for the latter. When looking at a work of Fontana's, they suggest, 'sight of the slashed painting invites a sense of empathetic movement that seems to coincide with the gesture felt to have produced the tear' (197). Empathy, here, is a mental simulation of motion that comes from an inner mirroring of external perception.

The tradition that moves from Titchener to Freedberg and Gallese is certainly significant, and Gallese is one of the most important neuroscientists of empathy today. But the limited definition of empathy here—which is more or less about the simulated mirroring of physical motion in the brain of an observer—is one part of a much larger history that draws out considerably different implications for empathy than, say, the claims Gallese makes about mirroring and intersubjectivity.[4] As a prefix, *Ein* suggests a movement from outside to inside—immersion into, moving beyond a barrier to penetrate within. *Fühlung*, which is derived from *fühlen*, means feeling. *Einfühlung*, however, was initially a concept proposed to describe how a viewer relates to a work of art, imported into German psychology to extend arguments about aesthetic experience to psychological states.[5] I want to suggest that the history of *Einfühlung* precedes our current discussions of the immersive potentials of videogames and digital art that employ VR technologies, arguing that the current articulation of immersion is deeply intertwined with assumptions indirectly derived from a particular moment in German aesthetic theory.

Einfühlung in German Aesthetics

Einfühlung first appeared in the writings of Romantic authors, such as Johann Gottfried Herder and Novalis, to describe the fundamental workings of beauty. In several books published in 1778, Herder argued that beauty was an effect of an inner transposition (or *Versetzung*) into the figure an observer contemplates. For Herder, contemplation allows us access to both the other and ourselves, as 'we can only feel ourselves inside others (*hineinfühlen*)' (Pinotti 2010, 93). Judgments of beauty, for Herder, come from an intersubjective relationship between observer and the observed, in which the boundary between the two is blurred as one enters the other and vice versa. While Herder and Novalis were clearly discussing art, it also seems that they elevate this aesthetic contemplation to a foundational aspect of all relations—something similar to other Romantic aesthetic theories, such as Friedrich Schiller's idea of the 'aesthetic state' and shared taste as a foundation for proper governance (see Chytry 1989). Judgments of beauty would become the grounds for moral and political unification.

It's worth noting this idea of in-feeling was subtly distinct from the Kantian thought of Schiller—at least the writings after Schiller's more philosophical turn. For Schiller, the intersubjective relations produced by shared taste required a balance between feeling and rationality, being about neither pure sensory experience nor dispassionate judgment. One does not feel into another to unify disparate subjects—at least not completely. As he notes in a letter to Prince Frederick Christian von Augustenburg, education in the service of training and refining aesthetic judgment,

> although it acquires for man no value in the moral world, renders him nonetheless viable in the physical world, making him capable of a regularity of conduct without which nature could never achieve its great aim of uniting men into a whole. But men are still a long way from being united when they have ceased disputing among themselves, and legality alone can prevent injustice tearing the bonds of society. To unite men truly and inwardly requires another, positive bond, that of social character, or the communication of sensations, and the exchange of ideas (2016, 167).

The unity of disparate subjects only emerges from a combination of feeling with the active discussion and communication of sensation—an idea that resonates both with Kant on communities of taste (cf. Friedlander 2015) and the importance of publicity, or the public sphere (*Öffentlichkeit*), in Enlightenment politics (e.g. Habermas 1989). If Schiller's aesthetic state required the making public of sensations, Herder and Novalis instead emphasized the access to the interiority of another via the penetrative sensation of *Einfühlung*. These two positions both necessitated the exteriorization of that which is interior, but they disagreed on the means for making public inner sensations and experiences for the purposes of aesthetic judgment. For Schiller, access to the feelings of another required communication and rational dialogue, while for Herder and Novalis, perhaps embodying a more Romantic spirit of experience and feeling, this access was innate in intersubjective encounter.

While Schiller's arguments were very influential, Romantic discussions of *Einfühlung* were inconsequential until nearly a century later, in 1873, when *Einfühlung* reappeared in Robert Vischer's dissertation *On the Optical Sense of Form: A Contribution to Aesthetics*. Drawing on the formalist aesthetic philosophy of Friedrich Theodor Vischer, his father,

Robert Vischer argued that form permitted the projection of the body into the object observed. The observer, he states,

> in responding to certain stimuli in dreams, objectifies itself in spatial forms. Thus it unconsciously projects its own bodily form—and with this also the soul—into the form of the object. From this I derived the notion that I call 'empathy' [*Einfühlung*]. Soon, however, I realized that this notion would only in part explain the symbolism of form, for the effect of light and color, the contour, and the pure line cannot be described by empathy. Here one can only assume a direct continuation of the external sensation into an internal one, a direct mental sublimation of the sensory response (1994, 92).

Visher, while he does not reduce all aesthetic experience to empathy, nonetheless proposes a particular movement from the observer into that which is experienced—a movement that has clear limits. Visher's arguments were influential for a number of art historians and aesthetic theorists writing around the same time, including Adolf Hildebrand, August Schmarsow, and Heinrich Wölfflin—the last of whom used *Einfühlung* to describe the aesthetic power of architecture. Eventually these theories were taken up by the psychologist Theodor Lipps, who expanded *Einfühlung* and aesthetic feeling by making it a general psychological principle of social relation. It is Lipps' psychology that Titchener was referring to when he described particular kinesthetic relations as empathetic, and it is Lipps who, unlike Visher, makes *Einfühlung* into a central psycho-aesthetic concept that grounds both aesthetic and intersubjective relations (Jarzombek 2000, 59; Jahoda 2005).

Juliet Koss (2009) has argued that *Einfühlung* be linked with the Wagnerian *Gesamtkunstwerk*, where the overwhelming, totalizing experience of Wagnerian opera in Bayreuth was intended to induce in the individual spectator a communal state that would emerge from feeling into others. While Wagner argued that the *Gesamtkunstwerk* be thought of as producing the organic unities that Schiller described in his writings on the aesthetic state (Smith 2007, 12), the emphasis on feeling and spectacle, rather than a balance between rationality and sensation, suggests an aesthetic politics more akin to Herder and Novalis, in which the emotional making-public of interiority is essential for the formation of political unity. As Koss argues, 'Wagner linked publicity to an idea of communism—as opposed to what he termed egoism, or individualism. The former described the public and communal experience, simultaneously political and aesthetic, of a group of people; the latter referred to the private experience of the solitary individual' (20).

Immersive spectacle would permit observers to lose themselves in the bodies of the performers and their fellow spectators, which Wagner described through the term *Mitfühlung*, or feeling with. Yet, this 'communism' or 'socialism' is more than a little problematic. Not only did Wagner describe 'communism' as emerging from and realizing the *Volk*, preceding the Nazi invocation of these same terms, the *Gesamtkunstwerk*, in its very architecture, required the fragmentation, separation, and isolation of bodies before they could be subsumed into a singular whole. As Adorno noted in his influential, trenchant critique of Wagner, this sensory and social unity is merely an ideology that covers over the extreme divisions implicit in the material form of the *Gesamtkunstwerk*:

> It is questionable, indeed, whether such a unity of sense experience has ever existed, dependent on it as Wagner's disillusioned mind may be. On the contrary, the senses, which all have a different history, end up poles apart from each other, as a consequence of the growing reification of reality as well as of the division of labour. For this not only separates men from each other but also divides each man with himself. It is for this reason that the music drama proves unable to assign meaningful functions to the different arts. Its form, therefore, is that of a spurious identity (1981, 91).

Rather than generate organic unities, Adorno sees the *Gesamtkunstwerk* as part of the modern fragmentation of the senses that characterized media culture in the late 1800s and early 1900s (cf. Kittler 1999), but with particular ends that required the exclusion of specific bodies from the unity generated by Wagnerian opera. Wagner's theater at Beyreuth, for instance, worked to make the orchestra invisible to the audience. The synthetic, immersive spectacle of the *Gesamtkusntwerk* could only occur if the fragmentation of its material base remained invisible. The unification of the senses could only happen if the fragmented physical labor that underpinned the sound and fury of Wagner was inaccessible, separating audience, performer, and musician. This theme is even foregrounded in Wagner's operas, which likewise reveal the parallels between the material structure of immersion and Wagner's anti-Semitism. As suggested by David Levin (1996, 131) and Matthew Wilson Smith (2007), the totality of the *Gesamtkunstwerk* can only be maintained by 'an Other whose purpose it is to *not belong* (…) the guarantor of totality who must be exorcised' (104), an Other who is, in Wagner's operas, Jewish. This structural logic is, of course, not limited to Wagner and the *Gesamtkunstwerk* (cf. Žižek 1989). But, for us, the immersion of the total work of art can only occur as long as *something* is deliberately excluded. 'Communism' can only come into being, here, through an empathy that relies on the deliberate erasure of particular Others.

Thus, if the immersion of the *Gesamtkunstwerk* is assumed to be a media-technical form that, through emotional spectacle, permits one spectator to transcend their own body and join with another, producing a kind of populist 'communism' through *Einfühlung*, then we can begin to see problems if this tradition is maintained. Namely, empathy becomes a function of media to separate bodies and recombine them—immersion into others becomes a function of individualistic isolation. And, additionally, this empathetic linkage is not truly totalizing, but rests on the necessary exclusion of the other. To return to the present, philosopher Kate Manne (2018) has argued that, today, we do not exist within a crisis of empathy. Rather, empathy is articulated only to specific bodies and individuals, bodies and individuals that correspond to those traditionally in power (Manne calls this 'himpathy' because the subject of empathetic investment, in various demands for emotional identification today, tends to be male). And even when marginal bodies become objects to empathize into, they do not exist beyond which can be assimilated (Bollmer 2017). Or, the problem is not a *lack* of empathy, but that any feeling into is directed towards the same forms of identification and relation that have long existed.

Conclusion

In foregrounding the history of *Einfühlung* I've attempted to describe how immersion and empathy are articulated today, and how this can be seen as a recursion of past debates about immersion and empathy derived from German aesthetics and the Wagnerian *Gesamtkunstwerk*, along with briefly noting how the political and aesthetic potential of VR in games and digital art ultimately leads us to questions of self-other relations. Any belief in a 'crisis of empathy' that can be solved through immersive, empathetic simulation assumes that, in effect if not in actuality, there is a generalized brain deficiency that can be corrected through moving images and sound, inducing in the brain of an observer the feelings of the one observed. But the logic of empathy via immersion has a long history, one that demonstrates not a perpetual deficiency of empathy, but that empathy and feeling are always directed at particular bodies and not others. The problem is not a lack of feeling, but with the barriers and oppositions that emerge from how feeling is articulated (cf. Ahmed 2015). Simulation and immersion necessarily involve the exclusion of something in order to make an experience 'whole' and to bind fragmented subjects together. The question should be to look not at these unities, but at that which is excluded—perhaps, when it comes VR and digital art, we find that the one excluded is the one the observer is supposed to empathize with, as the observer, ever fragmented and distanced from what they see and experience, absorbs the other into their own experience, their own body, into the same that denies the other.

Notes

1. What I'm arguing here bears some similarity to Katherine Guinness's critique of particular networked, artistic practices as framing the bodies of others as 'corpus nullius,' or empty bodies to be inhabited and colonized by others. Guinness sees this as the legacy of enlightenment humanism

when combined with digital simulation (see Guinness 2019). In this chapter, I hope to demonstrate how simulations designed to foster empathy perpetuate this logic.

2. The 'mechanical' metaphor additionally implies that the empathetic relation produced by media is automatic, and merely seeing particular images induces in the viewer a (neurological) reflex response that is independent of will. This is also implied by some cultural theories of 'affect' who draw on a similar psychological genealogy as theorists of empathy do for their claims (see Leys 2017; Bollmer 2014).
3. 'Empathy' can be found in print in English prior to Titchener's use of the term, by E. L. Hinman (1895), in a brief review of a German article on psychophysics written by Kurd Laßwitz, who is better remembered for his contributions to German-language science fiction. From the review, it's unclear what word Laßwitz actually used in his own writing, but Hinman uses empathy to describe a quantification of neurological energy that correlates to pleasure or pain—a definition that has little to no relation to its use since Titchener's lectures. Other uses of empathy in English from the early 1900s all appear to refer to Titchener's translation of *Einfühlung*.
4. Gallese is strongly influenced by phenomenology and, while I believe he misreads some of those he cites, like Husserl and Edith Stein, he affirms the separation of self and other in his definition of empathy. Mirroring, for Gallese, is not the same thing as a literal 'entering into' the body of another (see Gallese 2003). This leads me to wonder why he uses 'empathy' rather than 'sympathy,' as the latter affirms a fundamental separation between subjects. I would assume the distinction here—much like the distinction between 'affect' and 'emotion' in the humanities—is that empathy is thought neurological and pre-conscious while sympathy is, assumedly, based in conscious thought. As Ruth Leys (2017) notes about this distinction in affect theory, this unintentionally repeats a Cartesian split between mind and body in spite of other attempts to describe a 'materialist' theory of mind and brain. Additionally, my view is that empathy requires the assumptions of entering into another implied by the German *Ein* (see Bollmer 2017), which is also what differentiates empathy from other political theories of sentiment and sympathy, like that of Adam Smith (1759). While Gallese himself is clearly opposed to particular definitions of empathy, such as the 'theory of mind' defined by Simon Baron-Cohen (1995) or the more 'penetrative' definition of empathy described here, the uptake of his work on mirror neurons, especially by those in the humanities, often tends towards a penetrative definition of empathy rather than one that refuses a conflation of self and other.
5. Interestingly, *Einfühlung* was translated into French in 1869 as *sympathie symbolique*, suggesting that the French context may permit a completely different genealogy than the one I describe in this chapter. Andrea Pinotti, whom I draw on for some of the claims I'm making, has done much to outline the history of *Einfühlung* and empathy, and his history of empathy follows similar lines to what I outline here. Yet, while his tracing of the various debates about empathy and aesthetics is deeply nuanced, I nonetheless believe Pinotti defers to empathy as an innate human capacity, conflating it with intersubjectivity (as do those in neuroaesthetics), and also underempha-

sizes just how radically theorists like Edmund Husserl, Edith Stein, Max Scheler, and Wilhelm Worringer rejected the idea of empathetic relation rather than merely revising it in their definitions of intersubjectivity, which is what Pinotti seems to suggest. Rather than accept empathy and intersubjectivity to be similar or identical concepts—which is what Pinotti seems to claim, along with neuroscientists like Gallese—I argue that empathy must be understood an historically emergent discourse that was only called into being in the context of eighteenth- and nineteenth-century German aesthetics. See Pinotti (2010, 2016).

References

Adorno, Theodor. 1981. *In Search of Wagner*. Translated by Rodney Livingstone. London: Verso.

Ahmed, Sara. 2015. *The Cultural Politics of Emotion*. 2nd ed. New York: Routledge.

Baron-Cohen, Simon. 1995. *Mindblindness: An Essay on Autism and Theory of Mind*. Cambridge MA: MIT Press.

Bollmer, Grant David. 2014. “Pathologies of Affect: The ‘New Wounded’ and the Politics of Ontology.” *Cultural Studies* 28, no. 2: 298–326.

Bollmer, Grant. 2017. “Empathy Machines.” *Media International Australia* 165: 63–76.

Bollmer, Grant. 2019. *Materialist Media Theory: An Introduction*. New York: Bloomsbury Academic.

Bollmer, Grant, and Katherine Guinness. 2020. “Empathy and Nausea.” *Journal of Visual Culture* 19, no. 1.

Bradshaw, Tim. 2017. “VR Industry Faces Reality Check on Sales Growth.” *Financial Times*, February 28, 2017. https://www.ft.com/content/f7e231ee-fc84-11e6-96f8-3700c5664d30.

Chytry, Josef. 1989. *The Aesthetic State: A Quest in Modern German Thought*. Berkeley CA: University of California Press.

Freedberg, David, and Vittorio Gallese. 2007. “Motion, Emotion and Empathy in Esthetic Experience.” *Trends in Cognitive Sciences* 11, no. 5: 197–203.

Friedlander, Eli. 2015. *Expressions of Judgment: An Essay on Kant’s Aesthetics*. Cambridge MA: Harvard University Press.

Gallese, Vittorio. 2003. “The Roots of Empathy: The Shared Manifold Hypothesis and the Neural Basis of Intersubjectivity.” *Psychopathology* 36: 171–80.

Grau, Oliver. 2003. *Virtual Art: From Illusion to Immersion*. Translated by Gloria Custance. Cambridge MA: MIT Press.

Griffiths, Allison. 2008. *Shivers Down Your Spine: Cinema, Museums, and the Immersive View*. New York: Columbia University Press.

Guinness, Katherine. 2017. “Nuclear Utopias: Visibility and Indexicality from Hiroshima to Fukushima.” Paper presented at the *College Art Association Annual Convention*, New York, February 15–18.

Guinness, Katherine. 2019. “The Colonizer and Corpus Nullius.” *parallax* 25, no. 4.

Habermas, Jürgen. 1989. *The Structural Transformation of the Public Sphere: An Inquiry into a Category of Bourgeois Society*. Translated by Thomas Burger with Frederick Lawrence. Cambridge MA: MIT Press.

Han, Byung-Chul. 2018. *The Expulsion of the Other*. Translated by Weiland Hoban. Cambridge: Polity Press.

Hillis, Ken. 1999. *Digital Sensations: Space, Identity, and Embodiment in Virtual Reality*. Minneapolis MN: University of Minnesota Press.

hooks, bell. 1992. *Black Looks: Race and Representation*. Boston MA: South End Press.

Hinman, E.L. 1895. "Ueber psychophysische Energie und ihre Factoren by K. Lasswitz." *The Philosophical Review* 4, no. 6: 672–73.

Jahoda, Gustav. 2005. "Theodor Lipps and the Shift from 'Sympathy' to 'Empathy.'" *Journal of the History of the Behavioral Sciences* 41, no. 2: 151–63.

Jarzombek, Mark. 2000. *The Psychologizing of Modernity: Art, Architecture, and History*. Cambridge: Cambridge University Press.

Kittler, Friedrich A. 1999. *Gramophone, Film, Typewriter*. Translated by Geoffrey Winthrop-Young and Michael Wutz. Stanford CA: Stanford University Press.

Koss, Juliet. 2009. *Aesthetics After Wagner*. Minneapolis MN: University of Minnesota Press.

Kumparak, Greg. 2014. "A Brief History of Oculus." *TechCrunch*, March 26, 2014. https://techcrunch.com/2014/03/26/a-brief-history-of-oculus/.

Levin, David J. 1996. "Reading Beckmesser Reading: Antisemitism and Aesthetic Practice in *The Mastersingers of Nuremberg*." *New German Critique* 69: 127–46.

Levinas, Emmanuel. 1969. *Totality and Infinity: An Essay on Exteriority*. Translated by Alphonso Lingis. Pittsburgh PA: Duquesne University Press.

Leys, Ruth. 2017. *The Ascent of Affect: Genealogy and Critique*. Chicago IL: University of Chicago Press.

Lomas, Natasha. 2017. "This VR Cycle Is Dead." *TechCrunch*, August 26, 2017. https://techcrunch.com/2017/08/26/this-vr-cycle-is-dead/.

Manne, Kate. 2018. "Brett Kavanaugh and America's 'Himpathy' Reckoning." *New York Times*, September 6, 2018. https://www.nytimes.com/2018/09/26/opinion/brett-kavanaugh-hearing-himpathy.html.

Milk, Chris. 2015. "How Virtual Reality Can Create the Ultimate Empathy Machine." *TED: Ideas Worth Spreading*. https://www.ted.com/talks/chris_milk_how_virtual_reality_can_create_the_ultimate_empathy_machine.

Parikka, Jussi. 2015. *A Geology of Media*. Minneapolis MN: University of Minnesota Press.

Pinotti, Andrea. 2010. "Empathy." In *Handbook of Phenomenological Aesthetics*, edited by Hans Rainer Sepp and Lester Embree, 93–98. Dordrecht: Springer.

Pinotti, Andrea. 2016. *L'empathie: Historie d'une Idée de Platon au Posthumain*. Translated by Sophie Burdet. Paris: Vrin.

Ross, Miriam. 2018. "Virtual Reality's New Synesthetic Possibilities." *Television & New Media*.
Schiller, Friedrich. 2016. *On the Aesthetic Education of Man* and *Letters to Prince Frederick Christian von Augustenburg*. Translated by Keith Tribe. New York: Penguin Books.
Smith, Adam. 1759. *The Theory of Moral Sentiments*. Edited by Ryan Patrick Hanley. New York: Penguin Books.
Smith, Matthew Wilson. 2007. *The Total Work of Art: From Bayreuth to Cyberspace*. New York: Routledge.
Titchener, Edward Bradford. 1909. *Lectures on the Experimental Psychology of the Thought-Process*. New York: The Macmillan Company.
Turkle, Sherry. 2015. *Reclaiming Conversation: The Power of Talk in a Digital Age*. New York: Penguin Books.
Visher, Robert. 1994. "On the Optical Sense of Form: A Contribution to Aesthetics." In *Empathy, Form, and Space: Problems in German Aesthetics, 1973–1993*, edited by Harry Francis Mallgrave and Eleftherios Ikonomou, 89–123. Santa Monica CA: The Getty Center for the History of Art and the Humanities.
Watercutter, Angela. 2018. "A VR Movie Set In Space Just Landed a 7-Figure Deal at Sundance. Yes, You Read That Right." *Wired*, January 24, 2018. https://www.wired.com/story/vr-film-spheres-huge-sundance-deal/.
Žižek, Slavoj. 1989. *The Sublime Object of Ideology*. London: Verso.

Avatars
Shifting Identities in a Genealogical Perspective

Andrea Pinotti

Immersivity and Presence

In recent years a huge number of publications dealing with the experience of 'immersion' has appeared on the scene of contemporary research, both in the humanities and in media and computer science. According to this scenario, we are constantly immersed in a situation: in learning (Beck et al. 2018); in media (Lombard et al. 2015); in technology (Moser 1996); in power (Schatz 2009); in video games (Calleja 2011); in literature (Ryan 2011; Wolf et al. 2013); in theater (Frieze 2016); in visual arts (Grau 2003; Liptay and Dogramaci 2016).

Immersion is such a powerful concept that scholars have started to retrospectively apply it to past cultures, like in the case of nineteenth-century American literature (Jarenski 2015), or even of medieval times (Bleumer 2012). Operations of this kind should not be easily dismissed as anachronistic. The desire to enter an alternative world, another dimension parallel to the actual one, is very ancient, even ancestral, as attested in timeless myths.

The most famous immersive myth is the legend of Narcissus—a veritable proto-immersant—who fell in love with his image reflected by the water. As any other mythical narration, this legend knows different versions. The most popular is the version narrated by Ovid in the *Metamorphoses* (III, 339–510: 1971, 149–61): the young man, frustrated by his vain efforts to embrace his image, perishes on the bank, and the corruption of his body gives birth to the beautiful white and yellow flowers that carry the name of narcissus. In a less frequently cited version, adopted among others by Plotinus (*Enneads* l, 6, 8: 1989, 257), Narcissus dives into the water in order to join his image, and he drowns: a destiny that reveals the mortal risks connected to the nullification of the threshold separating image and reality and hints at the structural co-belonging of desire and fear.

Far from being limited to the Western cultural tradition, this desire of entering the image seems to be a transcultural one, as suggested by anecdotes regarding Chinese painting. In one of the most famous, the painter Wu Daozi disappears in the painting he has realized for the emperor's palace (Lindqvist 2012). This motif has been analogously varied in a literary form: trespassing on the threshold between the real and the fictional worlds, Wan Hu-Chen writes himself into his book as a character to join his beloved Li-Fan, whom he has already transformed into a literary figure (Balázs 2010).

Contemporary immersive virtual reality devices seem therefore to correspond to a very ancient—and possibly universally anthropological—desire, which needs to be understood from an archaeological and genealogical perspective. Media archaeologist Erkki Huhtamo has strongly warned against the risk of adopting a universalistic, 'continuistic,' and linear approach, considering the quest for immersion as an anthropological and a-historical structure. Nevertheless, his intention to address it from a historical perspective does not seem to be able to avoid this risk at all. Speaking of 'tracing and comparing some of its manifestations' (Huhtamo 1995, 161) always implies that a core is stable under the different historical concretizations, which share a Wittgensteinian family resemblance.

Over the centuries, the immersive drive has constituted a fundamental challenge to the very notion of 're-presentation' in literature, theater, and visual arts. It is a multifarious trend that seems to find a common denominator in the attempt to defy the traditional framing devices (be it the frame of a painting, the pedestal of a statue, the curtain of the theatrical stage, the screen of a moving theater or of a computer), operating through strategies which transform re-presentation into an experience of 'presence,' of 'being there' (Riva et al. 2003). Such 'unframedness' and 'presentness' go along with an effect of 'immediateness,' namely of transparency of the medium, which is paradoxically obtained through highly sophisticated mediatizations. As a result, these dynamics end up in what could be defined an 'environmentalization' of the fictional world: the representational sphere seems to exist in a seamless space-time continuity with the real spatio-temporal world of the user, it becomes an environment which offers affordances and allows agencies just like the real one.

Being aware of the fact that this desire has an ancient history does not mean denying that our times seem to be particularly involved in it, to an unprecedented extent and scale: a massive immersive drive, whose quantitative extension promises to produce qualitative effects as well. In the 1930s, Walter Benjamin had spoken in a truly prophetic way of a "tactile reception" (Benjamin 1935, 40) of the images, promoted by the invention of photography and cinema. Nowadays, digital natives grow up interacting with touch screens from their youngest age (what has been called the 'touch-screen generation': Rosin 2013). The experience with images is being radically transformed, because nowadays an image that cannot be modified (in dimension, orientation, detail zoom) by the action of our fingers is no longer able to attract the attention of children: digital (from Latin *digitus*: finger) images go literally hand in hand with finger users. But in relation to immersive virtual environments, these children are still 'immigrants' who need to acquire familiarity with the immersive transformations of sensibility. However, given the rapid pace of technological development and the huge amount of economic and scientific investment on a global scale, *immersive natives* are to be expected in the very near future. In a techno-enthusiastically inspired mood, Frank Steinicke has written:

> As an immersive immigrant and young father, I would love to welcome also future generations of immersive natives. I can only imagine how exciting it must be to grow up in a world in which the exploration of virtual environments or the seamless merging of digital information and physical reality is totally natural and unremarkable (Steinicke 2016, xiii).

It is highly disputable whether such environments will be able to 'contribute to a better world,' as he maintains: techno-apocalyptic thinkers would argue exactly the contrary. Keeping us far from both technophilia and technophobia, what is sure is that a critical understanding of such a process is urgently needed since it is likely to rapidly and radically transform our experience of iconic worlds and our experience in general.

Avatars: shifting-identity operators

In the dynamic of immersion and presence a crucial role is played by 'avatars': graphical representations functioning as digital proxies through which the users of the Internet, of a cyber-community or a computer-based interface (like in the case of video games), interact with synthetic objects or other avatars in the virtual world.

In its referring to a surrogate or representative of one's identity, the notion of the avatar belongs to a wider constellation of cognate concepts, which embraces analogous figures such as the 'double' (Rank 1914), the 'alter ego' (Cooper et al. 2007), the 'Doppelgänger' (Ascher Barnstone 2016), the 'hologram' (Johnston 2016). The functional spectrum of the avatar spans from the revelation of the self (a sort of truthful self-portrait) to the disguising deceit of the mask: under such an extremely wide range of possibilities, it acts as a powerful identity operator which allows for a virtually infinite number of negotiations of selfhood. The avatar allows for what has been called 'auto-empathy' (Tordo 2013): the possibility of empathizing with the other-in-me, of taking my perspective on the world as seen from an external point of view.

Far from being one-way access from the real into the digital domain, avatars constitute *two-way* mediators which permit interventions from the virtual onto the real world as well. As reported in recent neurocognitive research, avatars seem to be able to impact actual reality, modifying for instance gender and racial biases via the elicitation of a full-body ownership illusion: assuming an avatar of different gender (for instance in the case of domestic violence: Seinfeld et al. 2018) or putting oneself in a different skin color (Peck et al. 2013; Hasler et al. 2017) within the virtual environment can reverberate in the real world as well, helping to correct social stereotypes and prejudices through the adoption of an alternative perspective-taking and the experience of empathizing processes. Low-level bodily representations impact on high-level attitudes and beliefs. As a group of scholars intensely working on these issues brilliantly put it, 'Changing bodies changes minds' (Maister et al. 2015). Moreover, the recourse to avatars has proved to be particularly significant in clinical investigation of pathological syndromes, like in the case of assessing BID (Body Image Disturbance) connected to anorexia and other eating disorders (Mölbert 2018; Serino et al. 2018), or PTSD (Post-Traumatic Stress Disorder) symptoms (Myers et al. 2016).

In the domain of new media arts, the recourse to avatars massively involves both production and reception. This process goes evidently hand in hand with the mainstream tendency to immersivity: not only are exhibitions on immersive art organized (like *The Art of Immersion* at the Center for Art and Media in Karlsruhe, 2017–19, which displayed immersive digital artworks and performances), but exhibitions themselves turn to immersivity in their display modalities and presentation strategies, so that even Van Gogh can become 'immersive' (https://vangoghimmersion.com/ or www.imagine-vangogh.com/). Considering this trend, it is not surprising that avatars as well play a remarkable role in this scenario, which has been exemplarily modeled by the experiences conducted in the virtual world of Second Life (Doyle 2015).

Cao Fei is a Chinese multimedia artist who has documented in a video entitled *i.Mirror* (2007) the life of her avatar China Tracy in Second Life. In the same year and also in SL, Cao planned and developed RMB City, a virtual city offering a platform for creative activities and experimentations on the threshold between virtual and physical existence, and virtual art exhibits. This hybrid entity—both artwork in itself and institutionalized virtual space to display other digital artworks—was acquired by the Guggenheim Museum in 2008 (https://www.guggenheim.org/artwork/23251).

Also involved in SL, Canadian artist Jon Rafman defines himself as dominated by a 'flâneur-like gaze' while approaching via his avatar in his *Kool-Aid Man in Second Life* (2008–11: http://koolaidmaninsecondlife.com) the virtual subcultures proliferating in this parallel cyberworld 'as if I were a passing explorer or an amateur anthropologist' (Sanchez 2014). His recent work *Dream Journal* (2015–16: https://vimeo.com/jonrafman/review/179476655/1552a7d383)—a computer-generated imagery which aims to give visual expression to his oneiric fantasies—allows us to draw a crucial distinction: although one might be tempted for stylistic reasons to designate the digital beings crowding this CG animated film as 'avatars,' they are merely CG animated characters which lack the necessary correlation with a real being for whom they operate as proxies. Avatars in VR are a sub-class of CG figures, identified by their relationship with the real subject they represent.

However, such a fundamental correlation does not need to be conceived of as a human master vs. digital avatar relationship, nor in a univocal hierarchical direction, as if the controlling subject were necessarily the human master and the controlled entity the digital proxy: mixing computer-generated art with videogame stylemes, American artist Ian Cheng subverts both premises. Expressing his fondness for his pet dog Mars, he has introduced in his works canine avatars: of the Corgi breed, like in the case of *Bad Corgi* (a 2015 Serpentine Digital Commission, which has also become an app for smartphones: http://getbadcorgiapp.com), and of the Shiba Inu breed, like in the case of *Emissary Forks at Perfection* (2015–16), a work which comprises the 'Emissary' trilogy together with *Emissary in the Squat of Gods* (2015) and *Emissary Forks For You* (2016). Staging a sort of cyber-Hegelian master-slave dialectic, Ian Cheng inverts the conventional role and obliges the human users, equipped with Google Tango tablets, to follow the canine avatar in its wanderings through artificial ecosystems, thus reversing the conventional human-animal power relationship (Kerr 2016).

Negotiation of authorship can assume various forms: a multiple-use name, like in the case of the collective author of literary texts and musical works Luther Blissett; or a pseudonym to preserve the anonymity of an individual author, like in the case of the Italian writer Elena Ferrante or of the British artist Banksy. In none of these cases is the visual identity of the subject accessible. On the contrary, the virtual world can allow for both anonymity and visibility, like in the case of LaTurbo Avedon (https://turboavedon.com), an avatar operating since 2008 both as an artist and as a curator intensely active on social media platforms, whose screen name can at the same time be associated with a face depicted in different self-portraits and accurately protects the real identity of the human(s) behind it. The first virtual resident of the Somerset House Studios in London, LaTurbo Avedon, also paradoxically displayed her work in the solo exhibition *New Sculpt* at the real-life gallery Transfer in New York in 2013. After much investigation about the presence of the real in the virtual, this case calls for a study of the presence of the virtual in the real: 'Will you be "there" at the opening?,' asked an interviewer. And she replied, 'As much as I can be! I'll be chatting with everyone during the reception using various devices, but as you can imagine, I have my limitations' (Palop 2013).

Given the fundamental correlation between the production and reception of artworks, it is not surprising that avatarization impacts spectatorship as well. This process belongs to a wider phenomenon of virtualization of artistic institutions, thanks to which IRL (In Real Life) existing museums and galleries increasingly offer the possibility of online virtual explorations of their collections. A particularly massive case is represented by 'Google Arts & Culture' (https://artsandculture.google.com; formerly 'Google Art Project'), a kind of titanic on-line meta-museum bringing together hundreds of international institutions. Launched in 2011, and employing an indoor version of the Google Street View 360-degree camera system which allows the capturing of images at an ultra-high giga-pixel resolution, the project invites visitors to walk through the galleries, zoom in to focus on details at a microscopic level, and also create a personal virtual collection of artworks ('Be your own curator') selected from different locations: Malraux's *musée imaginaire* finds here its most sophisticated accomplishment.

In this context, a specific sub-class is the case in which the spectator is represented by its avatar performing as the visitor's surrogate during the virtual aesthetic experience. Once again, Second Life has been a pioneering space that has permitted the proliferation of a high number of exhibit locations (https://secondlife.com/destinations/arts). But what happens to our aesthetic experience when we go beyond the isolated individual on-line surfing and we get to share our emotions in co-presence and inter-communication with other avatars? Does this virtual setting specifically and qualitatively affect this particular lived event? This has been the object of a recent interdisciplinary experiment entitled *Art Distance Sharing* (Garnier et al. 2017), designed to evaluate the neural, cognitive, and perceptive processes elicited when several users, who are physically distant but present together in the digital space, share via their avatars the experience of artworks. While the data resulting from this experiment are still being analyzed and have not been published yet, the design of the experiment itself attests to the fecundity of this research field, which promises to open up new horizons in VR spectatorship studies as a particular chapter of what is being studied as the 'social life of avatars' (Schroeder 2002).

Notes for a genealogical approach

The list of avatarial artistic practices could be enriched by many other cases. But at this point it is perhaps more useful to adopt a genealogical perspective to understand the cultural and historical background of this phenomenon, whose roots can be traced back to very ancient times, starting from the very term 'avatar' itself.

Far from being a term coined in contemporary times, the notion of 'avatar' is rooted in the ancient Hinduist tradition: the Sanskrit term *avatāra* refers to the descent on earth, the material appearance, the sensible manifestation of a god or goddess, mostly Vishnu, who intervenes in earthly affairs to reestablish the cosmic order. Forms assumed by the deity can vary according to the circumstances: among those embodied by Vishnu we find the fish, the tortoise, the boar, the man-lion, the dwarf, the Buddha. Therefore, each *avatāra* is to be considered only a partial manifestation of the corresponding deity it visually exhibits. After accomplishing the task assigned, the *avatāra* merges back into its deity (Hacker 1960; Kinsley 1987). From a comparative point of view, the Hinduist idea expressed by the *avatāra* has been associated with other similar representations of the divine appearance, like the Christian 'incarnation' (Parrinder 1997; Sheth 2002).

Although its first occurrences in Western languages can be traced back to the eighteenth century, and although it appeared as the title of a short novel by Théophile Gautier in 1856, the fortune of the term is due to its massive employment in the domain of video games (starting from *Avatar*, developed in 1979), chat-rooms, social networks, and more generally in digital communication. The use of the word has been dramatically increasing in recent years, also thanks to the enormous success of James Cameron's film *Avatar*, released in 2009 and shot with sophisticated stereoscopic techniques (Depraz 2012).

Just as the immersive drive can be traced back to the a-historical dimension of myth, the avatarial status in the broadest sense of the term—namely intended as the surrogate presence of the image-maker/user immersed in their own image—can as well be detected as a strategy which has been in force since ancestral times, if we consider the case of the representation of hands during the Upper Paleolithic Period. The techniques vary: the hands could be either painted (mostly in red, white, or black pigment) and then applied to the surface of the cave (*positive* handprint), or the hand was placed on the rock and its silhouette was emphasized by spraying pigment through a tube, or by spitting the paint from the mouth, or even by painting around the hand's contours with a brush (*negative* hand stencil). The oldest stencil so far was found on the Indonesian island of Sulawesi, dating to 37,900 BCE (Aubert et al. 2014).

Although we do not know much of the ultimate reasons which motivated such manufactures, it does not appear far-fetched to read them as partial avatars of the prehistoric self, iconic representatives of the human body in its archaic identity which could allow self-recognition and at the same time self-duplication. As a sort of archaic prototype of Jacques Lacan's mirror stage (with which they share the lateral reversal, so that the right model-hand appears on the rock as a left picture-hand, and vice versa), handprints and hand stencils might be intended as figures that transform the opaque surface into a specular surface in which to reflect oneself. The flickering lights of the torches must have added the proper cinematic dynamization to the whole.

From such mirroring to the practice of self-portraits and selfies (Mirzoeff 2016) is perhaps not too much of a leap: it certainly is from the point of view of chronology, but it is just one step away from a conceptual perspective. One of the first self-portraits usually mentioned in the histories of the genre (Hall 2014) was realized in quartzite by the Pharaoh Akhenaten's chief sculptor Bak in ca. 1353–35 BC (at the Ägyptisches Museum und Papyrussammlung in Berlin): it represents the sculptor together with his wife, both watching us in a frontal pose.

Hans-Georg Gadamer claimed that we can instinctively tell two depictions of a human face apart, distinguishing which is a portrait and which is a simple representation of a generic human being, even if we do not know the person portrayed (Gadamer 1960, 140). Could we say the same for the distinction between portraits and self-portraits? This is a much harder claim to defend, but certainly it is the 'self' implied in self-portraiture which constitutes its avatarial status, its 'mineness': avatars necessarily need the reflection of the identity, its doubling, which allows both the split between the self and its representation on the one hand and their connection or even identification on the other. In this sense, frontality rather than profile seems to constitute the 'symbolic form' specific for an avatarial self-representation. As Meyer Schapiro pointed out in a brilliant essay,

> The profile face is detached from the viewer and belongs with the body in action (or in an intransitive state) in a space shared with other profiles on the surface of the image. It is, broadly speaking, like the grammatical form of the third person, the impersonal 'he' or 'she' with its concordantly inflected verb; while the face turned outwards is credited with intentness, a latent or potential glance directed to the observer, and corresponds to the role of 'I' in speech, with its complementary 'you'. It seems to exist both for us and for itself in a space virtually continuous with our own, and is therefore appropriate to the figure as symbol or as carrier of a message (Schapiro 1973, 38–39).

In the pre-virtual age, to establish an avatarial relationship with my iconic surrogate, the 'I' outside the image needs to address the other 'I' within the image in an 'I-You' connection to elicit a mirror-like self-recognition. This requirement has probably contributed to the statistically higher number of frontal or semi-profile self-portraits with respect to the profile ones: a constraint which the invention of photography made less impelling, and which has significantly been revamped by the practice of selfies, because the subject needs to frontally address the camera in order to control the framing.

A particularly interesting case is offered by the sub-genre of the role-playing self-portrait (Hall 2014, 233–95; Moran 2017, 82–132), in which the self is portrayed as an 'other': famous examples are the alleged self-portrait of Michelangelo depicted in Saint Bartholomew's flayed skin in *The Last Judgment*, Rembrandt's self-portraits as an oriental potentate or as the apostle Paul; in contemporary times, the series realized by Cindy Sherman and Yasumasa Morimura. This sub-genre reminds us of the ancipital nature of the avatar: while anchored on the one side in the real subject as its representative, it allows for shifting modulations of its very identity.

A remarkable exception to the frontal self-representation—or rather a complementary case—is represented by the so-called *Rückenfigur*: the figure seen from behind which has been made popular thanks to Caspar David Friedrich's romantic landscape paintings displaying characters depicted from the back, like the *Woman before the Setting Sun*, or *Wanderer above the Sea of Fog* (both painted in 1818), or the *Woman at a Window*, 1822 (Böhme 2006). This compositional choice has exerted a significative influence on many artists (for example on Edvard Munch or on Gerhard Richter). The origins of this *topos* are also very ancient: the presence of figures from the back in Paleolithic paintings is disputable, due to the high level of abstraction and stylization (Wilks 2005, 18); but it is incontrovertibly attested to in ancient mosaics, frescoes, and vase paintings (Koch 1965).

Performing a 'hinge function' (Prange 2010, 140), the figure from behind appears as a veritable dialectical entity, whose effects are paradoxical and oxymoronic: on the one hand, by turning its back on the beholder it seems to ensure the tight closure of the iconic space, separating it from the real space in which the observer stands; on the other hand, it makes permeable and passable the threshold between the interior fictional space of the image and the exterior real space, offering the possibility of a 'reflection' of the gaze of the external observer in the gaze of the depicted figure (Krüger 1995, 156), and inviting the spectator to empathetically merge with the depicted character and to look at the landscape from its vantage point.

By making full use of the technical possibilities offered by the photographic device, in 1971 Giovanni Anselmo (an Italian artist belonging to the 'Arte povera' movement) pointed his camera in the direction of a meadow, and adjusted the self-timer shooting in order to allow himself to run in front of the lens and literally 'enter the work' (as the title of this photograph suggests: *Entrare nell'opera*). In this way, being captured by the camera from the back, Anselmo realized a perfect combination of self-portrait and *Rückenfigur*.

As has been convincingly shown (Beil 2012, 131–70), the iconographic tradition of the *Rückenfigur* has deeply influenced the representation of the third-person perspective avatars in contemporary video games.

Avatarizing the gaze

Cinema has inherited many representational strategies from the history of visual arts, including the recourse to partial and full figures from behind, with different variations on this theme in diverse cinematographic traditions—classical, modernist, and realist (see Kirsten 2011; Thomas 2013). One might even conceive of the 'subjective' or 'point-of-view shot' (Branigan 1984, 103–21; Casetti 1985; Dagrada 2013)—in which we see exactly what the character's eyes see—as the result of the spectator's progressive approach to the *Rückenfigur* up to the complete incorporation of the former into the latter; an incorporation which allows a full identification of the three gazes: of the camera, of the character, and of the spectator.

The 'over-the-shoulder shot' (OTS), a shot taken from the camera angle established by the shoulder of a character (in which a part of the back and side of the head is included either at the right or at the left side of the frame and occupies more or less one third of it), constitutes an intermediate stage between the 'objective' and the 'subjective' shots. The partial head of the character operates as a partial avatar of the spectator, who is invited to assume the point of view of the character.

The embodying effect performed by the subjective shot is further intensified by what has been called the 'first-person shot' (Eugeni 2012): a shot made possible by technological innovations such as the Steadicam, hand, and helmet digital cameras, surveillance sensors, and increasingly fluid and realistic video games (especially the genres of the 'shoot 'em up' and the vehicle simulators). A paradigmatic example is offered by *Hardcore Henry*, a 2015 film directed and interpreted by Ilya Naishuller, which was entirely shot with a GoPro.

In the first-person shot recording the event, what is important is not so much the event itself: what matters is rather the fact of conveying the presence of the filmers in the scene, their neurophysiological reactions to the world, in what has been happily defined as a 'somatic image' produced by a 'body camera' (Bégin 2016). An invitation to somatically empathize is thus offered by this kind of shooting, which rests on the idea of a shared experience. The device itself undergoes a sort of 'avatarization,' becoming the substitute of the experiencer and generating a kind of mobile self-portrait, a dynamic *proxy*. Nick Paumgarten describes his reaction to the viewing of a GoPro shooting taken by his ten-year-old son while skiing as follows:

> I didn't need a camera to show me what he looked like to the world, but was delighted to find one that could show me what the world looked like to him. It captured him better than any camera pointed at him could. This was a proxy, of sorts (Paumgarten 2014).

The most radical version of such somatic first-person empathizing has been offered by *Strange Days*, a 1995 science fiction thriller film written by James Cameron and Jay Cocks and directed by Kathryn Bigelow. The story revolves around the SQUID, an illegal electronic head-mounted device capable of recording physical sensations directly from the wearer's cerebral cortex during the actual experience onto a MiniDisc-like support. The device subsequently allows playback so that the user sees and feels what was experienced by the recording subject, feeding a black market of scenes of violence, sex, rape, and murder. Although the setting of the movie was the city of Los Angeles in 1999, the integrally reproducible experience imagined by *Strange Days* seems to be picked up nowadays by the efforts accomplished in the domain of virtual reality and of the so-called 'expanded cinema' (according to the label coined by Youngblood as early as 1970) in order to obtain an increasingly multimodal and multisensory engagement with images by gradually involving the whole domain of human sensibility in order to mimic real-life perception. This process has led to the development of immersive sound systems (Török et al. 2015), to the integration of haptic stimuli (Paterson 2007), and to the involvement of smell and taste (Ischer et al. 2014; Ranasinghe and Yu-Luen Do 2016).

What is suggested by the above-mentioned cases—the subjective shot, the first-person shot, and the total-empathy shot—is that the experience of the avatar as the experience of an immersive presence of the self in the iconic environment is not necessarily linked with the perception of a visible figure representative of the self, but rather with a particular articulation of the gaze, of the perspective on the experienced world as representative of selfhood. Moreover, a GoPro can be installed on non-human entities: a bird, a drone, a boat (like in the paradigmatic case of *Leviathan*, a 2012 ethnographic documentary directed by Lucien Castaing-Taylor and Verena Paravel, and shot with GoPros in the North Atlantic, which revolves around the different gazes—of fishers, fish, and boat—intertwining on board a commercial fishing vessel: Pavsek 2015; Unger 2017). This simple fact induces the enlargement of the notion of gaze—and consequently the notions of selfhood and of the avatar, too—to comprehend also non-human entities.

Given the above-mentioned tendency to complement the audio-visual perception with multisensory and multimodal integrations, the notion of 'gaze' itself should be finally intended as an expanded one, eventually coinciding with the notion of 'experience' tout court. Correlatively, the subject connected to such integration is no longer a 'spectator' but rather an 'experiencer.'

Acknowledgements

This project has received funding from the European Research Council (ERC) under the European Union's Horizon 2020 research and innovation program (grant agreement No. [834033 AN-ICON]).

References

Ascher Barnstone, Deborah, ed. 2016. *The Doppelgänger*. Bern: Peter Lang.

Aubert, Maxime, et al. 2014. "Pleistocene Cave Art from Sulawesi, Indonesia." *Nature* 514: 223–27.

Balázs, Béla, ed. 2010. "The Book of Wan-Hu-Chen." In *The Cloak of Dreams. Chinese Fairy Tales*. Translated by Jack Zipes, 159–71. Princeton NJ: Princeton University Press.

Beck, Dennis et al. 2018. *Immersive Learning Research Network*. Cham: Springer.

Bégin, Richard. 2016. "GoPro: Augmented Bodies, Somatic Images." Translated by Timothy Barnard. In *Screens*, edited by Dominique Chateau and José Moure, 107–15. Amsterdam: Amsterdam University Press.

Beil, Benjamin. 2012. *Avatarbilder. Zur Bildlichkeit des zeitgenössischen Computerspiels*. Bielefeld: transcript.

Benjamin, Walter. 1935–36. "The Work of Art in the Age of Its Technological Reproducibility," 19–55. In *The Work of Art in the Age of Its Technological Reproducibility, and Other Writings on Media*, edited by Michael W. Jennings et al. Cambridge MA: The Belknap Press of Harvard University Press, 2008.

Bleumer, Hartmut, ed. 2012. "Immersion im Mittelalter." *Zeitschrift für Literaturwissenschaft und Linguistik* 42, no. 3.

Böhme, Hartmut. 2006. "Rückenfiguren bei Caspar David Friedrich." In *Caspar David Friedrich. Deutungen im Dialog*, edited by Gisela Greve, 49–94. Tübingen: Discord.

Branigan, Edward. 1984. *Point of View in the Cinema. A Theory of Narration and Subjectivity in Classical Film*. New York: Mouton.

Calleja, Gordon. 2011. *In-Game. From Immersion to Incorporation*, Cambridge MA: The MIT Press.

Casetti, Francesco. 1985. *Inside the Gaze: The Fiction Film and Its Spectator*. Translated by Nell Andrew with Charles O'Brien. Bloomington IN: Indiana University Press, 1998.

Cooper, Robbie, et al. 2007. *Alter Ego: Avatars and their Creators*. London: Chris Boot.

Dagrada, Elena. 2013. *Between the Eye and the World. The Emergence of the Point-of-View Shot*. Bruxelles: Peter Lang.

Depraz, Natalie. 2012. *Avatar "Je te vois". Une expérience philosophique*. Paris: Ellipses.

Dogramaci, Burcu, and Fabienne Liptay, eds. 2016. *Immersion in the Visual Arts and Media*. Leiden: Brill.

Doyle, Denise. 2015. *New Opportunities for Artistic Practice in Virtual Worlds*. Hershey PA: Information Science Reference.

Eugeni, Ruggero. 2012. "First Person Shot. New Forms of Subjectivity between Cinema and Intermedia Networks." *Anàlisi* 2–3: 19–31.

Frieze, James, ed. 2016. *Reframing Immersive Theatre. The Politics and Pragmatics of Participatory Performance*. London: Palgrave Macmillan.

Gadamer, Hans-Georg. 1960. *Truth and Method*. Translated by Joel Weinsheimer and Donald G. Marshall. London: Continuum, 2004.

Garnier, François, et al. 2017. "Art – Distance Sharing: A Virtual 3D paradigm for the study of the influence of co-presence through avatars on the emotional perception of fine arts." *VRIC '17. Proceedings of the Virtual Reality International Conference – Laval Virtual 2017*, no. 11. doi>10.1145/3110292.3110304.

Gautier, Théophile. 1856. "Avatar." In *Clarimonde and Other Stories*. Leyburn: Tartarus Press, 2011.

Grau, Oliver. 2003. *Virtual Art. From Illusion to Immersion*. Translated by Gloria Custance. Cambridge MA: The MIT Press.

Hacker, Paul. 1960. "Zur Entwicklung der Avatāralehre." *Wiener Zeitschrift für die Kunde Süd- und Ostasiens* 4: 47–70.

Hall, James. 2014. *The Self-Portrait. A Cultural History*. London: Thames & Hudson.

Hasler, Béatrice S., et al. 2017. "Virtual Race Transformation Reverses Racial In-Group Bias." *PLoS One* 12, no. 4. https://doi.org/10.1371/journal.pone.0174965.

Huhtamo, Erkki. 1995. "Encapsulated Bodies in Motion: Simulators and the Quest for Total Immersion." In *Critical Issues in Electronic Media*, edited by Simon Penny, 159–86. Albany NY: State University of New York Press.

Ischer, Matthieu, et al. 2014. "How Incorporation of Scents Could Enhance Immersive Virtual Experiences." *Frontiers in Psychology* 5: 1–11.

Jarenski, Michelle. 2015. *Immersive Words: Mass Media, Visuality, and American Literature, 1839–1893*. Tuscaloosa AL: The University of Alabama Press.

Johnston, Sean F. 2016. *Holograms: A Cultural History*. Oxford: Oxford University Press.

Kerr, Dylan. 2016. "Artificial Ecology: Ian Cheng on the Strange Art of Simulating Life, and the Conceptual Merits of Pokémon Go." *Artspace*, August 26, 2016. https://www.artspace.com/magazine/interviews_features/art-bytes/ian-cheng-interview-54128.

Kinsley, David. 1987. "Avatāra." In *Encyclopedia of Religion*. Second Edition, edited by Lindsay Jones, Vol. 2, 707–08. Farmington Hills MI: Thomson Gale.

Kirsten, Guido. 2011. "Zur Rückenfigur im Spielfilm." *Montage AV* 20, no. 2: 103–24.

Koch, Margarete. 1965. *Die Rückenfigur im Bild von der Antike bis zu Giotto*. Recklinghausen: Bongers.

Krüger, Klaus. 1995. "Der Blick ins Innere des Bildes. Ästhetische Illusion bei Gerhard Richter." In *Zur Eigensinnlichkeit der Bilder*, edited by Matthias Weiß et al., 155–202. Paderborn: Fink, 2017.

Lindqvist, Sven. 2012. *The Myth of Wu Tao-Tzu*. Translated by Joan Tate. London: Granta.

Lombard, Matthew, et al., eds. 2015. *Immersed in Media Telepresence Theory, Measurement & Technology*. Cham: Springer.

Maister, Lara, et al. 2015. "Changing Bodies Changes Minds: Owning Another Body Affects Social Cognition." *Trends in Cognitive Sciences* 19, no. 1: 6–12.

Mirzoeff, Nicholas. 2016. *How to See the World: an Introduction to Images, from Self-Portraits to Selfies, Maps to Movies, and More*. New York: Basic Books.

Mölbert, Simone C., et al. 2018. "Assessing Body Image in Anorexia Nervosa Using Biometric Self-Avatars in Virtual Reality: Attitudinal Components Rather Than Visual Body Size Estimation Are Distorted." *Psychological Medicine* 48, no. 4: 642–53.

Moran, Claire. 2017. *Staging the Artist: Performance and the Self-Portrait from Realism to Expressionism*. London: Routledge.

Moser, Mary Anne, ed. 1996. *Immersed in Technology. Art and Virtual Environments*. Cambridge MA: The MIT Press.

Myers, Catherine E., et al. 2016. "Beyond Symptom Self-Report: Use of a Computer 'Avatar' to Assess Post-Traumatic Stress Disorder (PTSD) Symptoms." *Stress* 19, NO. 6: 1–19.

Ovid. 1971. *Metamorphoses*, Books 1–8. Translated by Frank J. Miller. Cambridge MA: Harvard University Press.

Palop, Benoit. 2013. "What Kind of Art Does an Avatar Make? Q&A with LaTurbo Avedon." *Vice*, May 24, 2013. https://www.vice.com/en_au/article/bmdw4a/qa-with-laturbo-avedon.

Parrinder, Edward G. 1997. *Avatar and Incarnation: The Divine in Human Form in the World's Religions*. Oxford: Oneworld.

Paterson, Mark. 2007. *The Senses of Touch. Haptics, Affects and Technologies*. Oxford: Berg.

Paumgarten, Nick. 2014. "We are a Camera. Experience and Memory in the Age of GoPro." *The New Yorker*, September 22, 2014. http://www.newyorker.com/magazine/2014/09/22/camera.

Pavsek, Christopher. 2015. "*Leviathan* and the Experience of Sensory Ethnography." *Visual Anthropology Review* 31, no. 1: 4–11.

Peck, Tabitha C., et al. 2013. "Putting Yourself in the Skin of a Black Avatar Reduces Implicit Racial Bias." *Consciousness and Cognition* 22, no. 3: 779–87.

Plotinus. 1989. *Enneads*, vol. 1. Translated by Arthur H. Armstrong. Cambridge MA: Harvard University Press.

Prange, Regine. 2010. "Sinnoffenheit und Sinnverneinung als metapicturale Prinzipien. Zur Historizität bildlicher Selbstreferenz am Beispiel der Rückenfigur." In *Ambiguität in der Kunst. Typen und Funktionen eines ästhetischen Paradigmas*, edited by Verena Krieger and Rachel Mader, 125–67. Köln: Böhlau.

Ranasinghe, Nimesha, and Ellen Yu-Luen Do. 2016: "Virtual Sweet. Simulating Sweet Sensation Using Thermal Stimulation on the Tip of the Tongue." *UIST '16 Adjunct Proceedings of the 29th Annual Symposium on User Interface Software and Technology*, 127–28.

Rank, Otto. 1914. *The Double: A Psychoanalytic Study*. Translated by Harry Tucker Jr. Chapel Hill NC: University of North Carolina, 1971.

Riva, Giuseppe, et al., eds. 2003. *Being There. Concepts, Effects and Measurements of User Presence in Synthetic Environments*. Amsterdam: IOS Press.

Rosin, Hanna. 2013. "The Touch-Screen Generation." *The Atlantic*, April 2013. https://www.theatlantic.com/magazine/archive/2013/04/the-touch-screen-generation/309250/.

Ryan, Marie-Laure. 2011. *Narrative as Virtual Reality. Immersion and Interactivity in Literature and Electronic Media*. Baltimore MD: The Johns Hopkins University Press.

Sanchez, Gabriel H. 2014. "Jon Rafman Discusses His Show at the Contemporary Art Museum St. Louis." *Artforum*, July 3. https://www.artforum.com/interviews/jon-rafman-discusses-his-show-at-the-contemporary-art-museum-st-louis-47380.

Schapiro, Meyer. 1973. "Frontal and Profile as Symbolic Forms." In *Words and Pictures: on the Literal and the Symbolic in the Illustration of a Text*, 37–49. The Hague: Mouton.

Schatz, Edward, ed. 2009. *Political Ethnography. What Immersion Contributes to the Study of Power*. Chicago IL: The University of Chicago Press.

Schroeder, Ralph, ed. 2002. *The Social Life of Avatars. Presence and Interaction in Shared Virtual Environments*. London: Springer.

Seinfeld, Sofia, et al. 2018. “Offenders become the victim in virtual reality: impact of changing perspective in domestic violence.” *Scientific Reports* 8, no. 2692: 1–11.

Serino, Silvia, et al. 2018. “From Avatars to Body Swapping: The Use of Virtual Reality for Assessing and Treating Body-Size Distortion in Individuals with Anorexia.” *Journal of Clinical Psychology* 12: 1–10.

Sheth, Noel. 2002. “Hindu Avatāra and Christian Incarnation: A Comparison.” *Philosophy East and West* 52(1): 98–125.

Steinicke, Frank. 2016. *Being Really Virtual. Immersive Natives and the Future of Virtual Reality.* Cham: Springer.

Thomas, Benjamin, ed. 2013. *Tourner le dos. Sur l'envers du personnage au cinema.* Saint Denis: Presses Universitaires de Vincennes.

Tordo, Frédéric, and Caroline Binkley. 2013. “L'auto-empathie médiatisée par l'avatar, une subjectivation de soi.” In *Les avatars jouables des mondes numériques. Théories, terrains et témoignages de pratiques interactives*, edited by Etienne-Armand Amato and Etienne Perény, 91–109. Paris: Lavoisier.

Török, Ágoston, et al. 2015. “It Sounds Real When You See It. Realistic Sound Source Simulation in Multimodal Virtual Environments.” *Journal on Multimodal User Interfaces* 9: 323–31.

Unger, Michael A. 2017. “Castaing-Taylor and Paravel's GoPro Sensorium: *Leviathan* (2012), Experimental Documentary, and Subjective Sounds.” *Journal of Film and Video* 69, No. 3: 3–18.

Wilks, Guntram. 2005. *Das Motiv der Rückenfigur und dessen Bedeutungswandlungen in der deutschen und skandinavischen Malerei zwischen 1800 und der Mitte der 1940er Jahre.* Marburg: Tectum.

Wolf, Werner, et al., eds. 2013. *Immersion and Distance. Aesthetic Illusion in Literature and Other Media.* Amsterdam: Rodopi.

Youngblood, Gene. 1970. *Expanded Cinema.* New York: P. Dutton & Co.

Meeting Yourself in Virtual Reality
A Performative Experiment in Self-Compassion

Daniel Landau

Background

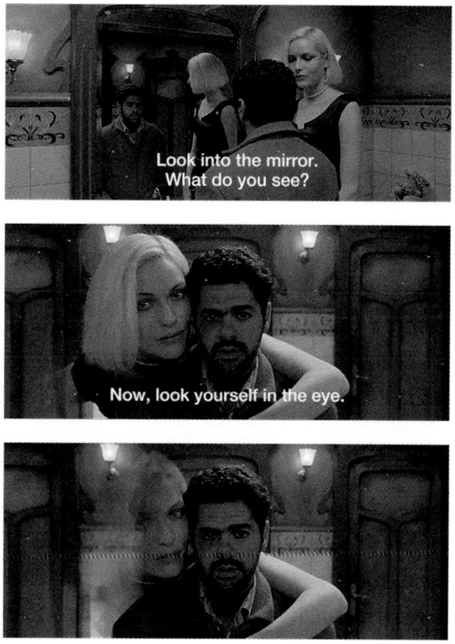

Screen shots from Luc Besson's film, *Angel-A* (2006).

Luc Besson's 2005 film '*Angel-A*' is a self-realization drama telling the story of an Angel '*Angela*,' sent down to teach the protagonist, '*Andre*,' to love himself. In a key scene, we see Angela and Andre standing in front of a mirror with Angela instructing him to look into his own eyes and say 'I love you.' Gradually the image of Angela disappears, and Andre merges his internal and external experiences of self in an intimate and meaningful moment.

Self-reflection is the capacity of humans to exercise introspection and the willingness to learn more about their fundamental nature, purpose, and essence. Between the internal process of self-reflection to the external observation of one's reflection runs a thin line marking the relationship between the private self and the public self.

From Narcissus's pond, through reflective surfaces and mirrors, to current-day selfies, the concepts of self, body-image, and self-awareness have been strongly influenced by the human interaction with physical reflections. In fact, one can say that the evolution of technologies reproducing images of ourselves has played a major role in the evolution of the Self as a construct. With the current wave of Virtual Reality (VR) technology making its early steps as a consumer product, we set out to explore the new ways in which VR technology may impact our concept of self and self-awareness. '*Self Study*' aims to critically explore VR as a significant and novel component in the history and tradition of the complex relationship between technology and the Self.

Introduction

Research in the field of compassion has shown that it is generally easier to express compassion to others than to oneself (Neff 2003). In our research, we use Immersive 360° stereoscopic video (360° 3D) to create an immersive virtual reality environment in which participants experience an illusion of meeting themselves. By using video editing techniques and a live performer to simulate touch, the participants are able to 'look into their own eyes,' to physically 'hold their own hands,' and hear compassionate words spoken to them by themselves. We achieve this interaction by meeting the participants in separate sessions. In the first session, each participant is filmed with a 360° 3D Camera system. During this session, he/she is asked to perform several tasks culminating in sitting in front of the 360° 3D camera, touching the hands of a confederate, and addressing him compassionately as if he were a close friend of the participant. This footage is then edited, and during the second session, which takes place a few days later, it is played back on a head-mounted display (HMD), with a live performer simulating physical touch in sync with the recorded virtual touch the participant sees in the HMD. Now the words that were meant for the close friend are addressing the participants themselves.

Method

We ran this set-up as an art performance at the 2017 B3 Biennial of the Moving Image in Frankfurt. We had 10 participants who attended both sessions. After the experience, we interviewed them using the semi-structured interview technique. Our research goal was to explore the impact of this encounter. In particular, we wanted to see whether this intervention can increase self-compassion and a sense of intimacy within oneself, improve perception of body image, and contribute to one's wellbeing.

In a follow-up study, we want to use a more empirical approach to measure the impact of meeting oneself. We will achieve this by using interpersonal distance as a measure for intimacy (see the appendix for a description).

Experiment procedure

PART 1: RECORDING SESSION

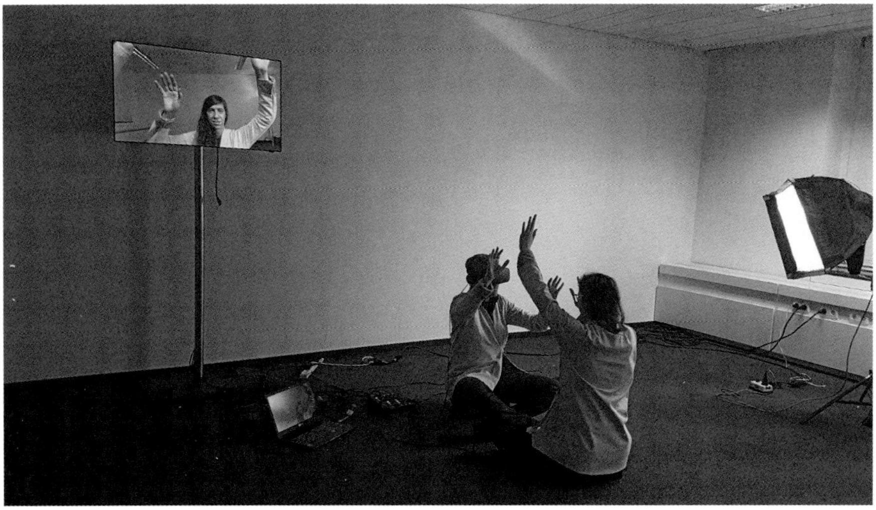

Recording session B3 Festival – November 2017, Frankfurt (Plate 1, p. 305)

1.1 Mirror-Game

The experiment begins with the participant entering the room and being greeted by the experimenter (Fig. 1). A confederate is seated on a mat behind an Immersive 360° stereoscopic video camera (360° 3D Camera). The camera captures the action in the room from the first-person perspective [1PP] of the confederate. The experimenter gives the participant a custom-designed jacket with extendable sleeves. Both the experimenter and the confederate are wearing similar jackets. The experimenter asks the participant to take a deep breath and then invites them to play the 'mirror game,' an improvisation technique in which one person performs a movement sequence and a person positioned in front of them mimics them. The sequence ends with the performer and the participant touching hands. The whole sequence is captured by the 360° 3D Camera.

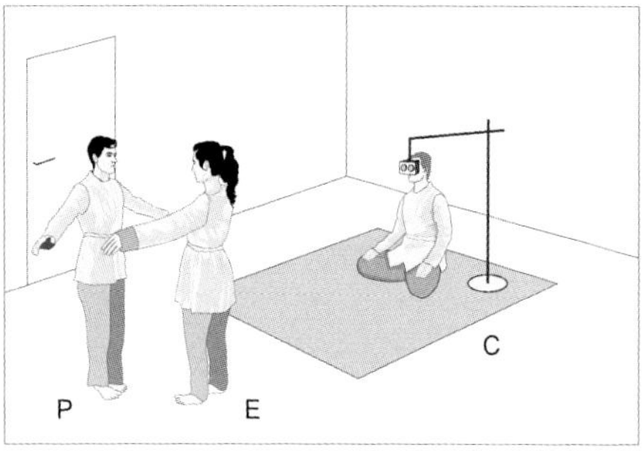

Fig. 1. Participant (P) playing the 'mirror game' with the experimenter (E). The confederate (C) is sitting behind the 360° 3D Camera wearing the custom-made jacket with its sleeves covering his hands. In the second viewing session, the confederate's body will seem to the participant like his own body, because they will be wearing the same jacket and his hands will be covered as well.

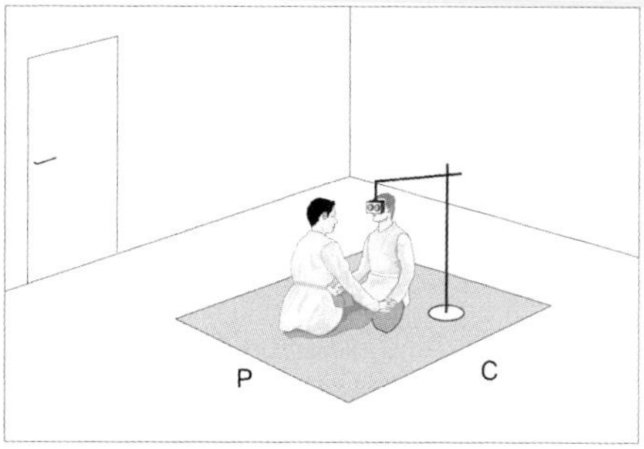

Fig. 2. Participant (P) sitting in front of the confederate (C), holding his hands, and looking into the camera, imagining a close friend, and saying all the good things he feels about his/her friend. This recording will be played back to the participant during the second session.

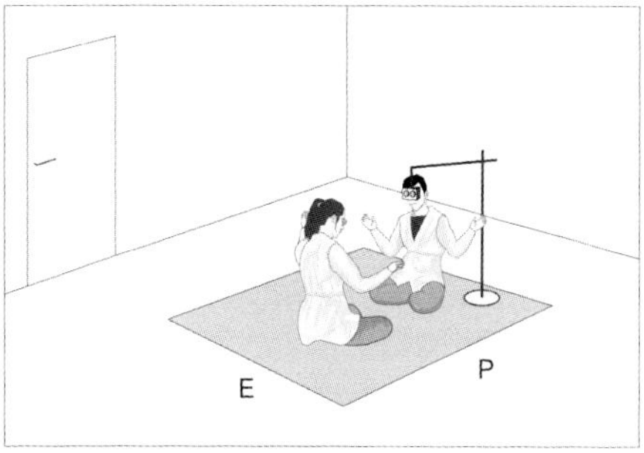

Fig. 3. The participant (P) is seated behind the 360° 3D Camera, playing the mirror game led by the experimenter (E). This recording will be played back in the second session and will establish the 1PP of the participant.

1.2 Compassion towards another

After the 'mirror game,' the participant is instructed to sit in front of the confederate and look into the lens of the camera (Fig. 2). The experimenter then gives the following instructions:

> *'Close your eyes. I want you to imagine that you are sitting in front of a person that you feel very close to. This person is going through a really rough time. Now open your eyes and look into the camera. Place your hands on the hands of the person in front of you. Take a minute, and tell the person that you were thinking of all the things you like about them.*

1.3 Embodying the participant in his own body

After the participant finishes talking to the confederate, he is requested to take a seat behind the 360° 3D Camera (Fig. 3). Once the participant is seated behind the 360° 3D Camera, the experimenter sits in front of him and leads the same movement sequence he did when he was standing. The sequence ends with the experimenter touching the hands of the participant. The purpose of this sequence is to establish a 1PP for the participant in his own body for the viewing session of the experiment. In the viewing session, he will wear an HMD and, looking down, he will see his own body from 1PP. The mirror game is used as a mean to establish agency—moving and seeing his/her own body in the HMD. The touch at the end of the sequence will be another sensory cue helping to intensify the illusion that the body seen in the HMD is the participant's own body.

PART 2: VIEWING SESSION

In the second session the participant views the footage filmed during the first session. This footage is edited, and dubbed with voice-over to achieve two manipulations:

- Body swap: enabling the participants to believe they are seeing themselves in both 1PP and 3PP.
- Interview swap: having the compassionate words they meant for another addressed to themselves.

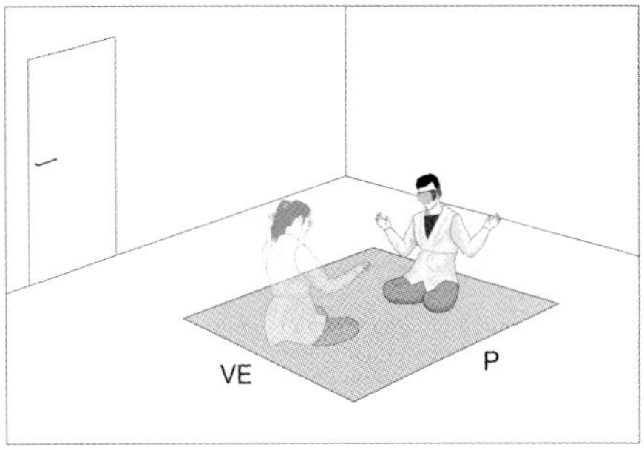

Fig. 4. The participant (P) sees the virtual experimenter (VE) in the HMD and follows her directions in the 'mirror game.' As he physically performs the movements, he will see his own hands in the HMD performing the same sequence as they were recorded in the first session.

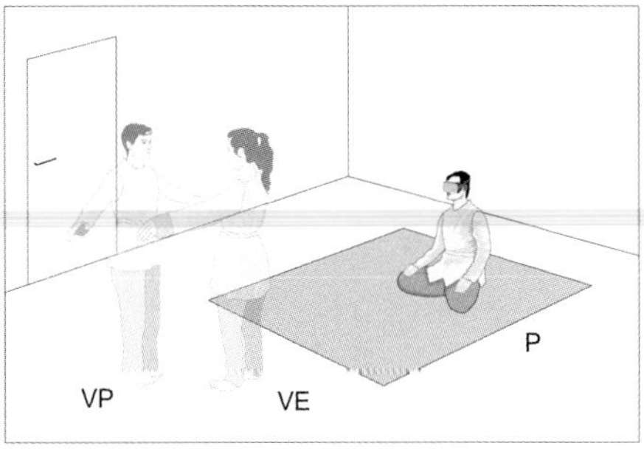

Fig. 5. The participant (P) sees his virtual self (VP) enter the room playing the 'mirror game' with the virtual experimenter (VE).

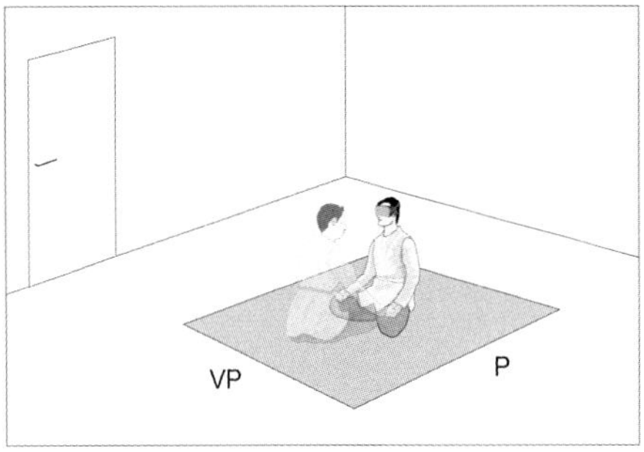

Fig. 6. The participant (P) watching his virtual self (VP) on the HMD. The virtual self is touching his hands while, simultaneously, the experimenter is physically touching the participant's hands, thus creating a visuo-tactile sensory integration.

2.1 Embodied in your own body

The second session begins with the experimenter greeting the participant and giving him the same jacket that he wore in the first session. Next, the participant is seated on the mat, at the same spot the confederate sat in during the first session, which is the perspective whence the entire scenario was filmed. The participant dons an HMD displaying the same room he is currently in. As the participant looks down, the HMD will show him his own physical body from 1PP—as it was filmed in the first session (Fig. 4). Next, the participant views the footage of the experimenter entering the room and sitting in front of him. In the video, the experimenter asks the participant to take a deep breath and to play the 'mirror game' with her (Fig. 4). The participant follows the recorded movements of the experimenter with the sequence ending with the hands of the experimenter and the participant touching. At this point, the live experimenter touches the hands of the participant in sync with the screened touch on the HMD, generating a sensory integration. The goal of this part is to fully establish the illusion of the participant being physically present in the room with visual, auditory, motor, and tactile information contributing to complete this illusion.

2.2 'I' enter the room

Next, the participant hears, via a headset, a voiceover asking him to close his eyes. The live experimenter then unfolds the sleeves of the jacket to cover the participant's hands. When the participant opens his eyes, we will not notice that the 1PP body is now a different body, wearing the same jacket, and since the hands are covered this body-swap is expected to remain unnoticed. The editing of the immersive video footage will enable the participant to see their virtual self entering the room and being greeted by the virtual experimenter (Fig. 5).

2.3 Compassion to self

After the 'mirror game' sequence, the participant will see his avatar sitting in front of him (see figure 6). He will see and feel 'himself' touching his own hands. This viewed touch will occur physically with the live experimenter touching the hands of the participant in perfect sync. Next, the live participant will hear the following voice-over text:

> *'Look at yourself. Look into your eyes. What do you see? Feel the touch on your hands… Now, think about times when you feel bad or are struggling.*
> *How do you typically respond to yourself in these situations?*
> *What do you typically do, what do you say?'*

At this point the pre-recorded self will be 'looking into the live participant's eyes'; the compassionate words that he previously intended for someone else are now addressing him.

Discussion

Our main conclusion is that meeting yourself in virtual reality using 360° 3D is an experience like nothing experienced before. Participants reported that the impact of meeting themselves through the immersive experience of the 'Self Study' intervention was fundamentally more powerful than that of the two related phenomena of (1) seeing oneself in a mirror and (2) watching oneself on video. In particular, participants reported that the human-touch part of the intervention produced a very profound and meaningful moment, especially while hearing their own voice addressing them. Participants reported that the experience felt like a 'real encounter' and they would want to further explore where it could lead. We also found a main effect of body image on how positive participants felt in the experience: The better the participants felt about their own body the more positive the experience felt. These finding are well worth exploring in the empirical follow-up study we describe in the appendix.

References

Banakou, Domna, and Mel Slater. (2014). "Body ownership causes illusory self-attribution of speaking and influences subsequent real speaking." *PNAS*, 111(49), 17678–83.

Blanke, Olaf (2012), "Multisensory brain mechanisms of bodily self-consciousness." *Nature Reviews Neuroscience*, 13, 556–71.

Clark, Andy (1997). *Being There: Putting Brain, Body, and World Together Again*. Cambridge, MA: MIT Press 1997.

Gavrilov-Jerković, Vesna, Veljko Jovanović, Dragan Žuljević, and Dragana Brdarić. (2014). "When less is more: A short version of the personal optimism scale and the self-efficacy optimism scale." *Journal of Happiness Studies*, 15(2), 455–74.

Hasler, Béatrice S., Spanlang, Bernhard and Mel Slater. (2017). "Virtual race transformation reverses racial in-group bias." *PloS one*, 12(4), https://doi.org/10.1371/journal.pone.0174965.

Lasher, K.P., and P.J. Faulkender. (1993). "Measurement of aging anxiety: Development of the anxiety about aging scale." *The International Journal of Aging and Human Development*, 37(4), 247–59.

Leary, Mark R., and June Price Tangney, eds. (2012), Handbook of Self and Identity. 2nd edition. New York: Guilford Press.

Tong, Eddie M.W., Barbara L. Fredrickson, Weining Chang, and Zi Xing Lim. (2010). "Re-examining hope: The roles of agency thinking and pathway thinking." *Cognition & Emotion*, 24(7), 1207–15.

Tsakiris, Manos. (2016), "The multisensory basis of the self: From body to identity to others," Quarterly Journal of Experimental Psychology, 70(4), 597–609. https://doi.org/10.1080/17470218.2016.1181768.

Appendix: Follow up study plan

Exploring Intimacy with Self, Friend, and Stranger using Immersive 360° video

Introduction

In 1963 Edward T. Hall developed the theory of proxemics, describing four distinct zones of interpersonal distances between people (Table 1): (1) intimate space, (2) personal space, (3) social space, and (4) public space. In this study, our goal is to explore the impact of meeting oneself by using Immersive 360° stereoscopic video (360° 3D). 360° 3D video viewed on an HMD generates sensorimotor contingency, producing a strong sense of presence for the viewer. Hence, using 360° 3D technology is an effective way to explore social interactions in general, and Intimacy in particular. In our experiment we will measure and compare the comfortable interpersonal distance (CID) (Nowicki, S., & Duke, M. P., 1972) between Friends, Strangers, and with Oneself. Specifically, we are interested in examining how body-image, body-esteem, and self-esteem may impact the CID with oneself.

Intimate distance (embracing, touching or whispering)	**Personal distance** (interactions among good friends or family)	**Social distance** (interactions among acquaintances)	**Public distance** (public speaking)
Close phase - less than 1 to 2 cm	Close phase - 46 to 76 cm	Close phase - 1.2 to 2.1 m	Close phase - 3.7 to 7.6 m
Far phase - 15 to 46 cm	Far phase - 76 to 122 cm	Far phase - 2.1 to 3.7 m	Far phase - 7.6 m or more.

Table 1. Hall's interpersonal distances chart

Method

Participants

Forty-eight Students from the IDC Herzliya will be recruited to participate in the study for credit. The participants will all be native speakers of Hebrew, and between 19 and 28 years of age, half of them female and half male, such that we will have twenty-four pairs of same gender friends. The female and male strangers will be nonIDC people.

Experimental design

This study will have a within-subject design. Each participant will be invited to two sessions.

In the first session ('recording session') the participant will be asked to fill in an informed consent form and the following questionnaires:

- Demographics (Age, Gender, BMI, Political Orientation, Religiousness)
- Intimacy question about the friend the participant came with
- Self-compassion
- Self-esteem (Rosenberg)
- Implicit Association Test (IAT) (Arab, Jewish IAT)
- Attractiveness rating of the IG and OG Stranger (post experiment)
- Body Satisfaction Scale (Slade 1990)
- Body Shape Questionnaire (Cooper 1987)
- Body-esteem (Franzoi & Shields, 1984)
- Stick figure CID test (post experiment)

We will use the Qualtrics system to run the questionnaires.

Next, the participant will be recorded walking three times towards a 360° 3D camera system, while looking directly into the left lens. In order to eliminate habituation each time, they will walk to the camera from a different angle. To ensure a slow and constant walking speed between participants, we will play a 40-beat-per-minute (BPM) metronome sound, instructing the participant to take one step every two beats. We will then record the same action with the participant's close friend of the same gender, and a same-gender total stranger. Additionally, we will record the experimenter walking towards the camera. This condition will be used as a practice run, before running the actual conditions.

Two weeks later, in the second session ('viewing session'), each participant will wear an HMD and view the video recording of his or her first session. The first clip will be the practice trial, with the experimenter walking towards the participant. The participant will see the experimenter walking towards him and he will be asked to report, by pressing a button on a controller, when the distance of the experimenter walking towards him becomes uncomfortable. The purpose of this practice is to familiarize the participant with the task, allow them to get used to the set-up, and eliminate the initial arousal, which might distort results. Next, the participant will view clips of the three conditions in counterbalanced order (Table 2). For every condition the participants will watch the three clips taken from different angles to avoid habituation. The order of the walking direction will be kept constant between conditions.

The viewing will take place using HMD. The nine clips will depict a person moving towards the subject (camera), three clips for each of the three conditions: Self, Friend, and Stranger. The participants will be asked to press a button on a controller as soon as they feel that the person coming towards them reaches an interpersonal distance that they are no longer comfortable with (CID).

1	Experimenter	Friend	Stranger	Self
2	Experimenter	Friend	Self	Stranger
3	Experimenter	Stranger	Self	Friend
4	Experimenter	Stranger	Friend	Self
5	Experimenter	Self	Friend	Stranger
6	Experimenter	Self	Stranger	Friend

Table 2. Order of viewing chart

Measures

Our measurement will be the mean comfortable interpersonal distance (CID) for each condition reported by the participants.

Technical Setup

Video recording system: Z-Cam camera rig with MK22 Lens
Virtual Reality playback system: Vive HMD with Tobii VR eye tracker
Tracker system: 2 Vive trackers

Application specifications

We developed two applications in Unity programming language, one for the recording session and one for the viewing session.

'Recording-Session' application specifications

Goal: recording the distance of the participant from the camera rig.
We will be using two controllers, one attached to the participant walking towards the camera, and the second attached to the camera.
Application flow:
01. Enter the participant's name and condition code.
02. Distance recording starts (temporal resolution of 1/10 of a second).
03. Sound trigger is played back to synchronize the distance record with the video footage.
04. Output file: Condition code (name of the participant), distance over time.

'Viewing-Session' application specifications

This app will playback the 360° 3D video file and will output the subject report of the CID.
Application flow:
01. Playback of video
02. Subject presses trigger on the controller that stops the video.
03. File output: Condition code, distance of CID.

Art and Presence
Investigating Embodiment in a Virtual Art Gallery

Hava Aldouby

The discourse on presence and embodiment in virtual reality has by now become pervasive both in the techno-commercial domain (Ijsselsteijn et al. 2000; Slater and Steed 2000; Riva and Waterworth 2014) and in historical, cultural, and phenomenological studies of immersion and VR (Calleja 2014; Dogramaci 2015; Dogramaci and Liptay 2015; Frohne 2016; Hansen 2006; Grau 2003; Kwastek 2015). Attention to the phenomenology of embodiment in VR (Garau et al. 2008; Pinotti 2019 [in this collection]) both meets with and deviates from the commercial race to increase the 'realness' of gaming and other entertainment media. But while VR prevails in entertainment media, the visual arts have until now retained a predominantly non-immersive aesthetic. Far from vanishing, or even becoming marginal, non-immersive visual art still engages wide audiences. At the same time, art created in both 'new' and 'old' media is evincing an intriguing *quest for presence*, as I propose to call it, albeit not through the same immersive qualities coveted by commercial agents of digital culture.

This chapter deploys VR to study born-digital, yet non-immersive art, asking whether the latter might still affect our sense of embodiment. A second fundamental question, raised by the project presented here, is whether VR might be put to use as an extremely versatile experimental platform for studies in experimental aesthetics, and whether it is applicable to the study of non-immersive artworks. To even begin probing these questions requires a multidisciplinary effort, bringing together VR research, image studies, and experimental aesthetics. This chapter presents a pilot trial that forms part of a larger art-based experimental project.[1] The experiment is conceived as the first building block in a multidisciplinary project designed to study born-digital art, both immersive and non-immersive, through participants' sense of embodiment, or Bodily Self Consciousness (henceforth BSC). To briefly elaborate, BSC depends on the integration of signals from different sensory modalities indicating the location of body parts and of the whole body in space, combined with body-related visual information (Blanke, Slater, and Serino 2015).

Building on the extensively studied effect of embodiment within VR (Maselli and Slater 2013, 2014), the present project focused attention on the responses of immersants to artworks embedded in a VR setting. In this experiment, the effect of non-immersive art on participants' awareness of their physical body was studied through the Rubber Hand Illusion paradigm (henceforth RHI: Botvinick and Cohen, 1998). Body ownership illusions like the RHI are widely accepted as an experimental paradigm enabling assessment of participants' sense of body ownership (Kilteni et al. 2015). In Botvinick and Cohen's seminal 1998 experiment, subjects watched an experimenter stroke a rubber hand which was placed on a table next to their own left hand. The real hand, resting behind a screen and thus hidden from view, was stimulated in synchrony with the rubber hand placed in front of the participant. In this condition, participants' sense of body ownership was extended to the artificial limb. Moreover, a consecutive experiment showed that participants who had experienced the illusion also manifested a sense of proprioceptive drift, expressed in erroneous assessment of the position of their (hidden) physical hand, drifting toward the fake hand. When asked, participants indicated that their real but invisible hand was closer to the rubber hand than it actually was.

The RHI experiment has since been replicated in VR with respect to a virtual hand, as well as demonstrating full body ownership in virtual reality (Ijsselsteijn et al. 2006; Slater et al. 2008; Sanchez-Vives et al. 2010). Our working hypothesis was that the selected art stimuli would affect participants' sense of embodiment, and that this could be productively investigated through the RHI experimental protocol, implemented in an adequately modulated VR setting. During the experiment, subjects were immersed in a virtual art gallery, and were asked to use their virtual hand to perform a task of picking and placing a virtual ball. The pick and place task was performed both before and after observing stimuli contributed by the media artist Ori Gersht.

A number of research projects involving artworks as stimuli have been carried out in the last two decades (Calbi et al. 2019; Freedberg and Gallese 2007; Mastandrea and Umiltà 2016; Sbriscia Fioretti et al. 2013; Umiltà et al. 2012). However, these studies, with the exception of Calbi et al. (2019), did not involve an artist in the research team. As in the experiment presented in Calbi et al. (2019), the present project benefited from the artist Ori Gersht's contribution of born-digital stimuli. Moreover, in an unexpected and intriguing turn—arising from the artist's experience of his own works in a VR environment—Gersht has begun experimenting with VR as a medium for creating immersive video art. The yield of this unforeseen 'VR turn' in Gersht's oeuvre will be further investigated in yet another iteration of the project, at IDC Herzliya's Advanced Reality Lab. In the conclusion of this chapter I discuss the significance of the art/science feedback-loop that emerged.

The project approaches aesthetic experience through its etymological roots in *aisthêsis* (Brudzińska 2010; Paterson 2012), generally emphasizing the embodied, sensory, or 'sensuous' nature of engaging with works of art. Particularly helpful here is the understanding that *aisthêsis* encompasses not just 'real' perceptual indications—received from the senses—but also 'the experience of the possible, [...] a consciousness in the mode of the as if' (Brudzińska 2010, 11). The present project relies on the important understanding of aesthetic experience as emerging from sensory experiences, whether these are physical, or anticipated through *the mode of the as if.* Informed by the phenomenological concepts of *haptic visuality* (Marks 2002) and *presence* (Gumbrecht 2004), we hypothesized that haptically effective art (i.e. art which suggestively activates our senses of touch) is capable of inducing the somatosensory state defined by Gumbrecht as *presence*, tapping a 'layer in our existence that simply wants the things of the world close to our skin' (Gumbrecht 2004, 106).

The phenomenological concept of aesthetic experience (Merleau Ponty [1945] 1993) implies that inducing or augmenting the sense of embodiment does not require actual physical stimulation, of the kind pursued by the entertainment industry (see for example the *ultrahaptics* technology).[2] Rather, aesthetic experience could involve 'as-if body loops' (Damasio 2010, 102–3), building on anticipated sensorimotor engagement regardless of whether the object of potential contact is materially available or is only suggestive of material presence. To corroborate this phenomenological assumption, it is productive to note that recent studies in cognitive neuroscience, investigating how the brain maps the space around the body, have shown the involvement of motor intentions and affordances in this process (Brozzoli et al. 2012). In other words, it is not merely neutral visual information that the brain-body system acquires and analyzes, as it orientates in its surroundings. Rather, the brain-body maps its environment in terms of physical proximity and the potential for physical interaction. We orientate toward *visual* stimuli (art objects included) as if preparing to physically interact or make contact, whether this is actually possible or not—for example in a prohibitive museum setting, or in VR.

Quite strikingly, it was shown that damage to the cortical circuits coordinating our motion toward objects not only impairs motion control, but also induces a deficiency in body awareness (Gallese 2015, 29). Thus, the way in which the mind-body system orientates towards the visual stimuli that surround it closely affects our very awareness of inhabiting a body in space. Based on this insight, the behavioral part of the present experiment consisted of subjective reports on bodily sensations while observing the artworks. The self-report questions focused on skin, touch, and limb location. Beyond the subjective reports, data pertaining to participants' hand movements were recorded during the pick and place task performance. Eye tracking will be added at future stages of the project.

As part of the desire to 'touch' in contemporary techno-culture, the visual arts are manifesting a growing concern with surfaces (Bruno 2014), textures, and skins, and altogether an increasing desire for tactile, or as-if tactile engagement. The present experiment was conceived as a first step toward investigating and characterizing this tactile turn as part of an urgent quest for presence, demonstrable through contemporary born-digital art. The general aim of the project is thus to cast a phenomenological, rather than political look at subtle negotiations of presence-related anxiety in early twenty-first-century art. Adopting what I propose to call a 'somato-cultural' viewpoint, this project makes a perhaps daring leap from the somatosensory dimension of aesthetic experience, accessible to empirical study, to the larger cultural domain. The growing concern with presence and the increasing desire to 'touch' are associated here with the increasing precariousness of place, identity, and body in this age of mass translocation and shifting ground. Where the experience of body, place, and self is radically destabilized, could haptically effective art activate the bodily senses responsible for our 'feeling of *having* (proprioception) and *moving* (kinaesthesia) *a body*'? (Paterson 2012, 476).

To emphasize, this project does not for a moment lose sight of Gumbrecht's caveat against conceiving presence naïvely, so to speak, as a primal or natural condition to be longed for nostalgically. Nor does the project's working hypothesis ignore the 'eventness' of aesthetic *presence effect*, which for Gumbrecht is inescapably 'mediated by clouds and cushions of meaning' (2004, 106–13). Also acknowledged is Janet Wolff's (2012) severe critique of 'the lure of immediacy' (3), and her skeptical assessment of the phenomenological turn represented by Gumbrecht and others, which threatens the primacy of analytic-discursive strategies in studies of culture and art. But, while Wolff emphatically argues against being 'seduced [...] into conversations about such mysterious things as "presence" or "immediacy"' (15), the present experimental project insists on asking whether an immediate haptic appeal may indeed be discerned in certain born-digital artworks. Furthermore, the project rests on the working hypothesis that such haptic appeal might momentarily increase viewers' sense of embodiment, thereby contributing to aesthetic pleasure by augmenting the experience of being present in *a body*. *Presence effect* (Gumbrecht 2004) is thus conceived here as offering momentary respite from millennial anxiety.

The stimuli: *New Orders* (2018) by Ori Gersht

Ori Gersht has long been experimenting with augmenting the sensory appeal and suggestive materiality of digital video and still photography. His oeuvre evinces a profound concern with presence, both as a cultural concept and as a phenomenological event, involving sensory and affective aspects. His reference to 'an ever-returning presence that is ultimately an eternal absence' (2013, 57) compellingly articulates this. In Gersht's works, especially those in which he reconstructed still life paintings by Jan Bruegel the Elder, Jean Simeon Chardin, or Henri Fantin Latour, the solid reality of the compositions collapses in the blink of an eye. In these instances, the destruction of the object (and image) by explosion or gunshot (Fig. 1) brings sheer materiality to the fore, driving strong anticipations of tactile contact with matter (see Aldouby 2019).

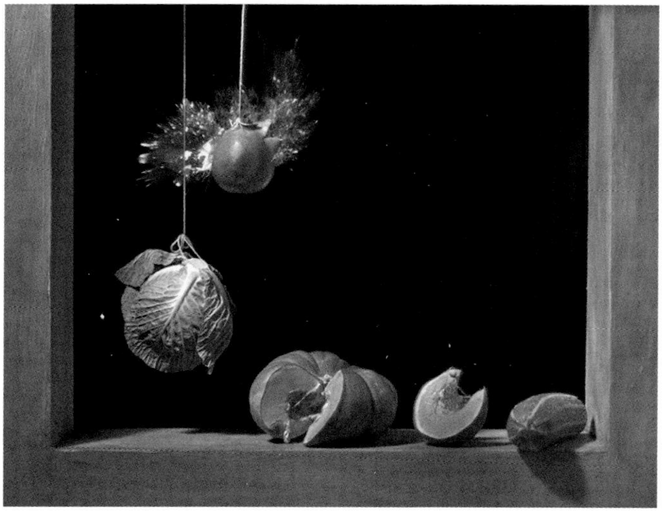

Fig. 1. Ori Gersht, *Pomegranate, Off Balance.* 2006. Archival ink print, 30 × 40 cm.
Image courtesy of the artist. (Plate 2, p. 305)

The underlying hypothesis of the present project is that the texture of the rubble left in the wake of destruction stimulates a sensorimotor response, as if preparing to physically engage. Gersht's works were chosen as stimuli for the present experiment based upon this hypothesis, assuming that their textural surplus would be compelling enough to induce sensations of embodiment and presence. *On Reflection* (Fig. 2) is a poignant example. Commenting about this triple channel video installation, and the series of stills that accompanies it, Gersht remarked:

> A work of art needs to have a body, a kind of physicality to it. Working with digital images, it's as though the work of art needs to be somehow fetishized as an object, in order for the viewer to get the full experience.[3]

Reflecting on surface and tactility in born-digital art, he went on to say:

> It interests me if you can create the same desire to touch, as in encountering a sculpture or a painting. To achieve this desire, there needs to be something to the surface, to the feel of it, that can only be experienced through the eye. But something in us really pushes to want to experience it in a much more sensual, physical way.[4]

The stimuli for the present experiment were selected from Gersht's series *New Orders* (2018; Figs. 3, 4, 5), which compellingly emulates a still life painting by the seventeenth-century

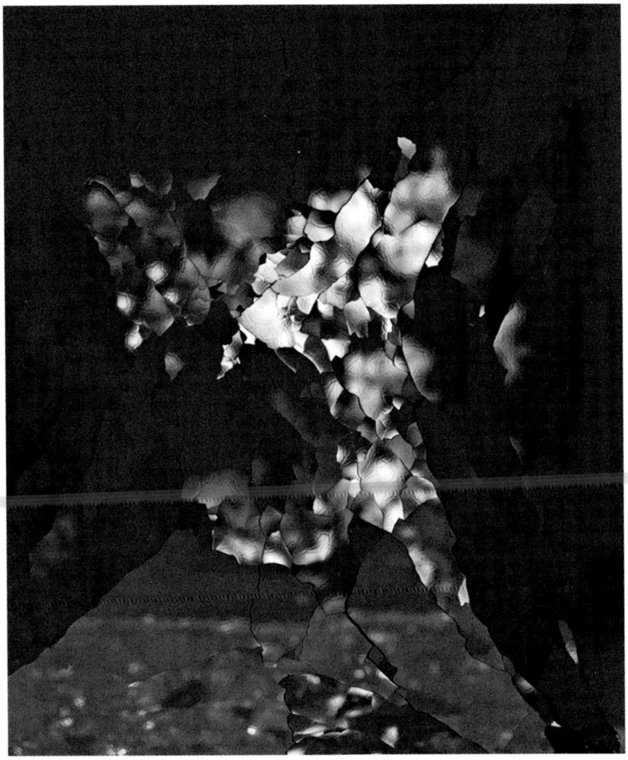

Fig. 2. Ori Gersht, *On Reflection, Material 11*, 2014. Still image from HD video. Image courtesy of the artist. (Plate 3, p. 306)

Spanish painter Francisco de Zurbarán, *A Cup of Water and a Rose* (ca. 1630). Gersht fractured the cup by gunshot, capturing the event in an extremely high frame rate so as to reveal increments of the occurrence that cannot possibly be perceived by the human eye (Fig. 4). These liminal moments border on the virtual, raising the ontological question of existence beyond the limits of human perception. Capturing increments of time on the verge of destruction, Gersht plays presence against absence, life against death.

For the experiment, three images from the series were selected: one in which the cup is still intact, and two other images featuring invisible micro-moments in which the shattered cup still holds the erupting body of water. In these images, the tension of pending collapse can be felt almost viscerally (Figs. 4, 5). Zurbarán's painting completed the selection. The four images were divided into two groups, tagged under 'texture' for the images of the fractured cup, and 'non-texture' for the unbroken cup.

The underlying hypothesis was that the cracked cups would induce strong anticipations of touch, whether active (i.e. intending or wanting to touch) or passive (i.e. expecting to be touched), owing to the suggestive sharpness of the fractured surface predicting

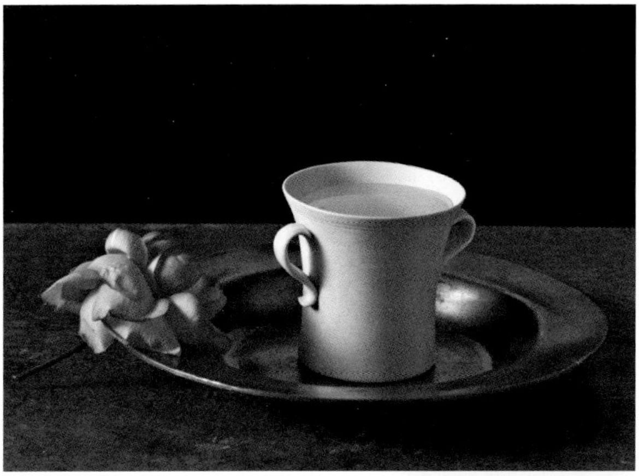

Fig. 3. Ori Gersht, from the series *New Orders* (2018). Image courtesy of the artist. (Plate 4, p. 306)

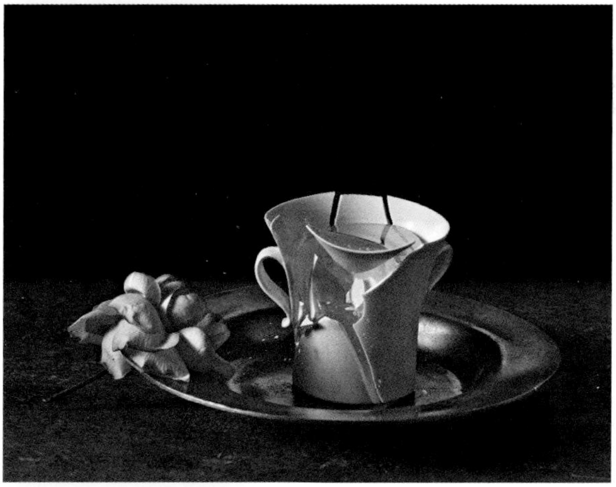

Fig. 4. Ori Gersht, from the series *New Orders* (2018). Image courtesy of the artist. (Plate 5, p. 307)

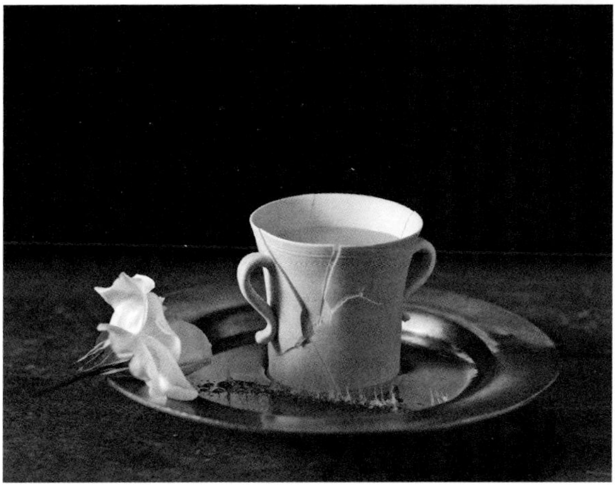

Fig. 5 Ori Gersht, from the series *New Orders* (2018). Image courtesy of the artist. (Plate 6, p. 307)

abrasive contact with the skin. Anticipation of tactile stimulation would, in turn, entail an altogether increased sense of embodiment. This assumption is supported by recent neurocognitive studies, showing that anticipation of touch is manifested in cortical networks that are also involved in actual touch (Ebisch et al. 2014; Carlsson et al. 2000). Haptically stimulating images may thus be expected to propel viewers into the type of prereflexive somatosensory engagement that, in Vittorio Gallese's terms, 'scaffolds' art's higher cognitive address (2015, 39).

Experimental procedure: observing artworks in a VR gallery

In order to deploy the Rubber Hand Illusion in VR, participants were immersed in a specially designed virtual environment, using a head-mounted display (HMD). Head tracking was provided by the HMD, and the participant's dominant hand was tracked using a Leap Motion depth camera. The physical set-up was designed to ensure optimal tracking conditions for the depth sensor, so that the virtual hand movements were congruent with the movement of

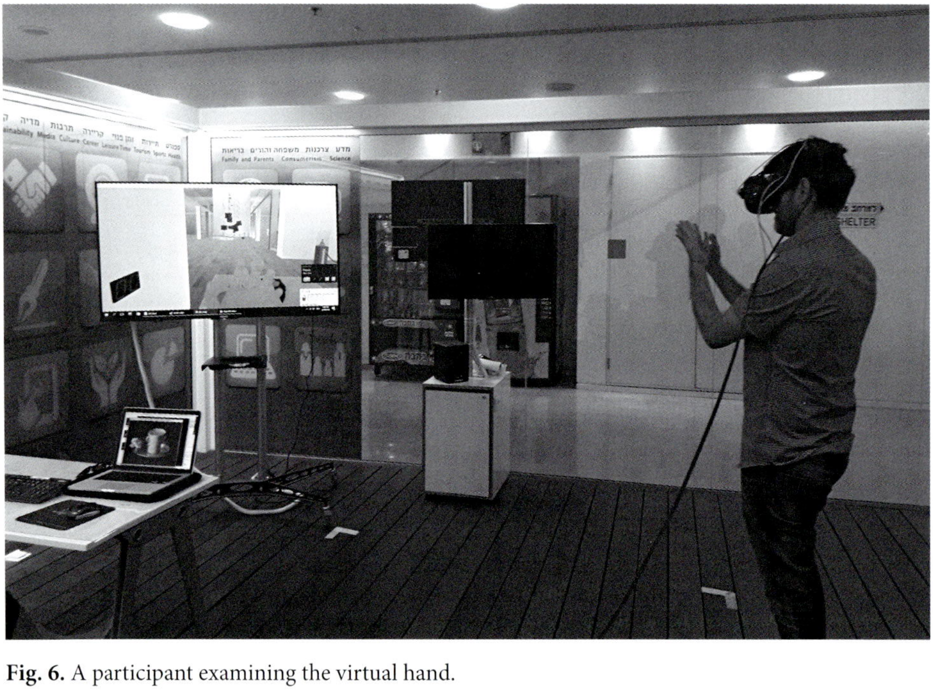

Fig. 6. A participant examining the virtual hand.

the participants' real hand. Maintaining visuo-motor coherence and a sense of agency in the virtual environment have been shown to constitute an important parameter in the production of presence in VR (Tsakiris et al. 2006). The experiment also included a 'drift' condition, where the virtual hand was represented at a slight shift from the position of the real hand. In this condition, the strength of the RHI was assessed through the extent of participants' conviction that the virtual hand and their real hand were still in an identical position in space.

Once the experimenter set up the VR equipment, participants were able to explore the virtual environment by moving their heads, and to test the virtual hand by moving it before the 'eye' of the Leap Motion sensor (Fig. 6).

The illusion was induced by requesting the participants to practice a pick and place task, moving a virtual ball above a stack of books, towards a predefined position on the other side of the shelf. At this stage, the task had the goal of creating a link between the real and the virtual hand, through the repetitive interaction task. The movements of participants' hands above the obstacle were recorded. The task was followed by a base-line self-report questionnaire referring to the SoE (Sense of Embodiment) of the virtual hand. Participants were requested to respond to the following statements by pressing numbered buttons on a panel embedded in the VR, grading their answers on a 1-6 scale (6 indicating full agreement with the statement):

(1) I feel that the virtual hand is my real hand.
(2) The virtual hand is positioned exactly where my real hand is.

The experiment then proceeded in four repeated cycles, each starting by viewing an art stimulus from the categories of either 'texture' or 'non-texture,' with the virtual hand presented in the condition of either 'drift' or 'no drift.' Thus, each participant experienced all four possible combinations. After observing the artwork for 30 seconds, the participants were again requested to respond to a short questionnaire by pressing buttons on a panel embedded in the VR. The questionnaire now consisted of the following four questions (responses graded on a 1-6 scale, 6 indicating maximal agreement):

(1) When I was observing the picture I could almost feel touch.
(2) When I was observing the picture I felt sensations on my skin.
(3) When I was observing the picture I felt a bit outside my own body.
(4) When I was observing the picture I became aware of the weight of my body.

After responding to the questions, the participants were asked to perform the same pick and place task as in the beginning. This time the impediment above which they had to

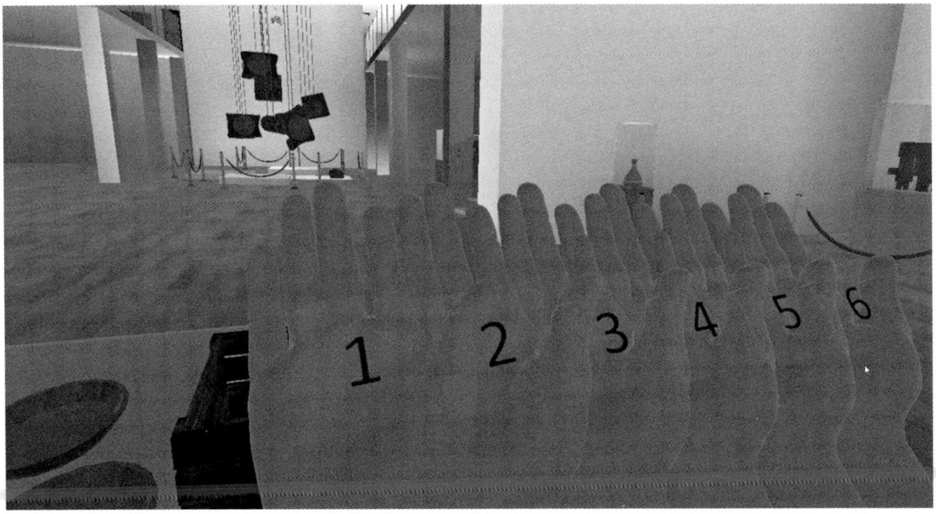

Fig. 7. Assessing the estimated position of the physical hand in relation to the virtual hand.

move the ball alternated between the neutral stack of books and a loud rotating chainsaw, suggesting potential threat to the virtual hand. The height and duration of hand movements above the obstacles were recorded, as an indication of participants' responses to the potential danger to their (virtual) hand. After completing each block, the participants were requested to respond to a subjective questionnaire, rating their sense of ownership over the virtual hand (responses graded 1-6):

(1) I feel that the virtual hand is my real hand.
(2) For a moment I was afraid that my real hand might be hurt by the chainsaw.
(3) My real hand is positioned exactly where the virtual hand is.

In addition, the participants were presented with six adjacent iterations of the virtual hand (Fig. 7) and were asked to pronounce out loud the number of the hand coinciding with the position of their own (physical) hand. The number was recorded by the experimenter.

Participants were thereby compelled to seek congruence between visual and proprioceptive information. The degree of interoceptive awareness, highly pertinent in this respect, was assessed for each participant by means of the MAIA (Multidimensional Assessment of Interoceptive Awareness) questionnaire (Mehling et al. 2012, 2018), which the participants filled in after completion of the experiment. General demographic details were collected in another questionnaire.

Discussion of preliminary results

The first experimental phase described here engaged 64 participants. An upcoming second phase will include eye tracking in addition to the hand movement recordings, and will increase the number of participants to 100. The primary picture emerging from the behavioral results obtained so far through the subjective questionnaires, and from certain correspondences between these results and the MAIA questionnaire of interoceptive awareness, appears to support the project's working hypothesis.

The first indication in this vein is a significant main effect for texture with respect to the following questions: 'when looking at the image *I could almost feel touch*,' and 'when looking at the image *I felt sensations on my skin*.' In other words, after observing stimuli designated as 'textured,' participants responded to the foregoing questions with significantly higher rates than after observing the 'non-textured' images. We interpret this result as an indication that the fractured cup, which we assumed would arouse anticipations of abrasive touch, was indeed more effective in arousing as-if tactile sensations than the unfractured cup.

In addition, the results indicate an interaction between participants' degree of interoceptive awareness (assessed through the MAIA questionnaire) and their susceptibility to images from the 'texture' category. The more interoceptively aware the participant, the stronger was the effect of textured stimuli in arousing sensations of skin and touch. Interoceptive awareness and image texture thus interacted towards a heightened sense of embodiment during contemplation of the more expressly textured stimuli. A synergy seems to be at work between the participant's body-awareness, indicated by the MAIA score, and certain aesthetic traits of the artwork (i.e. texture vs. smoothness), substantiating our working hypothesis—that haptic effectiveness strongly relies on texture. A further important insight arising from these results is the extent to which body-awareness participates in aesthetic experience.

The choice of a VR setting for exploring haptic effectiveness and embodiment was motivated by the assumption that it would offer participants a holistic experience, knitting together the virtual hand and the art-stimuli within the VR gallery space. The effect of embodiment in VR is induced by a congruency between proprioception, indicating the location of our own body and limbs, and the visible movements of the virtual hand, affording a sense of agency in the virtual setting. As it turned out, the sense of embodiment augmented participants' responses to the art images, making them robust enough for experimental detection. In our results, a mutual interaction emerged between the conditions of 'texture' and 'drift.' This interaction was expressed in the subjective reports on anticipation of touch and skin sensations and, most significantly, on the awareness of body weight. To explain: in the 'drift' condition (i.e. discrepancy between the positions of real and virtual hand), the textured stimuli had an even stronger effect on participants' awareness of the weight of their body than in the non-drift condition. In other words, gradings of the statement *I became aware of the weight of my body* in response to 'textured' vs. 'non-textured' stimuli differed more significantly in the 'drift' condition. Apparently, when subjects contemplated the fractured cup in the 'drift' condition, texture more strongly impacted their awareness of their body as a mass in space. This may be explained as an effect of increased arousal while attempting to reconcile incompatible proprioceptive and visual information. To get a grip on the situation, and bridge real and virtual space, somatosensory arousal is augmented and with it the susceptibility to texture. As it appears, VR is poised to tease out the haptic effect, and thus emerges as an effective experimental setting for studying the interface between aesthetic experience and embodiment.

Conclusion: VR as an experimental (art) medium

VR has played an intriguing role in this project's ongoing formation. An unexpected result of the pilot experiment was the opportunity given the artist to experience his own work in a VR environment. This had the radical effect of ushering in a 'VR turn' in Gersht's oeuvre. The artist is currently experimenting with virtual reality as a medium affording enhanced haptic experience. The feedback loop that emerged between art and research has not reached closure at this point, and is still open to new turns. Gersht's evolving VR art will be subject to further experimental studies at the Advanced Reality Lab through a complex experimental design that will immerse participants in the work itself, rather than in a virtual gallery hosting non-immersive stimuli. The pilot trial discussed here thus highlights the versatility of VR as a mediator of aesthetic experience, and as an experimental medium of art and research.

Notes

1. The project was initiated by the author, an art historian and image scholar. It was carried out in collaboration with VR and HCI researcher Doron Friedman, psychologist Béatrice Hasler, media artist Ori Gersht, and research assistant Tehila Nadav, at IDC Herzliya's Advanced Reality Lab. Publication of empirical results is forthcoming 2020. This research was supported by The Open University of Israel's Research Fund (grants no. 41139 and 37197)
2. https://www.ultrahaptics.com (last accessed 14 November 2019).
3. Ori Gersht, personal communication to the author, February 2015.
4. Ori Gersht, personal communication to the author, February 2015.

References

Aldouby, Hava. 2019. "Balancing on shifting ground: Migratory aesthetics and recuperation of presence in Ori Gersht's video installation On Reflection." *Crossings: Journal of Migration & Culture* 10:2, 161–181, doi: 10.1386/ cjmc_00001_1.

Barker, Jennifer. 2009. *The Tactile Eye: Touch and the Cinematic Experience*. Berkeley CA: University of California Press.

Blanke, Olaf, Mel Slater, and Andrea Serino. 2015. "Behavioral, Neural, and Computational Principles of Bodily Self-Consciousness." *Neuron* 88:7 (October), 145–66.

Botvinick, M., and J. Cohen. 1998. "Rubber hands 'feel' touch that eyes see." *Nature* 391: 756.

Brozzoli, Claudio, Tamar R. Makin, Lucilla Cardinali, Nicholas P. Holmes, and Alessandro Farnè. 2012. "Peripersonal Space: A Multisensory Interface for Body–Object Interactions." In *The Neural Bases of Multisensory Processes*, edited by M.M. Murray and M.T. Wallace, 1–17. Boca Raton FL: Taylor & Francis.

Brudzińska, Jagna. 2010. "Aisthesis." In *Handbook of Phenomenological Aesthetics*, edited by Hans Rainer Sepp and L. Ambree, 9–15. Dordrecht: Springer Netherlands.

Bruno, Giuliana. 2014. *Surface: Matters of Aesthetics, Materiality, and Media*. Chicago IL: University of Chicago Press.

Calbi Marta, Hava Aldouby, Ori Gersht, Nunzio Langiulli, Vittorio Gallese, and Maria A Umiltà. 2019. "Haptic Aesthetics and Bodily Properties of Ori Gersht's Digital Art: A Behavioral and Eye-Tracking Study." *Frontiers in Psychology* 10: 2520. doi: 10.3389/fpsyg.2019.02520

Calleja, Gordon. 2014. "Immersion in Virtual Worlds." In *The Oxford Handbook of Virtuality*, edited by Mark Grimshaw, 222–36. New York: Oxford University Press.

Carlsson, Katrina, Predrag Petrovic, Stefan Skare, Karl Magnus Petersson, and Martin Ingvar. 2000. "Tickling expectations: Neural processing in anticipation of a sensory stimulus." *Journal of Cognitive Neuroscience* 12(4): 691–703. doi: 10.1162/089892900562318.

Damasio, Antonio R. 2010. *Self Comes to Mind: Constructing the Conscious Brain*. 1st ed. New York: Pantheon Books.

Dogramaci, Burcu. 2015. "Water, Steam, Light: Artistic Materials of Immersion." In *Immersion in the Visual Arts and Media*, edited by Dogramaci, B., and F. Liptay, 21–39. Leiden: Brill.

Dogramaci, Burcu, and Fabienne Liptay, 2015. "Immersion in the Visual Arts and Media". In *Immersion in the Visual Arts and Media*, edited by Dogramaci, B., and F. Liptay, 1–17. Leiden: Brill.

Ebisch, Sjoerd J. H., Francesca Ferri, Gian Luca Romani, and Vittorio Gallese. 2014. "Reach Out and Touch Someone: Anticipatory Sensorimotor Processes of Active Interpersonal Touch." *Journal of Cognitive Neuroscience* 26(9): 2171–85. doi: 10.1162/jocn_a_00610.

Freedberg, David, and Vittorio Gallese. 2007. "Motion, Emotion and Empathy in Esthetic Experience." *Trends in Cognitive Sciences* 11(5):197–203. doi: 10.1016/j.tics.2007.02.003.

Frohne, Ursula. 2016. "Expansion of the Immersion Zone: Military Simulacra between Strategic Training and Trauma." In *Immersion in the Visual Arts and Media*, edited by Dogramaci, B., and F. Liptay, 215–50. Leiden: Brill.

Grau, Oliver. 2003. *Virtual Art: From Illusion to Immersion*. Cambridge MA and London: MIT Press.

Garau, Maia, Doron Friedman, Hila Ritter Widenfeld, Angus Antley, and Andrea Brogni. 2008. "Temporal and Spatial Variations in Presence: Qualitative Analysis of Interviews from an Experiment on Breaks in Presence." *Presence* 17, no. 3 (June): 293–309.

Gersht, Ori. 2013. "Somewhere Between." In *Ori Gersht: History Repeating*, edited by Al Miner, Yoav Rinon, and Ronni Baer, 57. Boston MA: Museum of Fine Arts.

Grimashaw, Mark (ed.). 2014. *The Oxford Handbook of Virtuality*. Oxford: Oxford University Press.

Gumbrecht, Hans Ulrich. 2004. *Production of Presence: What Meaning Cannot Convey*. Stanford CA: Stanford University Press.

Hansen, Mark B N. 2006. *Bodies in Code: Interfaces with Digital Media*. New York: Routledge.

Ijsselsteijn, Wijnand A., Huib de Ridder, Jonathan Freeman, and Steve E. Avons. 2000. "Presence: Concept, Determinants, and Measurement." Proc. SPIE 3959, *Human Vision and Electronic Imaging* 5, 520. http://dx.doi.org/10.1117/12.387188.

Ijsselsteijn W., Y. de Kort, and A. Haans. 2006. "Is this my hand I see before me? The rubber hand illusion in Reality, Virtual Reality and Mixed Reality." *Presence: Teleoperators and Virtual Environments* 15: 455–64.

Kilteni K., J.-M. Normand, M.V. Sanchez-Vives, and M. Slater. 2012. "Extending Body Space in Immersive Virtual Reality: A Very Long Arm Illusion." *PLoS ONE* 7(7): e40867. doi:10.1371/journal.pone.0040867.

Kilteni K., A. Maselli, K.P. Kording, and M. Slater. 2015. "Over my fake body: body ownership illusions for studying the multisensory basis of own-body perception." *Frontiers in Human Neuroscience* 9:141. doi: 10.3389/fnhum.2015.00141.

Kwastek, Katja. 2015. "Immersed in Reflection? The Aesthetic Experience of Interactive Media Art." In *Immersion in the Visual Arts and Media*, edited by Dogramaci, B., and F. Liptay, 67–86. Leiden: Brill.

Marks, Laura U. 2002. *Touch: Sensuous Theory and Multisensory Media*. Minneapolis MN: University of Minnesota Press.

Maselli, A., and M. Slater. 2013. "The building blocks of the full body ownership illusion." *Frontiers in Human Neuroscience* 7(83). doi: 10.3389/fnhum.2013.00083.

Maselli, A., and M. Slater. 2014. "Sliding perspectives: dissociating ownership from self-location during full body illusions in virtual reality." *Frontiers in Human Neuroscience* 8:693. doi: 10.3389/fnhum.2014.00693.

Mastandrea, Stefano, and Maria A. Umiltà. 2016. "Futurist Art: Motion and Aesthetics As a Function of Title." *Frontiers in Human Neuroscience* 10:201. doi: 10.3389/fnhum.2016.00201.

Mehling, W.E., C. Price, J.J. Daubenmier, M. Acree, E. Bartmess, and A. Stewart. 2012. "The Multidimensional Assessment of Interoceptive Awareness (MAIA)." *PLoS ONE* 7(11): e48230. doi:10.1371/journal.pone.0048230.

Mehling, W.E., M. Acree, A. Stewart, J. SilasJ, and A. Jones. 2018. "The Multidimensional Assessment of Interoceptive Awareness, Version 2 (MAIA-2)". *PLoS ONE* 13(12). https://doi.org/10.1371/journal.pone.0208034.

Merleau-Ponty, Maurice. [1945] 1993. "Cézanne's Doubt." In *The Merleau Ponty Aesthetics Reader: Philosophy and Painting.* Trans. and ed. Galen A. Johnson and Michael B. Smith, 59–75. Evanston IL: Northwestern University Press.

Paterson, Mark. 2012. "Movement for Movement's Sake? On the Relationship Between Kinaesthesia and Aesthetics." *Essays in Philosophy* 13(2): 471–97. http://dx.doi.org/10.7710/1526-0569.1433.

Riva, Giuseppe, and John A. Waterworth. 2014. "Being Present in a Virtual World." In *The Oxford Handbook of Virtuality*, edited by Mark Grimshaw, 205–21. New York: Oxford University Press.

Sanchez-Vives, M.V., B. Spanlang, A. Frisoli, M. Bergamasco, and M. Slater. 2010. "Virtual hand illusion induced by visuomotor correlations." *PLoS One* 5. https://doi.org/10.1371/journal.pone.0010381.

Sbriscia-Fioretti, B., C. Berchio, D. Freedberg, V. Gallese, and M.A. Umiltà. 2013. "ERP Modulation during Observation of Abstract Paintings by Franz Kline." *PLoS ONE* 8(10). https://doi.org/10.1371/journal.pone.0075241.

Slater, M., and A. Steed. 2000. "A virtual presence counter." *Presence: Teleoperators and Virtual Environments* 9(5): 413–34.

Slater, M., D. Perez-Marcos, H.H. Ehrsson, and M.V. Sanchez-Vives. 2008. "Towards a digital body: the virtual arm illusion." *Frontiers in Human Neurocience* 2: 6.

Tsakiris, M., G. Prabhu, and P. Haggard. 2006. "Having a body versus moving your body: How agency structures body-ownership." *Consciousness and Cognition* 15: 423–32.

Umiltà, M. Alessandra, Cristina Berchio, Mariateresa Sestito, David Freedberg, and Vittorio Gallese. 2012. "Abstract art and cortical motor activation: an EEG study." *Frontiers in Human Neuroscience* 6:311. https://doi.org/10.3389/fnhum.2012.00311.

Waterworth, John, E.L. Waterworth, F. Mantovani, and G. Riva. 2010. "On Feeling (the) Present: An Evolutionary Account of the Sense of Presence in Physical and Electronically-mediated Environments." *Journal of Consciousness Studies* 17 (1–2): 167–88.

Wolff, Janet. 2012. "After Cultural Theory: The Power of Images, the Lure of Immediacy." *Journal of Visual Culture* 11 (1): 3–19. https://doi.org/10.1177/1470412911430461.

Shared Objective Empathy in Telematic Space

Paul Sermon

A Proprioceptive Arts Practice

This essay discusses the juncture between empathy and presence manifest in my telematic artworks that converge space and distant audience participants in telepresent interactive video installations. The discussion will focus on three specific works that chart the kinaesthetic and proprioceptive nature of my practice, which includes 'Telematic Dreaming,' first shown in 1992, 'Peace Talks' from 2003, and 'Touched,' produced in 2017. All these works have involved networked videoconference technology, but range in contextual installation settings from the intimacy of a bed surface to a peace negotiations table and the keyboard of an online chat conversation. Through each of these works I will present and discuss the empathetic nature of the participant's interactions and experiences that I have encountered since the early 1990s, described through the conversations, observations, and participants accounts I have witnessed in the production and realization of these works.

My work in the field of telematic arts explores human relationships and the identification of the self through the emergence of shared objective empathy between remote participants who are brought together within merged telepresent environments. Through the use of live video cameras, projection mapping techniques, chroma-keying, and videoconferencing technology these divided audience participants are brought together to coexist in what Randall Packer terms the 'third space' (Packer 2018); the fusion of the physical and the remote in a networked place that can be inhabited by distant users simultaneously sitting, standing, or sometimes lying together within it. Their presence within the space is captured live on video and combined in real time with an identical camera view of another participant in a matching installation space; mixing two live video shots by projecting a person into the environment of another or by replacing a blue or green backdrop in one image with the image of the other. The two spaces are linked via a networked videoconference connection, making it possible to merge these telematic installations and their performing audiences between geographically distant global locations.

Telematic Dreaming

This research began in June 1992 with the development and presentation of 'Telematic Dreaming,' an experimental telepresence performance installation produced for the 'Koti' exhibition at the Kajaani Art Museum in Finland, curated by Marita Liulia and sponsored by Finnish Telecom. The summer exhibition entitled 'Koti' (home) was used in reference to Jean Baudrillard's dystopian vision of the 'telematic home' in his 1987 essay 'The Ecstasy of Communication.' The combination of these factors, and in particular access to the latest videoconference technology from Finnish Telecom, allowed me to propose and develop this first networked installation. Linked between the Kajaani Art Museum and the Finnish Telecom's Telegalleria in Helsinki, it originated a unique method of performing third-person experiences of telepresence.

This is an installation that exists within a telematic network between two installation locations, connected through customized videoconferencing technologies. In the first location a video camera picks up a bird's-eye view of a queen-size double bed beneath it—the exact same 3:4 aspect ratio as the PAL video format of the time (Fig. 1). The camera image of the person lying on the bed is sent to a geographically distant location and video projected down onto another double bed, where a second person is lying. A further video camera situated next to the projector captures the superimposed images of the projected person and second person together on the same bed. This combined telepresent image is then fed directly to two video monitors on either side of each bed via the videoconference connection, allowing the participants to physically interact.

Telematic Dreaming encountered all the 'jouissance' (Baudrillard 1988, 5) that Baudrillard was cynically advocating, and in doing so it became the antithesis of his argument—it was 'The Ecstasy of Communication' in practice, a certainty for extending the body beyond the screen to a distant simulation of it. Phenomenologically, the remote 'other' body is no less of a simulation than its carbon original. Its physical movements were taking place locally whilst its cause and effects were taking place remotely—simply extending the sensory inputs from the cognitive process.

As the participants begin to discover this telematic third space and the potentiality of this new immersive coexistence, the telematic technology (tools) and geographical distances involved are of least concern and essentially become invisible, and increasingly more so the longer the participants interact on the bed. Aided by the complexity of this public bed surface and the challenging invitation to share it, these kinaesthetic encounters are far from contemplations of the network hardware. In many ways this reflects the ontological claim that Martin Heidegger makes in 'Being and Time' (2001 [1953]) with the example of a 'hammer' (tool) (Heidegger 2001 [1953], 195–203) as having two states. The hammer is either an innate object when not in use, what Heidegger refers to as 'present-at-hand' (Heidegger 2001 [1953], 200) or a 'tool' when in use, the result of which is the primary focus of our attention, whereby the object itself disappears, in a state that is 'ready-to-hand' (Heidegger 2001 [1953], 200). The analogy goes even further when we consider the breakdown in the network technology of 'Telematic Dreaming,' which occurred on numerous occasions, returning the audience participant to the immediate individual state of their experience, reflecting on the technology around them and what has happened. A third state that Heidegger refers to as 'un-ready-to-hand' (Heidegger 2001 [1953], 103) is that when the ham mer/tool breaks whilst in use. In Heidegger's experiment the invisible hammer has essentially become an extension of the body, or what Maurice Merleau-Ponty calls the 'extension of the bodily synthesis' (Merleau-Ponty 2012 [1945], 154) when referring to the experience of our own body and its sensory shift from the surface of the skin to the tip of a blind person's stick, 'the stick is no longer an object perceived by the blind man, but an instrument *with* which he perceives. It is a bodily auxiliary, an extension of the bodily synthesis' (Merleau-Ponty 2012 [1945], 154).

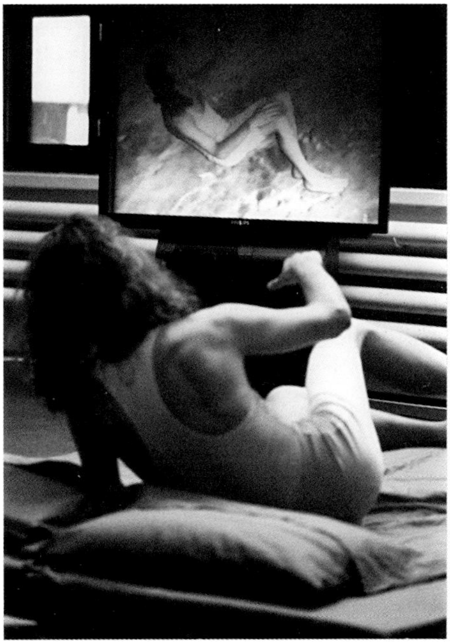

Fig. 1. Paul Sermon participating as a performer in 'Telemtic Dreaming' at the Telegalleria. (Plate 7, p. 308)

In 'Telematic Dreaming' the extension of the bodily synthesis occurs in the same way through our proprioception of the telepresent body image, like a 'phantom limb/body' where the sense of touch has shifted from our hands to our eyes, the blind person's stick in this case being the telepresent body at a distance that places our finger nerve endings within the auxiliary body image, where a shift of senses occurs—

touching with your eyes as if you are touching with your hands. In the same way a blind person will rely on the sensory input of touch in the absence of sight, the feeling of touch in 'Telematic Dreaming' is reliant on the sense of sight. Not unlike the visual sensory input of pain that is often stimulated prior to the momentarily numbed nerve endings of the tissue at the cause of it—the cognitive process of pain taking place via the eyes—regardless of where they are located.

From the observations I have made of 'Telematic Dreaming' in the 29 presentations of it since its development in 1992, participants were consistently initially hesitant about interacting on the bed surface due to its public context and the implications of it—reasons I can completely appreciate myself. An example of this apprehension occurred when I presented 'Telematic Dreaming' in Austria, linked between the Ars Electronica Center in Linz and the Kulturvermerke Symposium in Gmunden in 1996, when a man had been standing a meter away from the bed watching the projected body for almost twenty minutes, he took the plunge and dived onto the bed to embrace a new virtual participant who had just arrived, shouting 'mein frau, mein frau!' I later learned it was actually the man's wife, and earlier that day they had decided to attend two different gallery openings on the same evening; she wanted to attend an opening in Linz and he wanted to attend an opening he had heard about in Gumunden; little did either of them know, they were in fact attending the same telematic event. It was as if the surprise of seeing his wife 'broke the ice,' and the need to disclose this telematic coincidence provided him with the confidence he needed to enter the space. A similar response occurred in 2003 when I presented 'Telematic Dreaming' for the Sonar Festival at MACBA Museu d'Art Contemporani de Barcelona. On observing a long queue of people wanting to visit the installation I had to weave my way through a large crowd of people around the bed to see what was happening, only to discover they were all staring at an empty bed. The potential for what could happen was as compelling to them as the spectacle that was about to start.

However, once the participants overcame the initial awkwardness of entering the bed, they became far less concerned about the consequence of their immediate surroundings and corporal bodily existence, and far more attuned with their telematic bodily synthesis. So much so, that it became increasingly difficult to then exit from the bed and return to the original state, a disruption in their conscious experience—a state of being 'un-ready-to-hand' (Heidegger 2001 [1953], 103). Participants would frequently remain in the installation for sometimes up to an hour or more, and make return visits on numerous occasions.

In 1994 I presented 'Telematic Dreaming' in Amsterdam for the third time since it was first developed in Finland. The exhibition, entitled 'Ik + De Ander' (I + The Other), was shown at the 'Beurs van Berlage' (Amsterdam's former stock exchange building) from June 16 to August 14, 1994, curated by Ine Gevers and Jeanne van Heeswijk. For this particular group exhibition 'Telematic Dreaming' was linked via video cable between a public video projection space and a disclosed performance room within the same building for the three-month duration of the show. During this time I commissioned dancer, choreographer, and writer Susan Kozel to participate in the disclosed performance room for four weeks (Fig. 2). Completing her PhD at the time, Susan Kozel went on to write a vivid account of her experiences and reactions whilst performing in the installation, described in her article 'Spacemaking: Experiences of a Virtual Body' for the Laban Dance Theater Journal in 1994. Kozel's detailed description of varying encounters, from initial touch to physical intimacy and on some occasions even aggression, further evidences the phenomenological interpretations I have observed and experienced myself.

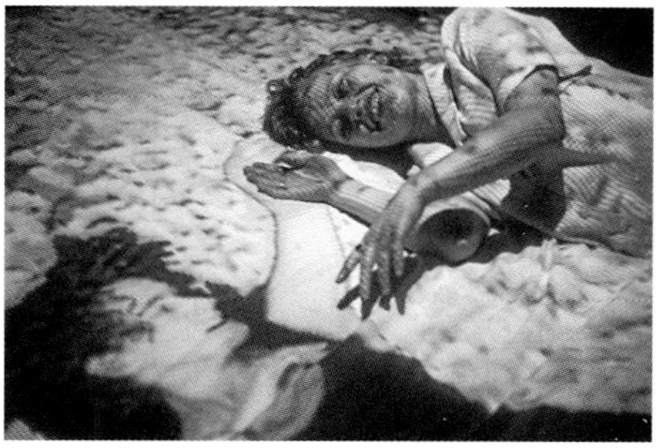

Fig. 2. Projected image of Susan Kozel with a gallery participant in 'Telematic Dreaming.' (Plate 8, p. 309)

> Movement usually began in a hesitant way with hand contact taking on excessive importance. The impact of slow and small movement became enormous (...). When the movement progressed from these early stages to a sort of full body choreography the piece became an emotional investment (Kozel 1994).

Beyond the initial 'electric shock' (Kozel 1994) of telematic touch Susan Kozel reflects on the level of empathy and trust built up between herself as the sole performer and the visiting public participants over time, and ultimately how this experience affected both her and those taking part.

> Many members of the public were overwhelmed by their experiences on the other bed. Some felt protective toward me, or stayed on the bed because they didn't want me to be alone in my virtual world. Others claimed to have been 'changed' by the experience. The installation was paradoxical for not only using technology to provide a forum for experiencing the basics of human intimacy, but for situating this private interaction within a public domain (Kozel 1994).

Returning to the complexity of exiting 'Telematic Dreaming,' Kozel observes a state of change and lasting effect. 'Yet what intrigues me is the return to the body which is implied by any voyage beyond it. Once plunged back into flesh, what has changed?' (Kozel 1994). This marked experience leaves participants with something learned and gained from the installation and their telematic extension of the bodily synthesis. This is an observation I share, from my own experiences as a performer in the first presentation of 'Telematic Dreaming' in 1992. Whilst I returned to a corporal state of existence, a lasting neurological effect from this shift of senses still remained through my learnt experience, in much the same way as the blind person will build on their experiences through the auxiliary extension of the stick.

Peace Talks

The findings from 'Telematic Dreaming' informed the development of further telepresence installations, explored through a range of domestic items of furniture following on from the bed. This included the sofa in 'Telematic Vision' in 1993 and the table in 'The Tables Turned, A Telematic Scene on the Same Subject' in 1997. The table interface was then further explored in the context of 'Peace Talks,' an installation commissioned by FACT (Foundation for Art and Creative Technology) for the BitParts Exhibition in February 2003, linked across the city of Worcester in England, using an internet videoconference connection between Worcester City Art Gallery and Worcester Guildhall.

Identical rooms containing a table and chair were located in the two separate exhibition spaces. The only difference was the colour of each room. The entire room and table at Worcester City Art Gallery were painted 'chroma-key green' (Fig. 3) and the complete room and furniture at Worcester Guildhall were painted 'chroma-key blue'. Using identically positioned cameras in each of the rooms it was possible to composite (or chroma-key) the two room views together, placing the separate participants at the same telepresent table. In addition a computer-generated 3D model of a peace negotiations room and table was mapped and composited within the image, thereby placing the remote participants into a simulated United Nations Peace Talks scene, appearing as if they were sitting on opposite sides of the same peace negotiations table.

Unique to the 'Peace Talks' installation, the participants observed the space and their shared existence within it via head-mounted video displays normally used for viewing DVDs—a technology originally developed by Sony[1] in the late 1990s. Importantly, this was not intended for a virtual reality experience; it simply pre-

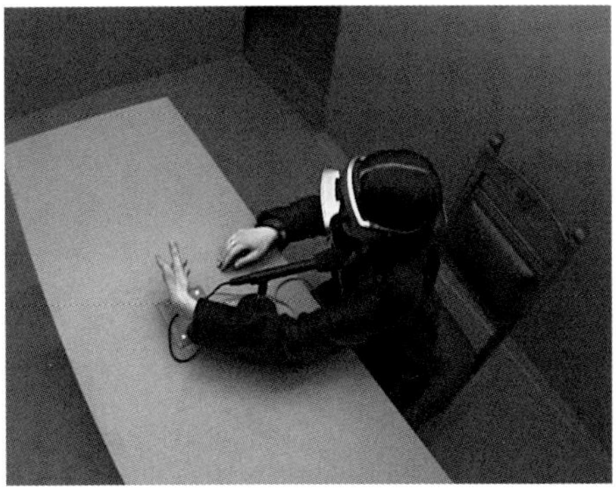

Fig. 3. Participant seated at the green table interacting and conversing in 'Peace Talks.' (Plate 9, p. 309)

sented 2D video in an immersive head-mounted display. By wearing a pair of these headset glasses in 'Peace Talks' the participants' view of themselves was no longer restricted to the location of the screens around them, as in previous telematic installations. Instead they could simply look in any direction and still have a complete view of the scene from the fixed camera position, in effect removing their eyes from their heads and locating them within the camera on which they relied to navigate within the space, resulting in a complete telepresent sense of immersion. The only downside was the image of themselves; in particular their faces were obscured by the head-mounted display (HMD) they were wearing and this was visually distracting for some participants, which is why they were not used in future installations. However, this did encourage them to remain seated at the table and engage in conversation with their telepresent participant sitting opposite, not only because of this increased sense of immersion, but because they could not easily stand up and walk about the room whilst being tethered to the HMD.

Put into context, 'Peace Talks' opened on February 1, 2003, just two weeks before one of the largest ever anti-war demonstrations took place in London, as well other cities across the world on February 15, 2003 in opposition to the imminent war in Iraq. According to BBC News, between six and eleven million people took part in protests in up to 60 countries. Much of the global opposition that preceded the demonstrations was of the opinion that diplomatic talks and peace negotiations had not been fully explored in resolving the situation and avoiding war. The 'Peace Talks' telematic installation was a direct response to this worldwide-felt concern.

As is often the case in previous telematic installations, the participants became amused by their self-image in this displaced telepresent experience, and whilst the installation retained this playfulness it simultaneously sparked a more urgent conversation when placed in the context of this United Nations (UN) Peace Negotiations backdrop. By simply augmenting the remote participants on a UN peace negotiations scene every public encounter that followed concerned the impending Iraq War and the inadequacies of politicians in resolving the situation. The 3D virtual model of the peace negotiations table and room was built using the 'Unreal Engine,' a 3D

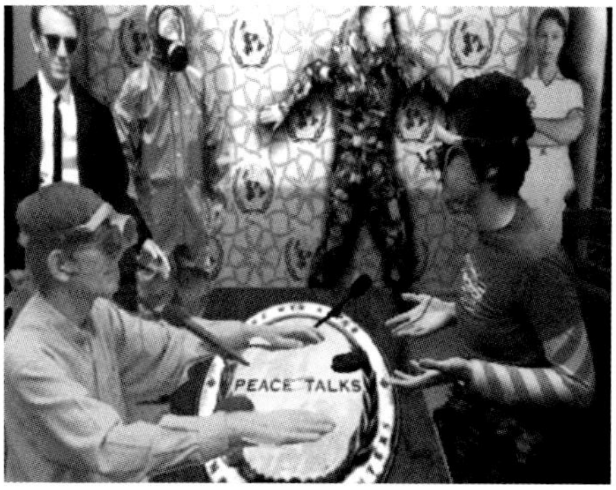

Fig. 4. Line-out video still of participants conversing in the virtual 'Peace Talks' environment. (Plate 10, p. 310)

modeling platform, primarily developed for first-person shooter games. Its digital aesthetic was clearly that of the computer game and included a typical peace talks room—an official table complete with UN insignia, papers of declaration, and microphones, as well as various military and security characters waiting and loitering in the background, ready to take up arms when necessary (Fig. 4). To add further confusion to the scene the two rooms were constructed using the 'Ames Room' technique, an optical illusion developed by American ophthalmologist Adelbert Ames, Jr. in 1946 (Dorward & Day 1997, 707–18). By using trapezoidal walls, floor, and ceiling the room still appears square from a fixed monocular perspective (camera position), but the person sitting on one side of the room appears much bigger than the participant sitting opposite.

The chaotic out-of-proportion scene depicted a political stubbornness and a futile attempt at holding peace negotiation talks, but simultaneously offered a viable telematic 'Peace Talks' alternative, which became more pronounced throughout the exhibition. Participants would talk almost exclusively on the subject of the looming war in Iraq more openly than they had appeared to previously. In order for 'Peace Talks' to function as an open system of interaction and conversation, audio communication had to be introduced. This was previously avoided in earlier telematic installations to encourage more expressive visual dialogues. However, in 'Peace Talks' speech was exploited to the point that participants began to forget the importance of time, place, and other appointments they might have had. Complete strangers were desperate to talk and share their feelings about the political situation for up to an hour on some occasions, as one participant exclaimed after 30 minutes of dialogue: 'I've just realised it must appear that I'm talking to myself here, people outside must be wondering what an earth is going on.'

I was struck by another participant's response when I asked how she dealt with differing points of view from her own, to which she replied, 'You kind of have the idea your going to come in here and negotiate or argue something out, and quite often you can be both on the same side.'

The telepresent view of the self in conversation was clearly affecting the objective view of the discussion. By viewing this conversation simultaneously as a third-person observer whilst also playing a first-person role directly within it provided each participant with a greater sense of empathy in wanting to resolve their differences, 'quite often you can be both on the same side' because you are observing the conversation from the same point of view. This is further expounded in the existential comments of another participant concerned about the consequences of their actions, 'I feel the image of myself viewed in this space talking to you is quite severer, I feel I could make some decisions here that would be viewed later as quite uncharacteristic.'

'Peace Talks' was a genuine proposal for the solution to the political stalemate of the time. It not only offers to overcome the geographical challenges involved in bringing heads of state together, it also meets the demands of body language, subtle gestures, and eye contact that are key to meaningful communication—that conventional videoconferencing falls short of achieving. While videoconferencing has become commonplace in business and domestic communication through applications such as 'Skype' or 'FaceTime,' peer-to-peer videoconferencing cannot replace a look them in the eye, the handshake that closes a deal, or a reassuring hug on the sofa with a relative. If you have ever used 'Skype' or 'FaceTime' I am sure you will have found yourself continually glancing at your own image in the smaller picture-in-picture window as much as the person you are videoconferencing with—and they will be doing the same. In effect this is a means of relayed eye-to-eye contact. Switching between the views of 'me looking at you' and 'you looking at me' for each participant is another way of converging these remote spaces. This is the first step towards creating the 'third space' that 'Peace Talks' and other telematic installations exploit by combining these views within the same image. The choreography of body movements, facial expressions, and hand gestures are key components of any conversation, often used unconsciously. 'Peace Talks' not only resolves the absence of body language, it offers more than our physical encounters permit. It allows the participants the opportunity to observe and reflect on the performed dialogue occurring in front of them whilst being directly responsible for it. A participant may be listening to the other, unaware of their hunched stance with arms crossed and head down, which is clearly seen as a stubborn confrontational posture by the other. But when the self is also the other, the participants are immediately aware of themselves both internally and externally, being confrontational to the other as well as themselves. Being in a position to reflect on the objective point of view they now find themselves within, they can adjust their pose and stance accordingly, having empathy with the other and the self.

On February 8, 2003 I participated in a seminar discussion at Worcester City Art Gallery on the theme of resolving conflict, involving humanitarian workers and other artists taking part in the BitParts Exhibition. One of the topics of the discussion concerned ways to bring communities of younger people together, with 'Peace Talks' as a feasible solution to connect younger participants who are divided by political conflict. The installation rooms were of shipping container dimensions, which could be feasibly fitted out with media and network technologies and deployed to make 'Peace Talks' possible across borders of longstanding conflicts, such as on either side of the Israeli/Palestinian West Bank, East and West Belfast, or between India and Pakistan etc. 'Peace Talks' is as much about play as it is about resolving conflict, and the role of play is at the core of its capability to establish greater empathy between remote participants. With younger participants in mind, the lasting effects of such encounters would impact the future of their generation through the legacy of this telematic intervention. The BitParts exhibition ended on March 1,2003, just 19 days before the Iraq War started.

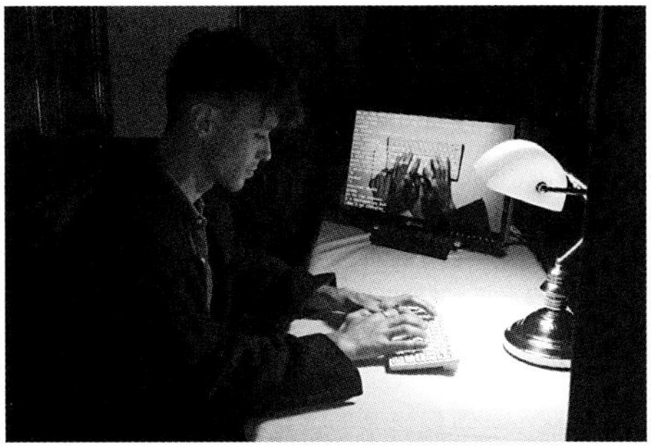

Fig. 5. Camera capturing image of participant's hands on the keyboard below in 'Touched. (Plate 11, p. 311)

Touched

Shortly after the development of 'Telematic Dreaming' I returned to Finland in 1993 to produce a new installation, 'The Telematic Séance,' for the exhibition 'Interactive Garden' curated by Erkki Huhtamo at the OTSO Gallery in Espoo, Finland. The installation was derived from the telepresent concept of 'Telematic Dreaming,' allowing two distant people the opportunity to coexist at the same séance table. What had explored a telematic extension of body scheme through a life-size digital other on a bed was used in 'The Telematic Séance' to new effect; echoing ideas of Spiritual Telegraphy in the 1900s that compared the invention of terrestrial telegraph to connections with the afterlife. Technically the installation functioned in exactly the same way as 'Telematic Dreaming,' but it was conceptually problematic for the participants involved. The title and séance context were too restrictive, limiting the potential agency within the piece and leaving the participants uncertain about what was being inferred. Twenty-five years on in 2017 I produced the installation 'Touched,' which technically worked in the same way as 'The Telematic Séance,' but in this new work the implications were more open. 'Touched' involved a computer screen and keyboard, where the projected hands of one person appear on the keyboard of another. The hands meet, touch, and type together on this keyboard surface, whilst also appearing as a live computer desktop image upon which a simple command line chat interface is overlaid (Fig. 5). This layering of keyboard, text, and image explores a new telematic experience of intimacy where the meaning of the type becomes dependent on the richness of touch.

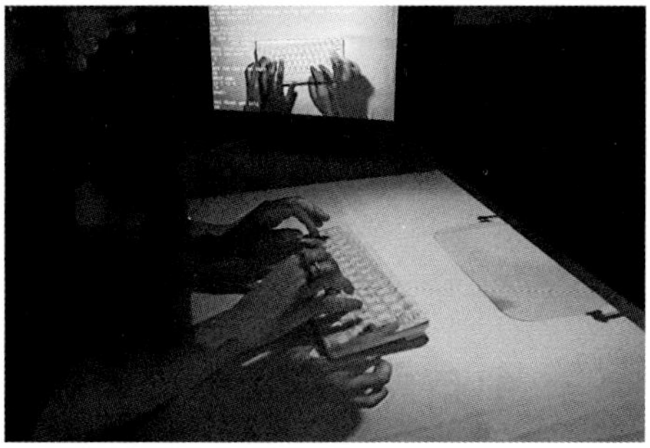

Fig. 6. Projected hands of the remote 'Touched' participant on the other participant's keyboard. (Plate 12, p. 311)

'Touched' was exhibited in the Digital Encounters Show at Horatio's Bar on Brighton Palace Pier for the British Science Festival in Brighton from September 5–9, 2017. Using standard video cables the installation was linked between two disclosed blacked-out booth spaces in the bar, approximately six meters apart. Horatio's Bar and its secluded ornate drinking booths, complete with tables and velvet seating, afforded an apt environment; its fake Edwardian interior assumed a context suggestive of a 'spiritualist' encounter of the 1900s, whilst its contrasting 1980s fabricated exterior provided a postmodern twist to the show.

The installation consisted of a keyboard and computer screen on a table with a video camera situated directly above it, picking up a bird's-eye view of the table top and keyboard below. This image was sent to a video projector and projected directly down onto another table and keyboard at exactly same scale and location as the original (see figure 6). A video camera situated next to the projector picked up an image of the projected keyboard and the aligned actual keyboard and sent it to desktop computer screens on the tables in each of the booth spaces. A Unix command line chat interface was then composited on top of the live video image, allowing the two separate participants to communicate via text over a Secure Shell (SSH) connection.[2] This provided a very simple raw text-messaging interface overlying the live video image, without any further distracting interface design. As each person began to type and chat his or her hands met and visually mixed on this telepresent keyboard surface where dialogues, games, and encounters unfolded.

Whilst on the one hand it is a completely simple set-up, it is on the other an entirely new physical encounter; echoing palmist conversations that suggest the convergence of virtual spirits and contemporary digital chat-room environments composited on screen. Our typing hands do not normally have any further role or function other than pressing keys, but 'Touched' turns our webcam chat attention away from our talking heads to focus on our hands, located on a shared telepresent keyboard, as a backdrop to a new space of remote intimacy. A space that was once the domain of textual exchanges now becomes the space of performance, abstract poetry, and hand-held encounters as a new sensation of being touched by text. The Brighton Pier venue and British Science Festival event attracted a diverse range of participants, from academics, local families, and teenage groups to seaside day-trippers and the occasional hen party. No one really knew what to expect; many participants were initially shocked and surprised, finding the opportunity to type messages a reassuring way of sharing their alarm. But playful encounters then unfolded. Participants played with aligning the keyboards, using them as props, as well other objects they might have on their person. One participant referred to it in a text message as 'keyboard ballet.' Hand clapping games and 'rock-paper-scissors' were also attempted, as well as the occasional rude hand gesture. Some people used their mobile phones to take and show photographs to each other. It also included putting their heads on the table, where profiles could meet and kiss. And it proved a good opportunity to discuss nail polish as well as a sense of proprioception. One member of a hen party turned up with a pack of Tarot cards; she moved the keyboard to one side and proceeded to make 'telepresent readings' for other members of the group. The table top alone provided an open platform to develop endless opportunities for engagement. The agency in 'Touched' was exploited in its secluded setting and contrasting context that was simultaneously artwork, invention, amusement, and fantasy.

In 1983 pioneering interactive digital media artist Myron Krueger wrote about his initial founding 'Metaplay' experiments in the late 1970s (Krueger 1991 [1983], 34) and the unexpected telepresent outcomes that led to his later VIDEOPLACE installations in the 1980s (Krueger 1991 [1983], 34–64). The encounter he described in those early experiments involving touching hands on a telepresent screen completely concurs with the phenomenological outcomes participants experienced in 'Touched.' Myron Krueger and his colleague were on the telephone discussing a waveform.

> After a few minutes of frustrating discussion, we realized that we had a far more powerful means of communication available. Using a two-way video link we turned the cameras on ourselves seated at our respective computers. As we did this, we used our hands to point to various features on the composite display

> (...). It was exactly as if we were sitting together at a table with a piece of paper between us. After a while, I realized that I was seeing more than an illusion. As I moved my hand to point to the data my friend had just sent, the image of my hand briefly overlapped the image of his. He moved his hand (...). I was struck with the thought that he was uncomfortable about the image of my hand touching the image of his (...). The inescapable conclusion was that the same etiquette of personal space and avoidance of touching that exists in the real world was operating at that moment in this purely visual experience (Krueger 1991 [1983], 34).

Both the awkward intimacy that Krueger experienced in 'Metaplay' and the converged images of hands and fingers playing, dancing, and caressing upon the keyboard in 'Touched' are clearly present in the findings of proprioceptive experiments in other discourses, namely the 'Rubber Hand Illusion' developed by Psychologists Matthew Botvinick and Jonathan Cohen from Pittsburgh in the early 1990s. This now classic 'body transfer' experiment was performed on me by a team of neuroscience researchers from the University of Manchester in 2003. I was positioned at a table so that one of my hands was out of sight under a small wooden bed tray, with a black tablecloth draped over it, from my shoulder to across the table. However my hand was still visible and accessible to the researcher sitting opposite me. A rubber hand was then placed on top of the tablecloth approximately 30cm to the left of my real hand, conceivably connected to my own arm under the draped tablecloth. The researcher then gently stroked both my hidden hand and the visible rubber hand simultaneously with a paintbrush in each of their hands. As I focused on the rubber hand in front of me the neuroscientist synchronously stroked the same parts of each hand, along each finger and across the back of my hand. After about ten minutes I was then asked to close my eyes and with my other hand point towards the index finger of my hand resting on the table. I pointed directly at the index finger of the rubber hand, which had clearly been incorporated into my own body scheme through this experiment. The same conclusive results also occurred extensively in the original 'Rubber Hand Illusion' experiment (Botvinick and Cohen 1998, 756). What Maurice Merleau-Ponty identified as the 'extension of the bodily synthesis' in 1945 (Merleau-Ponty 2012 [1945], 154) is now commonplace in what neuroscience and cognitive science refer to as 'body transfer' or 'body ownership' through experiments such as the 'Rubber Hand Illusion,' as well as discussions on virtual reality and presence research.

Sharing the same eyes

In Jacques Lacan's seminal paper 'The Mirror Stage as Formative of the Function of the *I* as Revealed in Psychoanalytical Experience' he describes the formation of the ego via the process of self-identification in the mirror, the 'self' being the result of identifying with one's own mirror image; a bodily wholeness constructed as if on stage in front of us (2003 [1966], 1–8). This concept has become increasingly significant in the development of my practice, as we can observe a similar process of identity construction through a sign and signifier relationship between performer and human-avatar in the digitally mirrored world. However, my work similarly concurs with Jean-Paul Sartre's provocation of the 'gaze' upon the other as confirmation of their presence; we perceive ourselves being perceived and come to objectify ourselves in the same way as we are being objectified (Sartre 2003 [1943], 245–326). Broadly speaking, Sartre and Lacan reach the same conclusion, but where they differ, in Lacan's view of the self and Sartre's gaze of others, my telematic practice simultaneously conflates their positions together. In my installations we share not only the reflection of the self but also the gaze upon the other from the same remote camera location. We are effectively sharing the same 'eyes'—the same point of view, where one's gaze of the other and view of the self are combined. The objectification of gaze is confronted on identical empathetic terms through this process of sharing our presence in a third space environment from a single viewpoint portal.

Notes

1. The Sony PLM-50 'Glasstron' personal LCD monitor was a head-mounted system that enabled individuals to enjoy realistic video equivalent to that of a 52-inch screen.
2. SSH, also known as Secure Shell or Secure Socket Shell, is a network protocol that gives users, particularly system administrators, a secure way to access a computer over an unsecured network.

References

Baudrillard, Jean. 1987. *The Ecstasy of Communication.* Semiotext(e), Translated by Bernard and Caroline Schutze. New York: Autonomedia.

Baudrillard, Jean. 1988. *Xerox and Infinity.* Translated by Agitac. London: Touchepas.

Botvinick, Matthew, and Jonathan Cohen. 1998. "Rubber hands 'feel' touch that eyes see." *Nature* 391(19): 756.

Dorward, Fiona M.C., and Ross H. Day. 1997. "Loss of 3D Shape Constancy in Interior Spaces: The Basis of the Ames-Room Illusion." *Perception* 26(6): 707–18.

Heidegger, Martin. 2001 [1953]. *Being and Time.* Translated by John Macquarrie and Edward Robinson. Oxford: Blackwell Publishers Ltd.

Krueger, Myron. 1991 [1983]. *Artificial Reality II.* Reading MA: Addison-Wesley Publishing Limited.

Kozel, Susan. 1994. "Spacemaking: Experiences of a Virtual Body." *Dance Theatre Journal* 11(3), November 7, 2018. http://www.art.net/~dtz/kozel.html.

Lacan, Jacques. 2003 [1966]. "The Mirror Stage as Formative of the Function of the *I* as Revealed in Psychoanalytical Experience." In *Écrits: a selection.* 1–8. Translated by Alan Sheridan. New York: Routledge.

Merleau-Ponty, Maurice. 2012 [1945]. *Phenomenology of perception.* Translated by Donald A. Landes. London: Routledge.

Packer, Randall. 2018. "The Third Space Network." *3SN*, November 7, 2018. https://thirdspace-network.com.

Sartre, Jean-Paul. 2003 [1943]. *Being and Nothingness: an essay in phenomenological ontology.* Translated by Hazel E. Barnes. London: Routledge.

Elaborating Mediation Through Media Arts

Wendy Coones

Volumes have been written on the development of museum exhibition mediation. Whether books or articles, specialist and interdisciplinary conference presentations or even TEDx talks, the dialogue on the mediation of content in a museum setting is ongoing and this article is only one humble piece in the larger puzzle. The research here does not attempt to develop the perfect exhibition strategy, nor list observations on Media Art's contributions to museums. The aim of this research is to connect a thread between two often discussed and rarely resolved aspects of museum exhibitions influenced by Media Arts in the early twenty-first century: first, the connection between emotions and mediation, and, second, an often implemented but rarely analyzed aspect of Media Art, namely the intentional maneuvering between medial manipulation and critical distance as a common artistic practice. Interpretive exhibitions have always made use of the most innovative media available. This has been and is still both lauded and criticized. The research in the next pages focuses on mapping examples across time spans and content to land at two contemporary mediated exhibitions influenced by media art practice. The common thread will also reference the connection to emotional engagement and the potential for creating an empathy-enabled environment.

Objects, Audiences, Media, and Art

Objects

The evolution from late eighteenth-century cabinet of curiosities (Wunderkammer) to late nineteenth-century natural history museums was centered on objects. Learning objects, remembering objects, exotic objects. Over time, museums developed into spaces where the audience stepped into the Wunderkammer and objects would thus become interpreted or (the current reference) mediated. Some previous mediation techniques could today be called immersive, since the media used brought the audience into very close, almost intimate proximity, or even direct interaction with the object. Exhibitions became objects themselves, and sometimes objects used in mediation balanced very close to the edge of spectacle or emotional manipulation. Current retrospection on the development of observer engagement in an emotional connection to museum content is exemplified by specifying 'Spectacle' in Rachel Poliquin's seven relationships to taxidermy: Wonder, Beauty, Spectacle, Order, Narrative, Allegory, and Remembrance (Poliquin 2012). We will visit spectacle in taxidermy shortly. We will also visit immersion and aura of a large-scale object, which becomes its own form of experiential spectacle.

Audiences

If the debate on audience engagement between Benjamin Ives Gillman and John Cotton Dana in the early twentieth century can be reduced to 'the Forum or the Temple' dichotomy (Duncan 1995 17, 70), then we see that theories about audiences' nature and motivation have developed in complexity over the past century as well. Jonathan Falk's research on 'types of visitors': Explorers, Facilitators, Professionals, Hobbyists, Experience Seekers, Rechargers (Falk and Dirking 2013), gives the audience varying degrees of autonomy, interaction, and responsibility. With the complexity of audience motivations, the mediation of content is equally complex.

> Discursive oppositions between science and spectacle, information and entertainment, immersion and distance, and passive and interactive spectators first articulated in relation to these visual technologies one hundred years ago have repeatedly resurfaced in contemporary debates over multimedia exhibitions in public museums (Griffiths 2008, 6).

Thus, any relevant dialogue on mediation as spectacle must examine the media literacy and expectations of the spectator.

Media and Art

There is a continual development and invention of new methods and materials to create images (images in the sense of the German word *Bild*, which can mean pictures moving, still, or 3-D, worldview images, or images from the mind's eye). Both art and industry stand at the forefront of the movement towards ever newer and novel interpretive media.

Although insights abound on the influence artists have to push the boundaries of the capacity of media, the words of media artist David Rokeby, active in the late twentieth and early twenty-first century are a cornerstone for this research:

> Many interactive artists create their own interfaces. Without the pressures and restrictions involved in getting a saleable and reliable product 'to market,' they are free to incorporate a richer complexity and ambiguity into these interfaces (Rokeby 1995, 150).

This richer complexity and ambiguity in mediation interfaces is a significant contribution Media Arts in the early twenty-first century can make to the museum exhibition strategies landscape.

Emotions, Manipulation, and Critical Distance

The engagement of the emotions through interaction in immersive settings has been well established by close examination of Virtual Art (Grau 2004). In tracing its historical predecessors and the reasons why we humans are so drawn to immersive environments, Oliver Grau's seminal research also pointed out the need for and important role of critical distance in allowing the observer to modulate the degree to which their emotions are manipulated by an immersive environment. The natural proclivity that artists have to critique culture and media manifests itself in the creation of media artwork's penchant to give critical ability back to the observer.

When museums moved from the touchable Wunderkammer of the Renaissance to the 'look but don't touch' of the Early Museum Age, the distancing between the object and the viewer began. This also continued the romanticization of the non-mimetic, non-haptic gaze. But as the museum evolved, so did the theories and practices of engagement. At the same time there were protagonists of the clean white cube where the works could affect the observer in visual stillness and silence, and there were also those dedicated to the teaching and study collections that were more constructivist and interactive in nature. As time progressed, the mediation techniques developed were ever-increasingly aligned with the design, craftsmanship, and technology available. As the eighteenth- and nineteenth-century collector of natural history brought back specimens to be prepared for display, and displays grew because they were found in large exhibiting institutions rather than in small private collections, the style of large dioramas with multiple animals in a life-like setting spread across the museum landscape. Institutions began to emulate one another's ever-higher degrees of suggestion with envy and ambition. The lifelike scenes of nesting and foraging birds, predators stalking prey, or the mating rituals of exotic fauna became the visual storytelling of natural history institutions. Museums used Magic Lantern presentations from the late eighteenth century well into the twentieth century to aid in the explanation of the objects, phenomena, and cultures presented (Griffiths 2008, 234). The late nineteenth and early twentieth centuries' predilection for Tableaux Vivants allowed audiences (predominantly women) with minimal or reduced geographical, educational, or financial access to museums to use the numerous 'how-to' manuals for the live creation of an iconic or popular artwork, allowing the participant's full-body engagement in an aesthetic experience.

This is the time when we see museums slowly begin to move in the two directions of the Forum and the Temple. The Temple was proselytized by Benjamin Ives Gilman, who wrote that museums should not come between the visitor and the work of art, whereas John Cotton Dana was calling for more interactive, experiential exchanges facilitated by the museum to take place between the observer and the work (Duncan 2009). During this debate, popular culture spectacles using immersive techniques like panoramas, peep-boxes, and other entertaining examples balanced on the precipice of overwhelming immersion that did not allow for enough critical distance (Grau 2004).

Distance in the classical sense vs. distance that allows critical thinking might be possible while immersed in an exhibition. The evolution of mediation techniques in exhibitions brings us to current practice, that now begs the following questions. Is it possible to have distance that is not critical? Is it possible to have a closeness and interactive immersion that also allows for critical thinking? Some of the current thinking on the use of new media exhibitions looks at cocreation and embodiment as bridges that connect immersion and critical examination by the viewer. The remediation of cultural narratives and experiences in a museum setting, rather than simply exhibiting collections, brings the audience into a remediated network.

> How can remediated networks serve the cultural institution sector? In the first instance, cultural institutions are concerned with authenticity and quality derived from interpretations of culture from a variety of sources. Remediated networks allow cultural institutions to be seen not as centers of knowledge, but rather as facilitators of networks of reliable, validated information (Russo & Watkins 2008, 156).

If we take into consideration Poliquin's differentiations (*Wonder, Beauty, Spectacle, Order, Narrative, Allegory, and Remembrance)*, but also the primary ways in which we make decisions, as demonstrated in the research by experimental psychologist and Nobel Prize winner Daniel Kahn, we see that humans go through the world making decisions as experiencing selves and remembering selves. One 'self' places value on the state of full experience above all else, and the other 'self' is most interested in the final impression that will be remembered. Just as these two selves decide whether we might choose the X or the Y in a laboratory experiment, these selves also react to the interpretation of museum content in exhibitions and the interaction with media artworks. The emotional considerations we employ in viewing, interacting with, and understanding exhibitions are largely beyond our control, and much happens at a purely intuitive level (Kahneman 2011). This sub-cognitive interaction with mediated content that artists do not shy away from should be further addressed and researched by the contemporary exhibition field.

Mediating Late Nineteenth-Century Natural Objects

In the year 1867, a taxidermy scene of four large figures: a man on a dromedary being attacked by two lions, was constructed partially from traditional taxidermy materials and partially from original body parts. The two lions are claimed to be Barbary Lions, now extinct, which at the time were a regular danger targeting travelers in Northern Africa. The work was presented in the Arts section of the Paris Exposition, where it won a gold medal. It was hailed as a masterpiece combining art and science, and exhibited in the Decorative Arts rather than the craftsmanship section, where taxidermy was typically displayed. The lions and dromedary in this work contain real bones from presumably the original animals. The skins are of course real, as is the usual practice in museum collections' prepared specimens and storytelling taxidermy. During a larger conservation action in 2017 that included deeper views through CT scans, it was discovered that the 'Arab' in the work *Arab Courier Attacked by Lions* also contains natural artifacts: a real human skull, its teeth having long been visible in the open cry for help. The dramatic/ized scene portrays the doomed man on a terrified beast of burden with a common dagger his only remaining weapon against the attacking male lion, as his damaged single-shot rifle lies across the lifeless body of the lion's mate. On display for 120 years at the Carnegie Natural History Museum in Pittsburg, which bought it from New York City's American Museum of Natural History in 1898, it has since been renamed to the translation of the original title, *Lion attaquant un Dromadaire.* (Fig. 1)

The work was bought by Andrew Carnegie for his new museum in Pittsburg after it spent 30 years in storage in New York, having been taken off display for being 'too theatrical' shortly after coming from France to America. It was sold for only US$ 50 at the time. This work's object biography is a demonstration of the unexpected heights realistic taxidermy traversed from the curiosity cabinets during the collection frenzy for exotic things, moving well beyond documentation and analysis by curators of Natural Science. The work's creators, the brothers Jules and Eduard Verreaux, were steeped in the art of their day with the likes of Mannerist-inspired contemporaries Eugene Delacroix (*Arab Horseman Attacked by a Lion* created 1849/50, lent to Paris by owner in 1864) and Antoine-Louis Barye (*Arab Killing a Lion*, 1860). The desire to bring pathos to the natural science specimen and raise it to an art form was being practiced and criticized in France, England, and Germany at the time. Already criticized in 1593 by Galileo, the emotionalization of 'scientific' objects remains even today under sharp dispute. Galileo perceived a great dichotomy

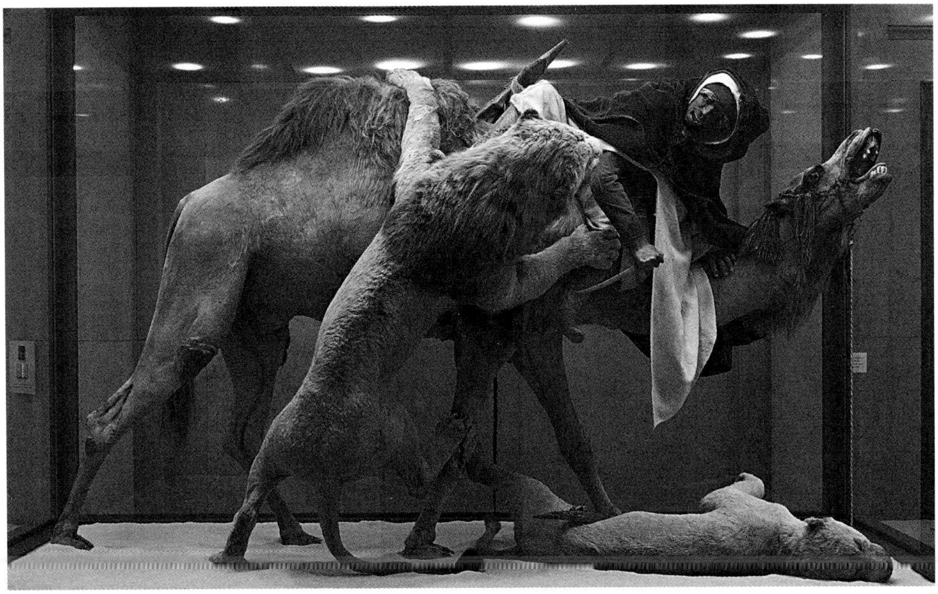

Fig. 1. Lion Attacking a Dromedary. Taxidermy scene constructed in 1867.
Photo: Ben Ledbetter (Wikimedia CC-BY-2.0).

between an insurmountable (…) classicist prejudice in favor of simplicity, order and *séparation des genres,* and against complexity, imbalance and all kinds of conflation (Panofsky 1956, 9–10).

We find it reiterated in ever-newer generations of artists, scientists, and presentation methods. Galileo's admiration for the classical and disdain of incongruence and 'curios' is found again in the nineteenth-century critique of the natural history dioramas as objects of voyeurism for spectacle-hungry general public visiting museums (Griffiths 2008). The opportunity to involve the audience in examination and discussion about emotionalization is currently being addressed through cocreation (Russo & Watkins 2008), and similar audience research critiques of 'theatricality' are being voiced about the un-reflective integration of simple spectacle-driven mediated techniques.

The opposite effect of *Lion Attacking a Dromedary* is reached by the classical aesthetic of collecting and displaying large-format objects with sparse to no interpretive techniques being used. The Pergamon Altar is such an example that would most likely conform to the opposing aesthetic taste: large-scale white marble with classical lines placed in a white cube that leaves nothing but the plain white object and the observer, for whom the object may or may not have emotional or intellectual meaning.

Immersion in an Original Object

The Pergamon Altar is a massive original object excavated from the Anatolia region of Turkey and relocated to Berlin. Excavation, transport, and eventual display began in 1878. The altar moved to its own house, the *Pergamon Museum*, in 1901, after two decades at the Altes Museum. The museological and imperialistic history of collecting Greek antiquities bears witness to the competition between the major capitals of Europe (Paris/Louvre, London/British Museum, Berlin/Altes Museum) in the race to possess a more impressive and significant exemplar of classical art from the Antique World, with the newly constituted German Empire as the youngest collector on the scene. The discovery, excavation, negotiated ownership, and display of the Pergamon Altar placed the new nation within the ranks of the European cultural elite. The identification of the new state with the Greek ideals represented in the altar structure and architecture, and the artistry of its elaborate frieze, was thus solidified for citizens of Germany.

The space where the object is housed has always had minimal interpretation. (Fig. 2) The audience is invited to experience the object in a way similar to a temple, but without visual, aural, or sensual disturbances. The focus of the emotional and haptic experience concentrates on the quietness of the space and the unmediated, sublime interaction with the object. The absence of mediation however means that the use and history of the object are not made readily available to the audience. When one sits on the cool marble steps to contemplate the aesthetics of Greek architecture, there is no smell of smoke from the altar's fires, no blood running down white stone, no cries of the sacrificial animals or bustle of the mingling Zeus and Athena devotees.

The interdisciplinary architectural historian Can Belsel coins the term 'regimes of the authentic' in reference to the phenomenon of tearing '"monuments" and "sites" away from the uses of everyday, creating a category akin to the sacred' (Belsel 2012). This enabling of a sacred-felt space in the architecture, accompanied by the aesthetic interaction with the Frieze sculptures, may eventually create an unbridgeable conflict between the overwhelming immersion in the original object and the unrealized critical distance that could enable the audience to understand the work(s).

> What the Pergamon Room has to offer is not merely an architectural backdrop to highlight and enrich the experience of the frieze itself. The feeling of the architectural sublime may overshadow, rather than assist, the experience of the original sculpture (Belsel 2012, 14).

An alternative mediation of the Pergamon Altar is currently on display, during the extensive renovations of the Pergamon Museum between 2014–23. A panoramic paint-

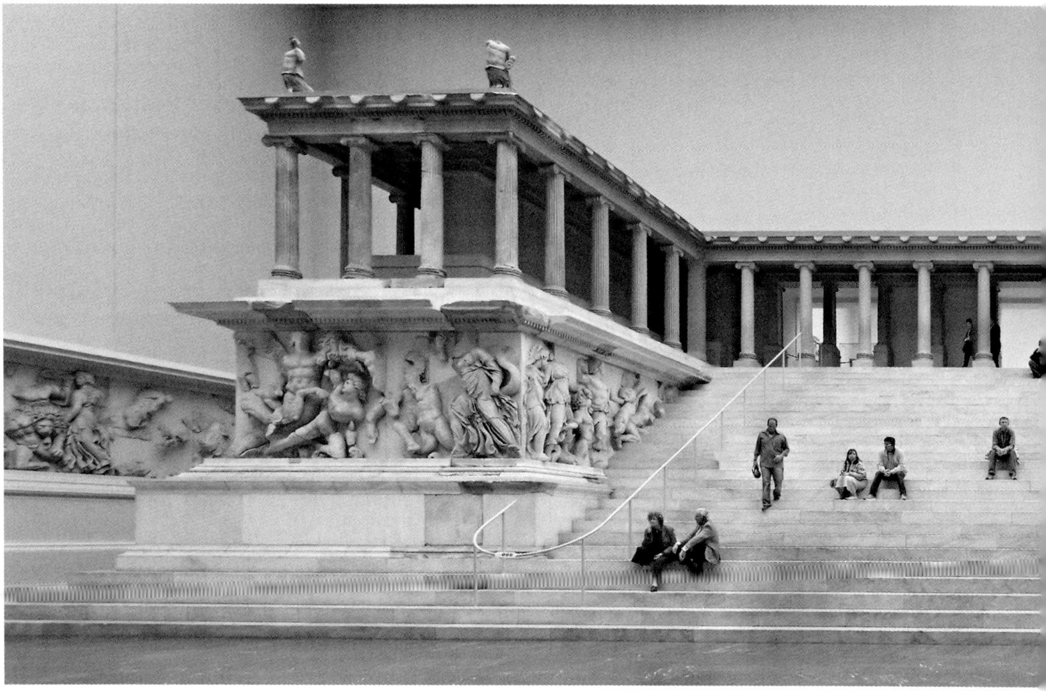

Fig. 2. Altar to Zeus in the Pergamonmuseum Berlin. Original temple built in the first half of the 2nd century BC in the ancient Greek city of Pergamon, currently in Turkey.
Photo by: Jan Mehlich (Wikimedia CC-BY-3.0).

ing by Yadegar Asisi enables the Berlin audience to experience a 360-degree visualization of the historical site as the entire landscape where the Pergamon Altar was located. The fully 'restored' image of the landscape integrates the 80 most significant artworks, reliefs, and sculptures recovered from the original archeological site and is combined with historical interpretations of the life and society of the time when the Altar was in use for its original purpose. The painting(s) depicts flora and fauna of the landscape, including the inhabitants, atmospheric elements like incoming rain clouds, rising smoke from the residential neighborhood, and many hidden details to be discovered. There are even background sounds and a soundtrack of music composed specifically for the panoramic experience.

The contemporary use of this historically laden mediation technique from the nineteenth/twentieth centuries inadvertently reverses the relationship to the earlier technique of placing an unmediated original object into a large neutral space. Perhaps somewhere between these two exhibitions lies a method for interpreting the Pergamon Altar that bal-

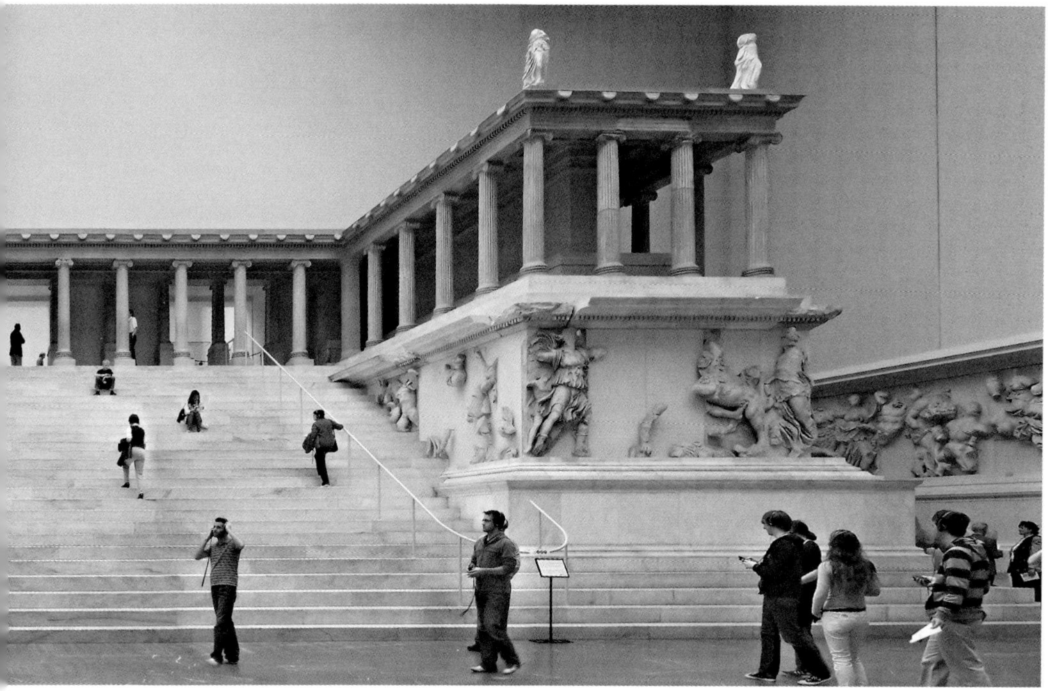

ances the sublime immersion in the immense object with a mediated critical distance to the historical and aesthetic interpretation.

We see in both the *Lion attacking a Dromedary* and the installation(s) of the Pergamon Altar how the evolution of mediation techniques in the pre-digital eras of the nineteenth and twentieth centuries set up a rubric of value systems and skill sets from the exhibiting institutions and the expectations and habits of their audiences. This basis of experience began to see dramatic changes in the late twentieth century based on factors of technology, society, and global interaction.

Interpretive Exhibition Strategies

The development of museums and the display of objects within them during the twentieth century is much too complex and multifarious to be adequately described here. Numerous books have been written on the subject, often focusing on a particular kind of museum, exhibition, or object type. Diverse interpretive styles add another layer to this variety. By the close of the twentieth century there could be found world-wide not only world fairs, festivals, and special exhibitions, collecting museums for natural history, archeology, culture, and the arts, but also thematical museums like the Hygiene Museum of Dresden, the Exploratorium in San Francisco, or the Rock-n-Roll Hall of Fame. Science Centers and Kunsthalle propagate across the globe, focusing not on collecting but on presenting concepts. During this birth of new genres of exhibition spaces and objects, the media arts were also going though radical changes. A new openness for materials, clearly embraced by the Dadaists, permanently opened up the gates of artistic creation to media beyond painting, drawing, marble, metal, and wood. Experimentation with all media by artists was possible and, as the digital age began in the 1960s, this new form of media also enthralled artists. Media artists were often pushing the boundaries of the genre to discover what it was physically and conceptually capable of. These works created as artworks would then later inspire spin-offs or distant relations that could be found in non-art-related exhibitions. The use of new media technologies as aids for interpretation in museum exhibitions gained ever-greater influence in the field. Though most often the new technologies were being adopted and developed in the larger budget institutions, even smaller houses created audio guides, interactive websites, or apps, authored intro videos, and interactive media explanations, as well as invited artists and creatives to co-develop exhibition elements. By the end of the twentieth century, transitioning into the beginning of the twenty-first century, companies specializing in the creation of new media interpretation strategies as well as in-house technology teams became the norm in all strata of collecting and exhibiting institutions. The questioning and critique of the spectacularization of knowledge presentation that has followed interpretive exhibitions since the beginning of the nineteenth century has lost some of its bluster in the twenty-first century. This is perhaps due to the fact that exhibitions must compete with alterative leisure time activities like film, video games, and theme parks, all of which have high levels of emotional engagement, but no obligation to self-examine the degree to which this might be misleading or detrimental to the audience. The pedagogical imperative of the museum exhibition automatically calls the pandering to the emotions of the audience into question. However, the competitiveness of the market for attention draws us back repeatedly to working with emotions or emotionally entwining media—simply because that is what people respond to. As soon as engaging Media Art techniques and technologies move to a popular landscape like a museum of natural science, the mandate not to involve the audience's emotion becomes much more difficult to adhere to, since the medium itself is emotionally engaging. Thus, new techniques are needed for critical distance that support engagement without getting so close as to render critique moot.

In searching for contemporary examples, two possible methods are found in the incorporate elements with potential for encouraging critical thinking. One method is related to Russo and Watkins opening up the opportunity for cocreation artifacts, to:

produce new types of cultural experiences where the audience is both the reader and the producer, and plays an active role in the remediation of knowledge (Russo and Watkins 2007, 161).

At the same time active knowledge creation is given to the audience, it takes a degree of authority and control away from the exhibiting institution. It is a dilemma well known to constructivists like George Hein and pragmatists like John Dewey.

The other method is activated through the interface itself. As Florian Wiencek makes clear in his ground-breaking research on mediation and museums, the affordances of engagement and distance can be given to the user through implementing either generous or ambiguous interfaces (Wiencek 2019, 259–62). Media artists regularly develop complex and ambiguous interfaces (Rokeby 1995) that allow interaction by the user that is both intellectual and critical. With these two methods in mind, we will now investigate contemporary examples bridging the gap between media manipulation and critical distance, between emotions and mediation.

Gundestrup Cauldron at Moesgaard Museum

The Gundestrup Cauldron found in northern Denmark is thought to have been created by multiple craftsmen with ties to Celtic traditions, and was made with materials originating outside Denmark, as far away as central Europe. The iconographical images on the cauldron are still being analyzed and debated as to their origin and the meaning behind the unusual combination of images most often associated with the different cultures known to have had contact at the time. The cauldron is one of the few existing examples of a highly decorated object that was created as a vessel for large festive meals most likely offered by someone of power. At the Moesgaard Museum just outside Aarhus one of the high-quality replicas is on display and an interactive media exhibition was created to help the visitor interact with the object's images and meaning. The object's meaning is primarily accessible through the interactive experience the exhibition developers chose to interpret through text, sound, and touch. The cauldron itself is on display with no text other than just the object information label. (Figs. 3-4)

The interpretive exhibition next to the object is a cozy, semi-dark circular room perhaps two or three meters in diameter, comfortably able to hold five to eight people. At the entrance there are four elaborate leather helmets hanging on a rack with instructions for visitors' use, including a diagram showing that the helmets cover the eyes and have headphones embedded in the earflaps that will accompany you during the short tour. Once activated, the text starts by inviting you to hold your hands out against the wall and follow the story

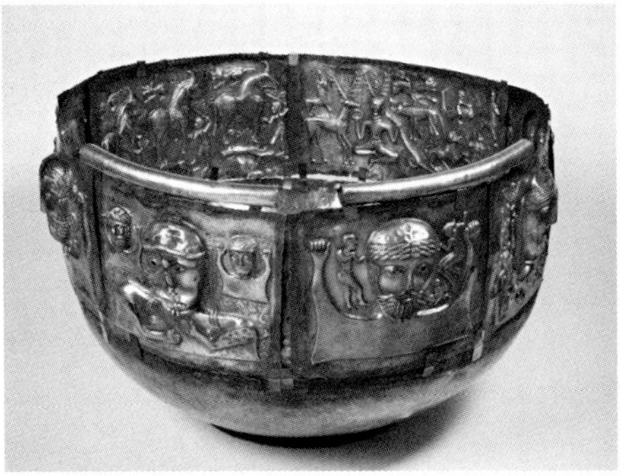

Fig. 3. Gundestrup Caldron. European Iron Age silver vessel assembled between 150 BC and 1 BC, found 1891 in Gundestrup, northern central Denmark.
Photo: Nationalmuseet (Wikimedia CC-BY-3.0).
(Plate 13, p. 312)

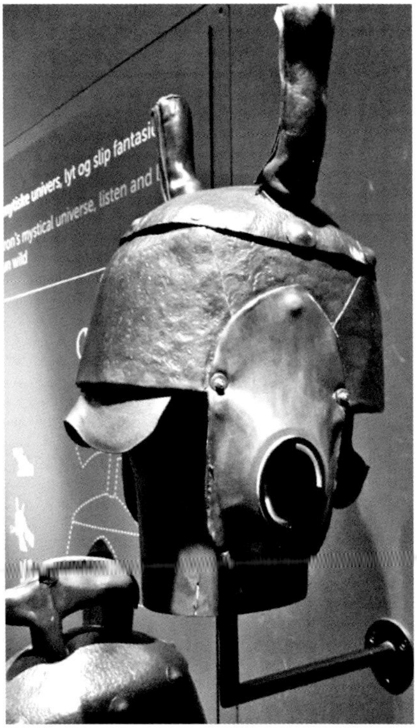

Fig. 4. Gundestrup Helmet. Leather "Viking style" helmet, which covers the eyes of the wearer and integrates headphones that play the exhibition's audio description of the relief wall.
Photo: Mie Ellekilde, with kind permission. (Plate 14, p. 312)

through your fingertips. The images that appear on the outside of the circular cauldron are recreated on the inside of the room. A kind of reverse embossing of the relief of the object is mirrored on the inner walls of the immersive space. The speaker lets you travel from one image to the next, telling you what it is that you are seeing (or rather touching) and what the meaning may be that is associated with the images that are found in the space. These range from images of fantastical animals that must have been part of the people's fantastical world of creatures, as well as humans interacting in both loving and violent ways. (Figs. 5-6)

The exhibit developer's choice to have the interaction with this object through the sense of hearing and the sense of touch is an interesting choice, especially in the context of the larger museum that is filled with objects and otherwise very visually stimulating. This one experience with one single, complex object is not interpreted through an audio guide or a video or a storyteller, but through an experience taking 10–15 minutes for an individual to move through. With their eyes covered and only touching the images, the user listens to the voice introducing each of the beings. Though this exhibition element is made as an interpretation of a historical object, the nature of that object is still and probably will ever remain under discussion regarding the meaning of its imagery and how this is connected

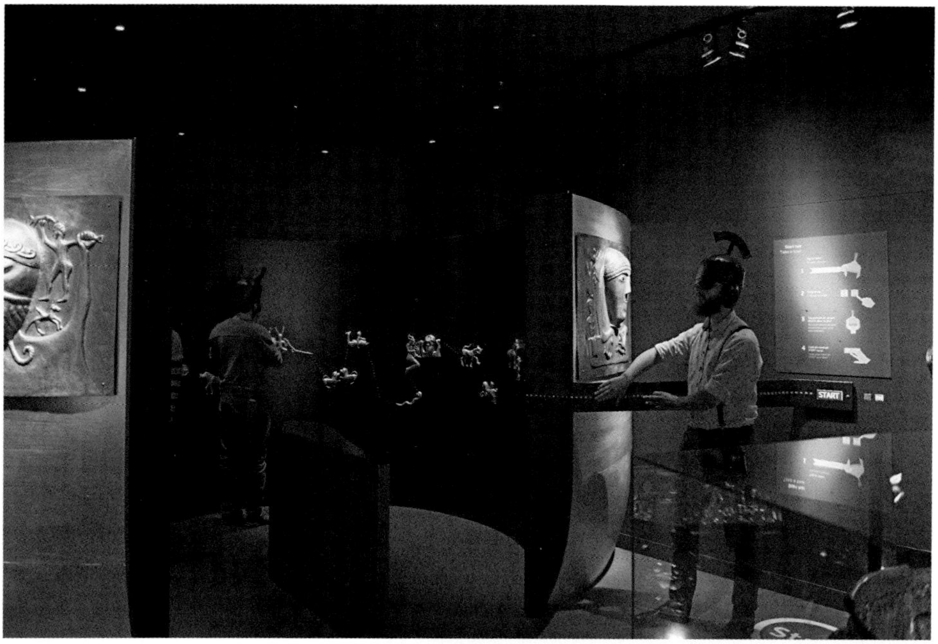

Fig. 5. Gundestrup Caldron installation. Part of the permanent exhibition at the Moesgaard Museum. Wide view of approach, entrance and inner caldron. Photo: Moesgaard Museum, with kind permission.

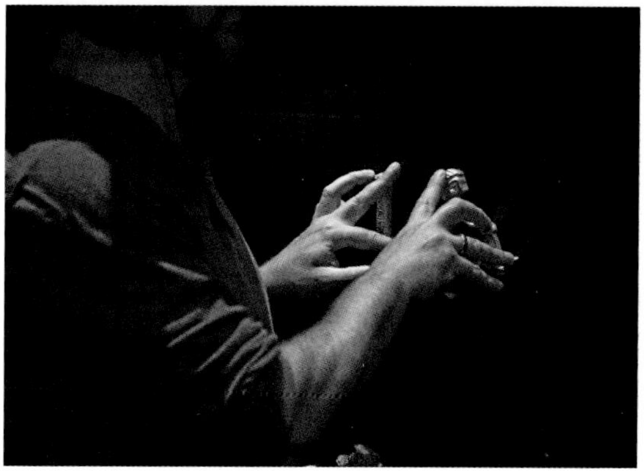

Fig. 6. Detail view of sensing relief figures of Gundestrup Caldron installation at the Moesgaard Museum. Photo: Moesgaard Museum, with kind permission. (Plate 15, p. 313)

to the lived experiences of the people of the time. The intricate details of the imagery are enlarged, so that being inside an oversized reverse-cauldron allows the chance to focus on specific imagery features and their stories. The exhibition is available for users to experience as they wish. It is possible to enter the room without a helmet and thus also without the audio commentary. The helmets that are available are large enough that one can 'cheat' and lift the front a bit to see the images quickly before returning to the touch/sound experience designed by the exhibition makers. This experience of a guided interactive sound experience has parallels in Media Art. The works of Janet Cardiff and George Miller invite observers to take a walk and experience a space under the guidance of an audio-experience artwork that overlaps the actual landscape with the virtual realm by influencing the listener to see, touch, and perhaps even perceive in a new way. The Gundestrup Cauldron exhibition is a similar experience, with the goal not to be an artwork itself, but to utilize techniques developed by early twenty-first-century media artists to engage the observer by offering the chance, and more importantly the choice, to immerse the senses into a story. The exhibit developers rely on the imagination and learned visualization capacity of their audience to fill in the gaps of the storytelling. Balanced somewhere between the dramaticism and decontextualization of *Lion Attacking a Dromedary* and the 'Object-Aura' of the Pergamon Altar in its museum setting, the Gundestrup Cauldron exhibition uses interpretative techniques that draw in the observer emotionally, while allowing for a critical distance that is open to and even encourages their own interpretation and enables a multisensory cocreation.

The interpretive technique-combination of immersing only one or two but not all the senses allows the observer to self-regulate the degree of immersion of the other senses. The Gundestrup Cauldron exhibition at the Moesgaard Museum creates an uncommon sensory experience of cutting off the sense of sight, and thus enhancing the sense of touch through the immersive technique of sound and text. Knowledge of the object itself is unclear as to the history and meaning of the imagery, setting up a precarious expectation of the exhibit developer to engage and inform the observer without introducing false or unproven information. In this exhibition, the observer is not only spared the manipulation of a full sensory interactive, but even given a small sense of what the scientists who study these objects must endure, namely, the scientist's uncertainty that comes with a limited amount of data available to understand and interpret such significant object's imagery is made to resemble the audience's limitation of observing it through only two senses. Not to mention the fact that this excludes the dominant sense of sight, which might echo the lack of information available about the living culture that created the object. Once the experience is over, the observer can return to the original object and see it again through the new eyes of someone who has haptically felt and aurally heard its contents.

The comments on the aims of the exhibition created at the Moesgaard Museum relay the intentions on emotional connection and critical distance:

> It [Gundestrup Caldron exhibition] strives to give experience of an opening to something that is not intellectual: the user enters a mythical universe where they must create their own images.
> MMEx – Center for Digital Interpretation in Museums http://mmex.dk (captured Jan 2019) (confirmed through telephone interview with exhibition designer November 2018).

This experience was created by a media designer and archeologist team, and gives a large degree of freedom to the observer to create their own interpretations and thus gauge their own internal flow of emotions opened up to the meaning making. Observers choose the level of critical distance and are even compelled to put some effort into the visualization of the images that their fingers feel. The use of the helmet as well insulates observers into their own realm of interpretation that cuts them off from the rest of the world and provides a closed-circuit connection between ear and finger and the reliefs of the Gundestrup Cauldron.

Pure Land 360 & Pure Land AR

Pure Land was created by the artist/museologist duo Jeffrey Shaw and Sarah Kenderdine, along with talented teams of technical support. Both of the primary authors have decades of experience with new media and cultural interpretation projects to draw upon. The Pure Land project has two primary outcomes, which continue to be displayed and expanded. It relies on the digitization of the Dunago Caves in China, whose documentation was embarked upon by the Chinese Government in 1944. The caves are covered in frescos spanning multiple dynasties and are located on an important stopping point along the Silk Road frequently traveled at the time the caves were decorated. The location is now remote, with few visitors, and (to protect the sensitive images) visiting highly limited. High-resolution scans of all surfaces in the Dunago Caves are part of ongoing research as to their imagery and meaning conducted by researchers with access to the data. Shaw and Kenderdine used the opportunity to create a new kind of mediated access to the images of Cave 220 in a deepened way. Through digital technology, they made multiple layers of analysis and interpretation possible. The images can come to life and also show views that are visualizations of scientists' theories of how it looked when created. Though it is technologically possible to create a fully immersive, total sensory experience of the original caves that would recreate the feeling of being there, this is not always the choice of the creators of Pure Land. There is one less exhibited version (*Pure Land Real*) that allows full 3-D Panoramic access via VR head-mounted display, which is not a primary focus here.

The two primary developments used as examples allow for interacting with the contents of the cave in a surrounding real architectural space. *Pure Land 360* is built for the Avi system, (Figs. 7-8) which is a surround projection and sound system that reacts to the user, pulling up and zooming in on information, images, and videos imbedded in the system. *Pure Land AR* (Augmented Reality) is built into a space the same size as the real caves. On the walls are drawn (not projected) the dots and lines typical of a 3-D camera's basis imaging. It feels like a cross between an old architectural blueprint and a starry sky.

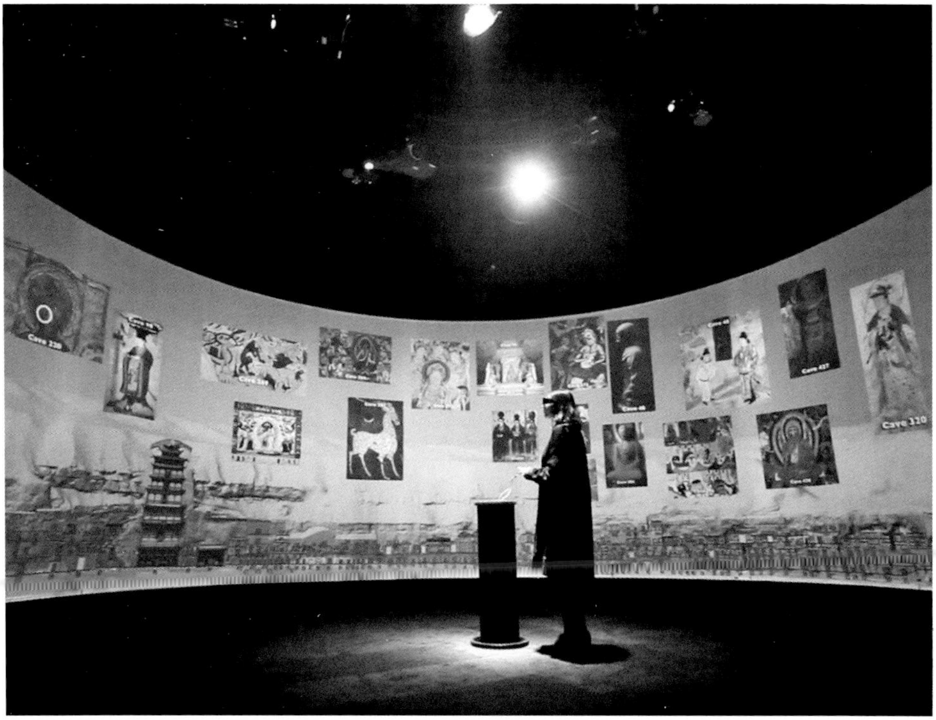

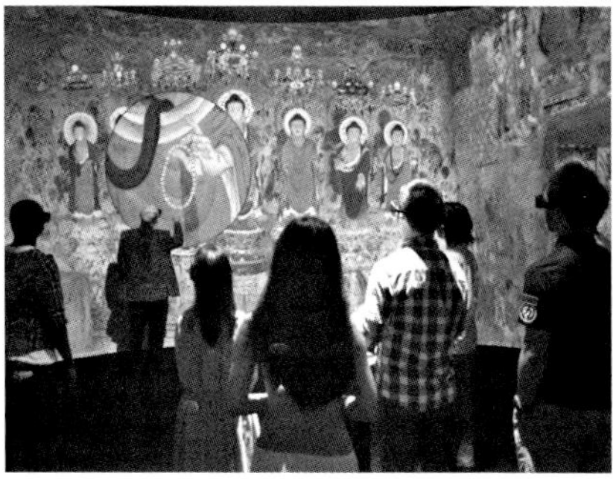

Fig. 7. *Pure Land 360*. Wide view. Jeffrey Shaw, Sarah Kenderdine, Cédric Maridet (sound), Grand Opening of Run Run Shaw Creative Media Centre, City University of Hong Kong, Hong Kong, China (Production: ALiVE/ACIM, School of Creative Media, City University of Hong Kong) (Documentation at Archive of Digital Art.)

Fig. 8. *Pure Land 360*. Close-up view highlighting 3-D interface with detailed information. Jeffrey Shaw, Sarah Kenderdine, Cédric Maridet (sound). With kind permission of the artists. (Plate 16, p. 313)

The interaction with the images happens through an iPad set-up with motion sensors so that the screen becomes a window through which the digital recordings of the originals and the interpreted visualizations and theories about the images can be seen, even while moving around in the physical space. The user takes the window into this world through the space with them. While *Pure Land 360* functions a bit like a surround network, giving

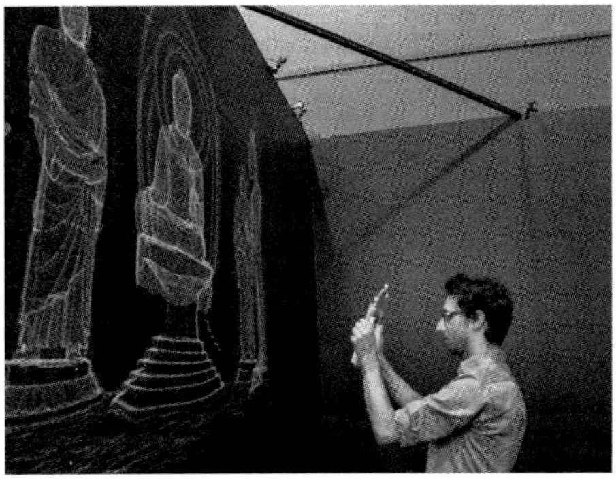

Fig. 9. *Pure Land AR.* Jeffrey Shaw, Sarah Kenderdine, Leith Chan, Art HK 2012, Hong Kong Convention and Exhibition Centre, Hong Kong, China (Production: ALiVE/ACIM, School of Creative Media, City University of Hong Kong)

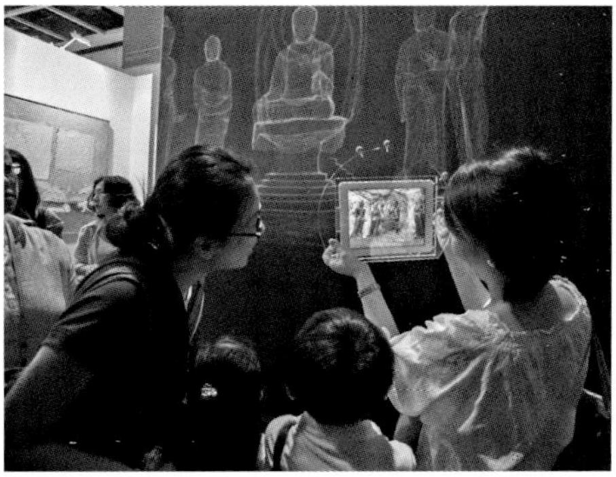

Fig. 10. *Pure Land AR.* Close-up and view of images and content as seen by the viewer via tracked ipad. Jeffrey Shaw, Sarah Kenderdine, Leith Chan,
(with kind permission of the artists.) (Plate 17, p. 314)

the observer full 360 degree access to the entire space of the cave, as well as the interactive ability to zoom in, concentrate, and pull up more detailed information, *Pure Land AR* gives the same information in a smaller and seemingly low-media space. (Figs. 9-10) The reduction of the total media presence in the immersion allows for more concentration on the information, though perhaps at the loss of the 'wow' effect that *Pure Land 360* offers.

It is clear that as Media Art production matures from the late twentieth to the early twenty-first century, many fields that grapple with the need for critical distance and must deliberately address the balance between media immersion and media control will benefit from the constant evolution of Media Arts entering non-art realms. The critical thinking involved in art making, when turning its energies upon technologies, creates a latent benefit for the other fields using new media technologies as part of their interpretive efforts.

> Ethical considerations are obvious if one considers that virtual reality tends to encourage users to suspend belief and render the interface 'invisible.' Terry Harpold (2001), researcher in new media culture, warns of 'direct manipulation' (possible when there is the illusion of user immersion) pronouncing it 'among the most privileged methods of psychic and political coercion of the post-enlightenment period.' He goes on to suggest, however, that we may—through understanding the history of contested spatiality of science, art, and politics away from the computer—'provide strategies of design that break the epistemic confines of direct manipulation' through the computer (Kenderdine 2007, 315).

This earlier writing of the *Pure Land 360* and *Pure Land AR* co-author Sarah Kenderdine demonstrates the basis for interpretive discussions on her later works. During the time that technology and parameters for the *Pure Land* project were in development, Kenderdine and Shaw were considering how to allow for immersion without direct manipulation, to suspend disbelief without coercion. An in-depth description and analysis of these works, in conjunction with the artists' own considerations, must be saved for another endeavor beyond this brief research. After both *Pure Land 360* and *Pure Land AR* have been displayed in numerous environments and undergone revisions and updates, Kenderdine's reflections almost a decade later on the influence of Media Arts and artists on the future of mediation technologies in museum contexts are significant.

> The history of experimental interfaces for cultural heritage materials dates back to the 1990s in a series of works by influential media artists. In 2015, the powerful nature of these experiences is now recognized by industry and will no doubt become the basis for further developments in screen(less) technologies and immersive environments (Kenderdine 2016, 37).

When seeing the three outcomes of the discrete endeavors, the Pure Land iterations provide contrasting examples of presenting the same content with different yet similar digital technologies. The 360-degree fully immersive single-user view of *Pure Land Real* leans more towards an intimate spectacle, while *Pure Land 360* feels like stepping into a personal interactive library with fully mediated access to imagery and research visualization. *Pure Land AR* feels like the dimensions of the real cave, but the compelling interface encourages exploration and rewards the user who seeks to discover invisible treasures through the view window on the subtly notated walls. Where *Pure Land Real* would be an example of a generous interface, both *Pure Land 360* and *Pure Land AR* have complex and ambiguous interface elements that give back critical distance to the audience while maintaining immersion.

Learning Mediation from Media Art

> As (exhibiting) institutions were reconsidered, the ways in which museums were used would change, developing deterritorialized models of museum environment. (...) In the deterritorialized space of remediated networks, ascribing meaning to space requires audiences to engage in a type of escape from reality and transport themselves from the physical space within the city to a potential space (Russo and Watkins 2008, 157).

There is an inverse relationship between the deepening refinement of the powers of engagement addressed in the Media Arts and the expansion of the definition of museum and its role in society. As museums and other institutions exhibiting cultural content adopt ever-increasingly new and more complex technologies for methods of interpretation and mediation, the debates over best practices will continue and evolve. Media Arts of the early twenty-first century are deeply concerned with actively addressing issues about engagement of media with the user. Critical thinking manifested as experiments or conceptual questioning in media artists' practice is being viewed and utilized by exhibition makers to encourage new kinds of collaborative meaning-making. The integration of ambiguous interfaces, intentionally restrained interactive engagement, and cocreation exhibitions can leave mental gaps in the story that give the audience the critical distance to be part of the storytelling. The influences of media arts also develop a 'deterritorialized space,' inviting the audience to enter a new 'potential space' where they are cocreators of the knowledge, experiences, and memories in the content.

Perhaps because new media are more readily capable of allowing criticism of the content presented, it is possible to afford a critical gaze without requiring distance. It is possible to be reflective and apprehensive of the experience while immersed in it. Perhaps it is also dependent on the degree to which the creator uses manipulative versus empowering techniques to overwhelm the audience, alongside the degree to which the audience can construct and insert their own perceptions and thinking. Perhaps under the example and guidance of Media Arts in the twenty-first century, museums and exhibitions can achieve critical distance *within* mediated immersion, in a new potential space that allows the audience emotional engagement while participating in a mediated meaning-making process.

References

Belsel, Can. 2012. *Antiquity on Display: Regimes of the Authentic in Berlin's Pergamon Museum.* Oxford: Oxford University Press.

Duncan, Carol. 1995. *Civilizing Ritual: Inside Public Art Museums.* London: Routledge.

Duncan, Carol. 2009. *A Matter of Class. John Cotton Dana, Progressive Reform, and the Newark Museum.* Pittsburg PA: Periscope Publishing.

Griffiths, Alison. 2008. *Shivers Down Your Spine.* New York: Columbia University Press.

Erwin Panofsky. 1956. "Galileo as a Critic of the Arts: Aesthetic Attitude and Scientific Thought." *Isis* 47(1): 3–15.

Falk, John H., and Lynn D. Dierking. 2013. *The Museum Experience Revisited.* Walnut Creek CA: Left Coast Press.

Grau, Oliver. 2004. *Virtual Art: From Illusion to Immersion.* Boston MA: MIT Press.

Kahneman, Daniel. 2011. *Thinking Fast and Slow.* New York: Farrar, Straus and Giroux.

Kenderdine, Sarah. 2016. "Embodiment, Entanglement and Immersion in Digital Cultural Heritage." In *A New Companion to Digital Humanities* edited by S. Schreibman, R. Siemans, and John Unsworth, 22–41. Arundel: John Wiley & Sons, Ltd.

Poliquin, Rachel. 2012. *The Breathless Zoo: Taxidermy and the Cultures of Longing*. University Park, PA: Pennsylvania State University Press.

Rokeby, David. 1995. "Transforming Mirrors: Subjectivity and Control in Interactive Media." In *Critical Issues in Electronic Art*, edited by S. Penny, 133–58. Albany NY: State University of New York Press.

Russo, Angelina, and Watkins, Jerry. 2007. "Digital Cultural Communication: Audience and Remediation." In *Theorizing Digital Cultural Heritage* edited by Fiona Cameron and Sarah Kenderdine. Boston MA: MIT Press.

Wiencek, Florian. 2019. "Digital Mediation of Art and Culture. A Database Approach." Published dissertation: IRC Library, Information Resource Center: Jacobs University Bremen.

II

Subjects of Surveillance

....

Reclaiming Human Agency in the Big Data Universe

Qualculative Poetics

An Artistic Critique of Rational Judgment

Derek Curry and Jennifer Gradecki

All online content is subject to surveillance by software-assisted intelligence analysts for the production of open-source intelligence (OSINT), or intelligence created from publicly available information. Systems for OSINT surveillance are designed to facilitate, and sometimes automate, rational scientific judgments. These judgments have been identified as a type of rationalized control that can be utilized for social domination. A framework for public oversight of OSINT systems has not yet been developed, nor have the critical conceptual tools needed for an effective critique. Our solution has been to situate the process in the material aspects of the rational decision-making process and to apply techniques from visual studies and critical theory as a way to both understand and critique secretive surveillance practices.

While rational decision-making is often seen as an innate human ability, some theorists have argued that it is the result of a set of materials and conventions. In his discussion of madness, Michel Foucault explains that the distinction between reason and nonreason was the result of a process of othering and that the scientific rationale for this distinction was applied afterward (Foucault 1988, ix-x). Herbert Marcuse traces the initial split between rational and nonrational to the emergence of formal logic, which distinguishes what is calculable from the incalculable. The calculable is deemed to be 'objective' and 'universal,' while the subjective, noncalculable is either ignored or reduced to a calculable form (Marcuse 1964, 137–38). Michel Callon and John Law have used the neologism 'qualculation' to describe the reductive process necessary to make judgment-based calculations. Qualculation de-emphasizes the distinction between rational and nonrational while revealing the unacknowledged array of materials and set of practices that combine to constitute rational judgment. Qualculation foregrounds how rational, scientific meaning depends on material objects in spatio-temporal arrangements. Since the production of non-rational meaning in art and poetry also depends upon the arrangement of visual and symbolic elements, homologies can be found in the way meaning is constructed in both rational and nonrational judgments.

While intelligence analysts tend to conceive of analysis and analytical tools as empirical and scientific, judgments made through OSINT systems depend upon visual representation, and therefore have an aesthetic, nonrational dimension. We will argue that OSINT surveillance tools are part of a socio-technical system that does not inductively reveal hidden facts to analysts; rather, these tools help to construct meaning through a complex, creative process where data are collaged together from disparate sources.

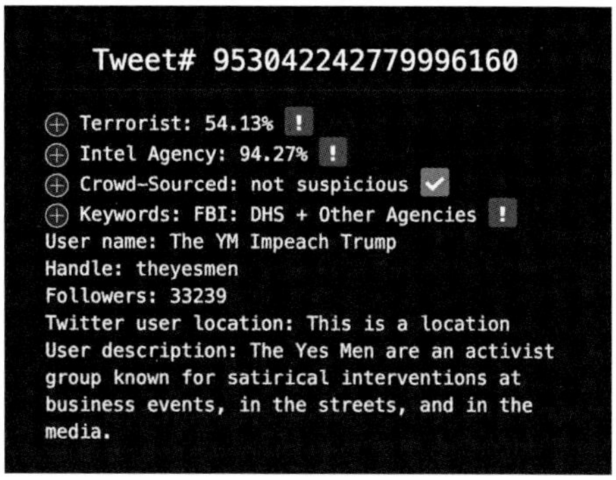

Fig. 1. Metadata in the CSIA Watchlist.

We have been exploring the role of qualculation in OSINT systems through our artistic research project, the Crowd-Sourced Intelligence Agency (CSIA). The CSIA is a partial replication of an OSINT surveillance system created by using information from leaked documents, technical manuals, ethnographic reports, and Freedom of Information Act (FOIA) files. Through the CSIA, we have been investigating how the arrangement and display of data in OSINT systems produces meaning, specifically in relation to social media. Users of the CSIA make decisions about whether posts from the microblogging site Twitter are threats to national security, and must base their decisions on the display and arrangement of information in the OSINT interface. For example, when a tweet is viewed in the Watchlist, users see the tweet's metadata, which include the user's name, handle, number of followers, location, and user description, as well as evaluations of the tweet made by multiple predictive policing machine-learning classifiers, and any keywords from the US Department of Homeland Security's (DHS) Analyst's Desktop Binder that have been found in the tweet (Fig. 1). Based on an isolated tweet's metadata and content, which are reframed within the OSINT interface, users are asked to determine whether or not the tweet represents a threat to national security. Users see the results of the machine-learning classifiers displayed as a percentage, so a tweet may have an 83% likelihood of being made by a terrorist. The classifiers are trained on different corpuses of data; for example, the CSIA's terrorist classifier contains posts from known terrorist groups such as the Islamic State of Iraq and Syria (ISIS) and Boko Haram. A tweet may be rated as a threat to national security, highly likely to be made by a terrorist, and simultaneously highly likely to be made by an intelligence agency, demonstrating how classification in one category does not exclude the possibility of classification in other categories, as well as the potential for false positives. The aesthetic experience inherent in making decisions about national security using these tools allows users to see that decision-making is not solely a rational, calculative process. We call this experiential understanding a visceral heuristic (Gradecki and Curry 2017). Replicating an OSINT system has allowed us to examine the types of decisions encountered by analysts without gaining access to the classified information produced by these systems.

'Qualculative poetics' is a conceptual tool we have been developing for resisting the domination of technological rationality, to enable artists and cultural theorists to analyze, critique, and potentially undermine OSINT surveillance processes (Knouf, Pederson, Curry, and Gradecki 2015). Qualculative poetics can add a much-needed counterpoint to the dominant rationalist narrative, providing a dialectic and a space for creative thinking. With this concept we hope to enable artists and cultural critics to parse and analyze practices that are both hidden and within their domain of expertise.

OSINT

Open-source intelligence has been used in the United States since at least 1939, when the US State Department formed the Joint Committee on Importations to secure access to foreign, primarily German, research publications amid the wartime embargos (Richards 1981, 253). This was followed in 1941 by the Foreign Broadcast Monitoring Service (FBMS), a precursor to the Office of Strategic Services (OSS), which was tasked with monitoring the radio and print news from Germany, Italy, and Japan. The FBMS was renamed the Foreign Broadcast Information Service (FBIS) in 1942, and placed under the administration of the Central Intelligence Agency (CIA) when it was created in 1947. Until the widespread use of the internet, the primary role of an OSINT agent was to collect, translate, and summarize foreign publications (Williams and Blum 2018, 1). This type of OSINT collection was not only uncontroversial, it was actually supported by academics, who have historically been among the CIA's harshest critics. For instance, because cultural studies departments of many universities in the United States were dependent upon information from the FBIS, when the CIA's budget was slated to be cut in 1997 to reflect a post-Cold War reality, the Federation of American Scientists (FAS), which was founded to promote nuclear disarmament, mounted a successful campaign to save the FBIS from budget cuts (Kempster 1997). In the current system of US intelligence, the collection and analysis of OSINT data are distributed across multiple agencies. The Defense Open Source Council (DOSC) oversees defense-related OSINT for the Department of Defense (US DoD 2010); geographical OSINT, such as satellite imagery, is under the purview of the National Geospatial-Intelligence Agency; and the CIA is the executive agent for the Open Source Enterprise (OSE), the successor to the FBIS. Even though OSINT is, by definition, compiled from public information, intelligence produced through OSINT systems often receives a classified designation (Williams and Blum 2018, 19).

With the rise of the Internet, new sources of information, such as real-time data feeds and user-generated content, have become available, necessitating a shift in OSINT gathering and processing. RAND researchers Heather Williams and Ilana Blum describe this shift in OSINT practices as 'second generation OSINT' because it represents such a significant break from previous practices (Williams and Blum 2018, ix). The role of an Open Source Officer has expanded over the past decade to include military operations and the surveillance of individuals and groups. For instance, the creation of weapons for cyber warfare became part of the OSE's purview when it was incorporated into the Directorate of Digital Innovation in 2015 (Brennan 2015). Automated translation programs have replaced human agents as the primary method of translation, and Open Source Officers must now be able to collect and process data across multiple, sometimes incompatible formats, contextualize both the source and the information produced, and, in some cases, protect the privacy of a source. With the development of Web 2.0, OSINT has come to include content produced by social media users, such as Twitter or Facebook posts, YouTube videos, images from photo-sharing sites like Instagram, or user-generated maps created on sites like Google Maps, Wikimapia, or Waze. These networked platforms have no standardized method for accessing their data; each platform has its own API and conventions for sorting and storing data. To deal with the difficulty posed by working across media, platforms, and data types, intelligence agencies typically use commercial 'off-the-shelf' tools for processing social media, and may choose to purchase data from a commercial aggregation service (Williams and Blum 2018, 17). These services were designed for commercial applications and are not optimized for national security objectives. To complicate matters, intelligence analysts are discouraged from using social media themselves (Williams and Blum 2018, 3), making them unfamiliar with the colloquialisms of social media communications.

The introduction of social media posts and other real-time datasets created by nonprofessional content creators means that qualitative judgments are now part of an OSINT analyst's job. Despite the central importance of interpretation in their practice, analysts tend to view their decision-making process as primarily inductive and empirical, and are skeptical of philosophical epistemological discussions and theoretical models of their practice.[1] This could be due to the fact that analysts receive very little training in methods of critical thinking (Herbert 2006, 667) and tend not to reflect on how the analytical process itself may affect or alter the information they are judging (Cooper 2005, 6). The disavowal of nonrational forms of knowing by the intelligence community is discussed by Jeffrey R. Cooper, who warns intelligence analysts that their mythologized way of thinking 'fails to recognize the important role of less "rational" and less "scientific" elements, such as imagination and intuition,' which are integral to intelligence analysis (2005, 27). He calls this narrow view of science 'scientism.'

What Cooper calls scientism is a view of science that is heavily influenced by the philosophy of positivism. For the positivists, the only true knowledge is that of empirical science, which can be instrumentalized or operationalized for human progress. Donald Schön refers to this 'harnessing [of] science to create technology for the achievement of human ends' as 'Technical Rationality' and traces how this positivist view of science became embedded in the modern university of the late 1800s (1984, 31). By the late 1800s, Positivism had already come to dominate the universities of Germany and the United States. Technical Rationality came to be considered conventional wisdom, science became 'the only source of positive knowledge of the world,' and metaphysics, or abstract theory, came to be seen as 'pseudo-knowledge' (32). The wars of the early twentieth century demonstrated how useful operationalized science could be for military purposes. The discipline of operations research was founded from this ethos, and the nuclear weapons developed from the Manhattan Project became a symbol of its success (37). The means-end rationality of the positivist paradigm is reflected in the tendency of intelligence analysts to be drawn to scientific knowledge to improve their practice, as well as their tendency to disavow the metaphysical aspects of their practice, such as deduction, creativity, and intuition, which cannot be scientifically rationalized.

Technological Rationality

Herbert Marcuse proposed that technology, which is the result of the means-end rationality of operationalized science, embodies the ideology of positivism. In a society that depends upon technology for its wellbeing, 'technological rationality' is how the scientific principles embodied in that society's ubiquitous technology come to be seen as reasonable, logical, and even natural. In this way, the technology will engender specific types of decision-making while discouraging others. Marcuse introduced the term in 'Some Social Implications of Modern Technology' (1941) and expounded on it in *One Dimensional Man* (1964). Technology is science operationalized for domination. Initially technology was used to dominate nature, then techno-scientific control was extended to people. The scientific management of Taylorism is an obvious example. Technology itself is politically indifferent, however, and the social control it embodies is easily mobilized by the dominant forces of any society for its own political ends. Marcuse explains that 'an electronic computer can serve equally a capitalist or socialist administration; a cyclotron can be an equally efficient tool for a war party or a peace party' (1964, 154).

Inherent in technological rationality is the belief that both nature and human actions are perfectly calculable using scientific tools (Marcuse 1964, 158). It is the result of Aristotelian formal logic, which takes specific instantiations of objects in the physical world, including people, and abstracts them into idealized forms that can be quantified, compared, measured, and mathematized. When this mathematical 'ideational' reality can be correlated with empirical reality, it is believed to be absolute truth (Marcuse 1964, 162–63). It is through this process that technological rationality turns nature and people into abstract elements that can be quantified for the domination of nature or society. Because these abstractions are needed to create technologies and are efficiently performed by technology, Marcuse calls technology 'the great vehicle of *reification* (…) in its most mature and effective form' (Marcuse 1964, 168–69).[2] Not only does technology allow workers to be interchangeable with one another, it can reduce people to quantifiable and operational data points to serve a variety of ends.

In the case of an OSINT surveillance system, communications of social media users, YouTube videos, images, online news reports, blogs, and global positioning data are all harvested and turned into measurable and quantifiable elements, often referred to as raw data. This reification is the first step towards attempting to make rational scientific decisions about intelligence. The next steps are analysis and exploitation, which can involve quantification, gaussian distribution, statistical analysis, and a number of other mathematical-scientific procedures. If the result of these processes is information that correlates to a plausible conception of reality, the information is included in a report and may be combined with classified intelligence to produce 'actionable' intelligence that allows the 'customers' of intelligence agencies to implement 'explicit responses' (Office of the Director of National Intelligence 2013, 71). The implementation of the intelligence cycle through OSINT systems is a process in which complex social interactions and events are reduced to data points, rationalized, and made into idealized forms that can be quantified, arranged, and visualized to allow analysts to make judgments and produce intelligence.

The ability to monitor entire populations as both individual data points and aggregated data is technological control par excellence. For US intelligence agencies,

OSINT from social media was the primary source of real-time information for the protests that followed the 2009 presidential elections in Iran (Williams and Blum 2018, 2) and has been used to gauge support and opposition for ISIS (Bodie-Baron, Helmus, Magnuson, and Winkelman, 2016, 5–11). Researchers for the Swedish Defence Research Agency believe that social media surveillance can be used to detect individual acts of violence, or 'Lone Wolf Terrorism' (Cohen, Johansson, Kaati, and Mork 2014). These same technologies have also been used by many governments to monitor and disrupt domestic movements pertaining to issues such as economic justice, civil liberties, and fair elections (Owen 2017, 688–97).

Qualculation

Sociologists Michel Callon and John Law argue that what is typically referred to as rational decision-making is really a type of judgment-based calculation which they refer to as 'qualculation'. Just like numerical calculation, qualculation is not a natural or innate quality of humans: to move from estimation[3] or intuition, both require an often unacknowledged set of rules and material devices (Callon and Muniesa 2005, 1230). Qualculation is 'a process in which entities are detached from other contexts, reworked, displayed, related, manipulated, transformed, and summed in a single space' (Callon and Law 2005, 730). Qualculation depends on material arrangements, systems of measurement and standardization, methods of displacement, and a spatio-temporal frame (Callon and Law 2005, 730). Callon and Law argue that what is typically believed to be a distinction between the rational and nonrational is actually a distinction between the ability and inability to qualculate (2005, 729). Nonqualculability is the foundation for religious and postmodern discourses, as well as philosophical skepticism, which all tend to be considered unscientific and nonrational (Callon and Law 2005, 732). Using the concept of qualculation to understand how rational judgments are made possible facilitates an examination of the typically unacknowledged processes and materials that constitute rational judgment.

Callon and Law appropriated the term 'qualculation' from sociologist Franck Cochoy, who coined the term to help describe different types of decision-making by consumers while shopping. For Cochoy, qualculation and calculation are distinct forms of judgment, rather than qualculation being an expanded form of calculation. For Cochoy, qualculation refers only to quality-based rational judgments, and calculation is restricted to numerical judgments. The two forms of judgment work in conjunction with each other in addition to 'calqulation,' which refers to the process of collective decision-making where different models for judgment are adjusted to one another. For example, if a family is shopping together, a mother and father may have differing opinions about how much should be spent for a product, while their young child may want a product because of a toy included with the item and will not consider price at all in her judgment. The process of adjusting these different calculative and qualculative judgments to make one decision for the whole family is calqulation. Cochoy was specifically interested in how the equipment used for shopping affects cognition and behavior. He found that the shopping cart modifies consumers' decision-making by transforming budgetary constraints into a volumetric consideration while creating a social scene or frame for calqulation (Cochoy 2008, 17–27).

The process of detaching entities from their initial context and reworking them within a set of material arrangements is intrinsic to any system that involves qualculation, whether it is a supermarket, court of law, or OSINT surveillance system. Just as a shopping cart creates a buffer (in the form of a storage zone) which temporarily dissociates product choices from payment so that monetary constraints become volumetric constraints (Cochoy 2008, 20), an OSINT system must dissociate pieces of information from their original context so they can be analysed in relation to national security. In an OSINT system, many of the decisions take place in the realm of visual representation. What is presented in a new context could be text, image, or video. The original associations of the material might be lost as new meanings are created. Some of this process takes place outside the analyst's control and may even be automated. Analysts are not trained to critique the role of visual representation in the process of analysis, and since the selection, arrangement, and display of elements in a visual space impacts how OSINT analysts make decisions, they may be unaware of the role played by aesthetics in qualculative judgment.

Qualculative Poetics

The visual metaphors that intelligence analysts use reveal a great deal about how they view their practice. Metaphors such as 'jigsaw puzzles,' 'connect the dots,' or 'mosaics'[4] indicate that they believe they are assembling a static image from a series of pieces. Data and information are regarded as discrete entities of raw material, pieces of a puzzle, that can be discovered, gathered from the world, and then combined to create a complete picture. By contrast, artists usually understand that meaning is not found in the world, but something that is created between a constructed object and a viewer.[5] What images and text mean is not fixed and will vary according to placement and context. If there is an apt visual metaphor for how OSINT systems help analysts assemble a picture, it would be building a collage where the selection of images and text as well as how they are arranged can create a multiplicity of meanings. As images and text are combined, new meanings are created through juxtaposition and recontextualization. The numerical judgments provided by machine learning algorithms, in spite of their computational origin, do not function as a type of calculation. Instead, they work as labels that potentially change the meaning of what they are attached to. Aspects of this collaging process may even be automated by the system. Analyzing how images, icons, text, and other media create meaning and inform judgment is the purview of artists and cultural critics. Because these techniques are standard practices in the visual arts, tools developed for critiques of representation allows us to read OSINT systems differently.

'Qualculative poetics' is a term we have been using to describe the *creative* arrangement of elements needed to make qualitative rational judgments. As designers of video games and other interactive media are well aware, decision-making itself is an aesthetic experience, which makes the arrangement of elements within a spatio-temporal frame a creative medium. Part of the balancing act in game design is providing the player with just enough components to solve a problem without giving them so many that the game is no longer challenging. The popular framework for game design, MDA (or Mechanics, Dynamics, and Aesthetics), defines aesthetics as 'the desirable emotional responses evoked in the player when she interacts with the game system' (Hunicke, Leblanc, and Zubek 2004). Similarly, artists have experimented with how altering or removing tools for making decisions or measurements changes how objects are interpreted. Claes Oldenburg's sculptures of everyday objects presented at a gigantic scale or Ben Grosser's Facebook and Twitter Demetricator projects, which hide numerical values from the user, are two examples of projects that defamiliarize everyday objects. When a viewer confronts a familiar object at an extraordinary scale, they must renegotiate their relationship to that object, which makes the viewer aware of the role that scale plays in their phenomenological experience of the world. Likewise, Facebook users are accustomed to making judgments based on the number of 'likes' or 'friends' a given post or user has. Removing these metrics alters how the user makes judgments about what they see on the platform. In their work, artists and designers must contend with the fact that perception and judgment are produced through the selection and arrangement of elements. Just as poets can undermine the analytic use of words and produce new significations through creative arrangement and use, rearranging and using data in different ways can produce new meanings.

Cultural and visual studies theorists have described how captions alter the way images are perceived. In *Ways of Seeing*, John Berger demonstrates how words attached to an image can change the meaning of that image, using the example of Van Gogh's *Wheatfield with Crows*. The viewer is initially presented with only the image. On the following page, the image is accompanied by handwritten text that says 'This is the last picture Van Gogh painted before he killed himself' (Berger 1973, 28). While art historians have debated whether or not this was actually Van Gogh's last painting, Berger points out that the inclusion of the text has changed the image, even if it is difficult to describe exactly what those changes are. Likewise, in 'The Work of Art in the Age of Mechanical Reproduction,' Walter Benjamin discusses how photography was seen as 'standard evidence for historical occurrences' (1936, 8). Photographs that were regarded as evidence 'acquire[d] a hidden political significance' (8). Because these photographic images-as-evidence were politicized, Benjamin argues that picture magazines began to 'put up signposts' for the viewer, providing 'directives' to shape how viewers looked at and interpreted them (8). In a similar way, in the US intelligence community, data are now seen by many as evidence of behaviors and even as a means to predict future behaviors. In the context of an OSINT system, much like a magazine, keywords, agent reviews, and machine learning predictions can function as labels and captions, providing signposts that direct analysts to see the posts and their authors as threatening or not, changing and shaping the meaning that is produced.

Juxtaposition, an artistic strategy that involves placing two or more images or objects next to one another, is used to emphasize certain qualities via contrast and comparison. Berger argues that 'the meaning of an image is changed according to what one sees immediately beside it or what comes immediately after it' (29). The arrangement of images or data-objects in an OSINT interface can produce meaning based on placement and the associations produced by juxtaposition. This can be seen in the CSIA when an analyst sees a tweet in the context of other tweets, some of which clearly represent threats. Visual similarities between social media posts can invite associations that lead analysts to judge them as threatening or nonthreatening.

Closely related to juxtaposition, recontextualization is an artistic technique in which meaning is created by placing an image or object in a different frame of reference. This could be simply 'positioning a familiar image in relationship to pictures, symbols, or texts with which it is not usually associated' (Gude 2004, 9), or placing an object in a new situation to give it a different meaning. It was through recontextualization that Marcel Duchamp was able to claim that the ordinary found objects that he called 'readymades' were art when he placed them in a museum or gallery. Likewise, when the CSIA is placed into an art context, users experience OSINT surveillance with the aesthetic contemplation reserved for art objects. When users monitor their own Twitter posts or the posts of their friends with the CSIA, they can experience how OSINT interfaces recontextualize social media posts, potentially impacting how an intelligence analyst could interpret their posts. Because users

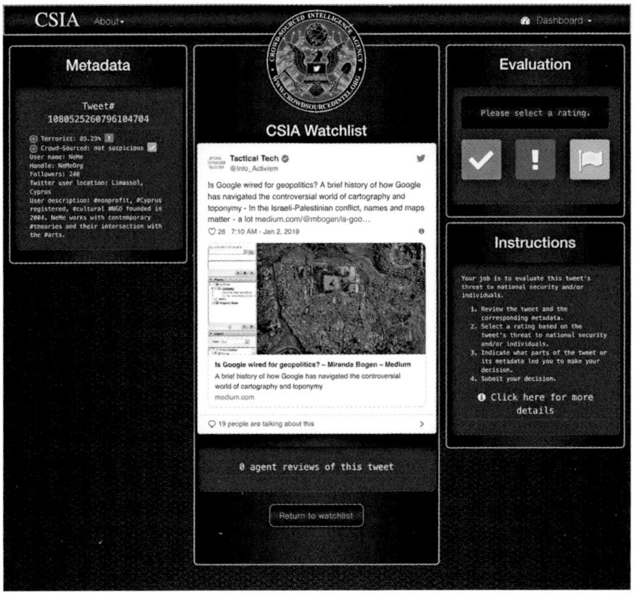

Fig. 2. The CSIA Watchlist.
(Plate 18, p. 314)

have intimate familiarity with the posts in their original and intended context, they can experience how OSINT processing reinterprets and reframes them as potential threats to national security. When a user sees their own post in the CSIA, it can appear alongside other posts which could actually be threatening. That juxtaposition, coupled with their tweet's metadata, DHS keywords that were found in the post, threat ratings from several machine learning classifiers, and reviews of the post by other users of the CSIA might all influence the user's judgment. For example, a user may find that a post they made contains a DHS keyword and the CSIA's terrorist classifier is displaying a 92% chance that the tweet indicates a terrorist threat. The user can then see first-hand how their tweet, which did not seem ambiguous within the context of their own Twitter feed, can take on a new meaning in the socio-temporal frame of an OSINT interface. To give a specific example from the CSIA Watchlist, a post made originally by Tactical Tech[6] and retweeted by NeMe Art Center links to a news story about the politics behind Google failing to label Palestine on Google Maps, and is rated as 85.23% similar to posts made by the CSIA's terrorist classifier (Fig. 2). When the post was retweeted by the NeMe Art Center, it appeared alongside an announcement for a panel on the environmental impact of computation, an open call for artists to exhibit at Ars Electronica, and a tribute to the artist and curator Alex Adriaansens, who had recently passed away. But when the tweet appears in the context of a surveillance interface, it is situated alongside posts made by Hamas and other organizations that are classified as terrorist. When the post was initially made, the focus was the geopolitics of technology, but when placed alongside posts made by organizations advocating for the violent liberation of Palestine, and given a machine learning classification of terrorist with a high degree of certainty, the focus of the post may appear to be less about the implications of technology.

Conclusion

The CSIA is not alone in its artistic critique of surveillance techniques and technologies. Other contemporary artists have used their work to reveal the inner workings of the surveillance state while simultaneously engaging in a critique of those practices. For example, Grayson Earle's *NSA Haiku* (2014) is a Twitter bot that randomly generates haikus with DHS keywords intended for OSINT analysts, and David Huerta's *Begin PGP Mixtape* (2014) is a 'soundtrack for the modern surveillance state' that he encrypted and sent to the NSA (Huerta 2014). Critiques of rationality are pervasive in Modernist art movements such as Dada and Surrealism. The ability of art to critique and offer alternatives to a rationalized existence was recognized by Marcuse, who characterizes art's freedom from rationality as both a privilege (Marcuse 1964, 228) and a threat to the established reality (Marcuse 1964, 238). Art's ability to define new possibilities is both '*validated by and functioning in the scientific-technological transformation of the world*. Rather than being the handmaiden of the established apparatus, beautifying its business and its misery, art would become a technique for destroying this business and misery' (Marcuse 1964, 239).

Cultural critics have used techniques of visual analysis as a way to dissect complex political and social issues, providing ways to examine subjects often shrouded from debate. Gilles Deleuze's analyses of films were often a firm rebuke of capitalism, Roland Barthes used a magazine cover to discuss race relations in postcolonial France, and for Susan Sontag photography was a vehicle to critique politics and society as well as art. Because they are often classified, surveillance techniques and technologies are largely inaccessible. Using tools developed for critiques of representation allows us to read OSINT systems differently to gain insights into the impact that technological rationality has on judgments made in the OSINT generation process and the domination it entails. Using the visual metaphor of puzzle-solving to represent judgments made with OSINT systems suggests that a predetermined, static image is being assembled. Adopting the metaphor of collage-making helps to illustrate that a multiplicity of images can be created through OSINT systems. Each of these images can be interpreted in multiple ways, leading to numerous actionable results that all appear rational. Qualculative poetics demonstrates that the difference between rational and nonrational judgment may hinge on visual arrangement. Because art is free from the constraints of rationality, it can reveal how rationality itself is manufactured by an OSINT system, and provide ways to critique or circumvent domination through rationality.

Notes

1. There are numerous accounts of the analytical process that reveal its philosophical nature despite focusing on empiricism, including: Kent's influential 1968 article 'Estimates and Influence' which describes the analytical process as based on 'solid information,' while it simultaneously requires deduction and interpretation, Heuer's *Psychology of Intelligence Analysis* (1999) which details analysts' decision-making processes, Herbert's "The Intelligence Analyst as Epistemologist" (2006), and the SIGINT Philosopher's intranet column, featuring essays like "Descartes Would Have Been a Lousy SIGINT Reporter" (2012), which self-consciously begins with a critique of 'abstract' philosophy.
2. Marcuse intends reification to be used in multiple senses of the word. In the Marxist sense, workers become objectified; in the mathematical-scientific meaning, reification is the turning of physical objects into mathematical representations.
3. Intelligence agents often use the term 'estimate' to mean 'predict.' Callon and Muniesa are using the term to mean an approximate judgment.
4. See Hollywood et al. '"Connecting the Dots" in Intelligence Detecting Terrorist Threats in the Out-of-the-Ordinary' (2005) for one use of the connect-the-dots metaphor; Heuer's *Psychology of Intelligence Analysis* (1999) for one critique of the mosaic metaphor; and Treverton's 'Making Sense of Transnational Threats: Workshop Reports' (2005) for an argument in support of the jigsaw metaphor.
5. For example, in Duchamp's 1957 speech 'The Creative Act' he argues that 'the creative act is not performed by the artist alone; the spectator brings the work in contact with the external world by deciphering and interpreting its inner qualifications and thus adds his contribution to the creative act.'
6. Tactical Tech is a Berlin-based nonprofit focusing on digital security, privacy, and the ethics of data.

References

Benjamin, Walter. 1936. "Art in the Age of Mechanical Reproduction." In *Illuminations*, edited by Hannah Arendt, 217–51. New York: Schocken.

Berger, John. 1973. *Ways of Seeing*. London: Penguin Press.

Bodine-Baron, Elizabeth, Todd C. Helmus, Madeline Magnuson, and Zev Winkelman. 2016. "Examining ISIS Support and Opposition Networks on Twitter." Santa Monica CA: RAND Corporation. https://www.rand.org/pubs/research_reports/RR1328.html.

Brennan, John. 2015. "Unclassified Version of March 6, 2015 Message to the Workforce from CIA Director John Brennan: Our Agency's Blueprint for the Future." March 6. https://www.cia.gov/news-information/press-releases-statements/2015-press-releases-statements/message-to-workforce-agencys-blueprint-for-the-future.html.

Callon, Michel, and Fabian Muniesa. 2005. "Economic Markets as Calculative Collective Devices." *Organization Studies* 26(8): 1229–50.

Callon, Michel, and John Law. 2005. "On Qualculation, Agency, and Otherness." *Environment and Planning D: Society and Space* 23(5): 717–33. doi:10.1068/d343t.

Cochoy, Franck. 2008. "Calculation, qualculation, calqulation: shopping cart arithmetic, equipped cognition and the clustered consumer." *Marketing Theory* 8(1): 15–44. doi:10.1177/1470593107086483.

Duchamp, Marcel. (1957) 1996. "The Creative Act." In *Theories and Documents of Contemporary Art: A Sourcebook of Artists' Writings*, edited by Kristine Stiles and Peter Selz. Berkeley CA: University of California Press. 818–19.

Foucault, Michael. 1988. *Madness and Civilization* (Richard Howard, trans.). New York: Vintage Books.

Gradecki, Jennifer, and Derek Curry. 2017. "The Crowd-Sourced Intelligence Agency: Prototyping Counterveillance." *Big Data & Society* 4(1): 1–7. DOI: 10.1177/2053951717693259.

Gude, Olivia. 2004. "Postmodern Principles: In Search of a 21st-Century Art Education." *Art Education* 57(1): 6–14.

Herbert, Matthew. 2006. "The Intelligence Analyst as Epistemologist." *International Journal of Intelligence and CounterIntelligence* 19(4): 666–84.

Heuer, Jr., Richards J. 1999. *Psychology of Intelligence Analysis*. Center for the Study of Intelligence, CIA.

Hollywood, John S., Diane Snyder, Kenneth N. McKay, and John E. Boon, Jr. 2005. "'Connecting the Dots' in Intelligence Detecting Terrorist Threats in the Out-of-the-Ordinary." Santa Monica, CA: RAND Corporation. https://www.rand.org/pubs/research_briefs/RB9079.html.

Huerta, David. 2014. "Why the NSA Can't Listen to My Mixtape…and why I mailed it to them as a reminder." *Medium.com*. May 24. https://medium.com/@huertanix/why-the-nsa-cant-listen-to-my-mixtape-739c147d9c04.

Hunicke, Robin, Marc Leblanc, and Robert Zubek. 2004. "MDA: A Formal Approach to Game Design Research." In *Proceedings of the AAAI Workshop on Challenges in Games AI* 4(1): 1722.

Kempster, Norman. 1997. "Academia Mounts Fight to Save a CIA Program." *Los Angeles Times*. January 14.

Kent, Sherman. (1968) 1994. "Estimates and Influence." In *Sherman Kent and the Board of National Estimates: Collected Essays*, edited by Donald P. Steury. Washington DC: Center for the Study of Intelligence.

Knouf, Nick, Claudia Pederson, Derek Curry, and Jennifer Gradecki. 2015. "Poetic Disruption in a Time of Surveillance." In *Proceedings of the 21st International Symposium on Electronic Art ISEA 2015: Disruption*, Vancouver. http://isea2015.org/proceeding/submissions/ISEA2015_submission_176.pdf.

Marcuse, Herbert. 1964. *One Dimensional Man; Studies in the Ideology of Advanced Industrial Society*. Boston MA: Beacon Press.

Marcuse, Herbert. (1941) 1985. "Some Social Implications of Modern Technology." In *The Essential Frankfurt School Reader*, edited by Andrew Arato and Eike Gebhardt, 138–62. New York: Continuum.

Office of the Director of National Intelligence. 2013. *US National Intelligence: An Overview 2013*. April 9. https://www.dni.gov/files/documents/USNI%202013%20Overview_web.pdf.

Owen, Stephen. 2017. "Monitoring social media and protest movements: ensuring political order through surveillance and surveillance discourse." *Social Identities* 23(6): 688–700. DOI: 10.1080/13504630.2017.1291092.

Richards, Pamela Spence. 1981. "Gathering Enemy Scientific Information in Wartime: The OSS and the Periodical Republication Program." *The Journal of Library History* (1974–87) 16(2): 253–64. http://www.jstor.org/stable/25541193.

SIGINT Philosopher. 2012. "Descartes Would Have Been a Lousy SIGINT Reporter." *SID-Today*. November 10. https://www.documentcloud.org/documents/2259079-descartes-would-have-been-a-lousy-sigint-reporter.html.

Schön, Donald. 1984. *The Reflexive Practitioner: How Professionals Think in Action*. New York: Basic Books.

Treverton, Gregory F. 2005. "Making Sense of Transnational Threats: Workshop Reports." Santa Monica, CA: RAND Corporation. https://www.rand.org/pubs/conf_proceedings/CF200.html.

US Department of Defense. 2010. *Department of Defense Instruction* 3115.12, August 24. https://fas.org/irp/doddir/dod/i3115_12.pdf.

Williams, Heather J., and Ilana Blum. 2018. "Defining Second Generation Open Source Intelligence (OSINT) for the Defense Enterprise." Santa Monica CA: RAND Corporation. https://www.rand.org/pubs/research_reports/RR1964.html.

Google, *Shmoogle*
A Critical Twist on Search Engine Ideology

Ziv Neeman and Tsila Hassine

Introduction

For the past few years, the Internet's dominant companies such as Facebook and Google have come under criticism as drivers of the Internet's negative impact. For the past decade, Google has controlled just above 90% of worldwide online searches. Its search engine has become the start or homepage for many users, serving as their central conduit into the Internet and a central gateway to the world. In other words, the billions of websites and pages Google indexes have become the *de facto* horizon of what users believe the Internet comprises More importantly, the top sites Google provides for each query have a critical impact on users' choices of sites, news, and other information providers (Heffernan 2017; StatCounter 2018).

Since the advent of the Internet in the early 1990s, there has been a sharply expanding mutual influence between the online world and the 'real' world. For some time now, this 'real world' has been dubbed online 'IRL' (short for 'In Real Life'). As the interfaces between the online world and IRL have multiplied, diversified, and intensified, and as the number of global users continues to grow, the boundaries between them are becoming harder to determine. As a result of these dynamics, there is a growing influence of the online world on IRL. In turn, Google's position as the monopolistic collator of online information and its highly curated and controlled allocation of information to users—due to its highly secret and complex search algorithm—has become dangerously outsized, not merely in the online realm, but in IRL as well. Hence the term 'Google Goggles' proposed by Tsila Hassine in the context of her digital artwork *Shmoogle* (2005-present), which suggests that above and beyond all the online (and subsequently offline) echo chambers, Google's search engine constitutes the most omnipresent and influential *filter bubble* of our times (Pariser 2011, 83–85).

Shmoogle

Shmoogle (2005-to present), an artwork by Tsila Hassine, is a search engine that serves all of Google's results, but in random order (Search Engine Watch, 2005; Rogers, 2018, 5). It is actually simpler than it sounds: once the user types a query into the *Shmoogle* search engine, *Shmoogle* retrieves all the results Google makes available for this query (Google very rarely returns more than 1,000 results). Once *Shmoogle* obtains these results, it overturns Google's PageRank hierarchy, presenting them in a long, randomly ordered list, tagging each result with its original position in Google's hierarchy. Moreover, *Shmoogle* avoids Google's portioning of results into separate pages, and prefers the 'long scroll down' approach. As Richard Rogers notes, '*Shmoogle* seeks to intervene in the hierarchy of source credibility suggested by [Google's] rankings' (2018, 5; see Fig. 1).

As we will explain in this paper, extensive research shows that Google's highly monopolistic position in the search engine market has markedly influenced for the worse people's sense of online information and its mediation with reality (Galloway 2017, 124–51). Searching the web on any topic, users have grown used to restricting their search to the first results found on Google Search (Search Engine Watch 2018). *Shmoogle* mimics Google's function, but with a minor, yet significant distortion that renders it—in terms of efficiency and utility—inefficient. However, it is precisely this

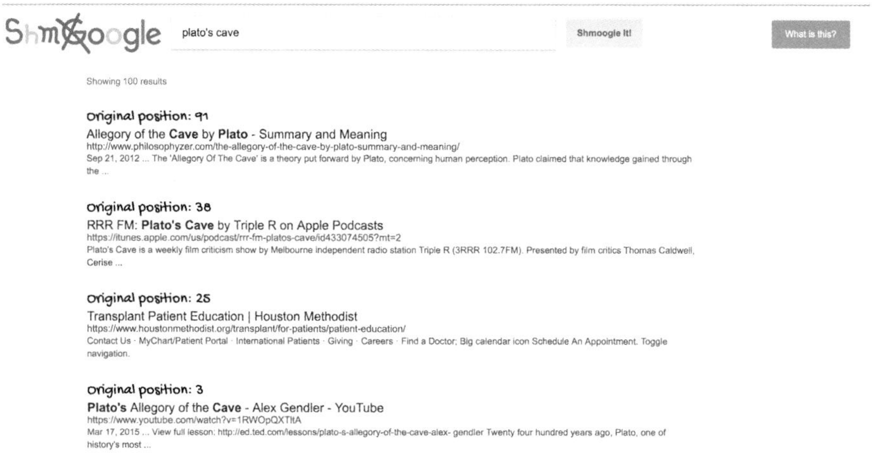

Fig. 1: Screen capture from *Shmoogle*. Captured on XXXXXX.

'inefficiency' that lets *Shmoogle* create a 'temporary autonomous zone' which allows the user to get a broader view of the mediated world beyond Google's first results (Bey 1985). By expanding the list of presented results, *Shmoogle* amplifies users' agency in deciding which search results are relevant for them and which are not. This disordering of Google's results critiques and questions the growing role of corporate-owned algorithms in society's mediation of reality. Moreover, as we shall discuss, *Shmoogle* occupies a nexus of critical artistic practices and perspectives, among them: culture jamming, institutional critique, and the particular form of mimetic representation referred to as the 'hypericon.'

A Very Brief Introduction to Information Society

During the past few decades the world has witnessed vast interrelated politico-economic and technological changes, and with them substantial changes in the dominant ideological formations and technologies. These include the rise of neoliberalism and the widespread adoption of digital technology. As David Harvey explains, global neoliberal expansion:

> requires technologies of information creation and capacities to accumulate, store, transfer, analyze, and use massive databases to guide decisions in the global marketplace. Hence neoliberalism's intense interest in and pursuit of information technologies (...) leading some to proclaim the emergence of a new kind of 'information society' (Harvey 2014, 3–4).

In this context, the Internet provides the most prominent and influential example. One can, indeed, look at the interaction between the rise of neoliberal globalism and digital technology as the basis for the ideology of information capitalism, comprising its own values and norms (Castells 1996–2010; Poster 1990; Zuboff 2019).[1]

The Dominance of Google's Search Engine and Its Quest for Efficiency

The highest ideological value of global neoliberalism's digital sphere is efficiency at all costs, as efficiency is a necessary condition for profits, and in the digital world it is achieved, above all, through algorithmization. The paradigm of efficiency is so engrained that it has become the benchmark for hardware and software development.

In this context of a neoliberal culture dominated by information technology and massive databases, it is only logical to seek the most efficient solutions for data indexing and retrieval. In what follows we focus on Google's search engine, because of its overwhelming dominance, and because it has 'long [been] considered the "gold standard" in [online] search[es]' in terms of algorithmic efficiency (Search Engine Land 2017). Note that the measure of search engine efficiency is based on a number of combined parameters that include speed, accuracy, relevance, and personalization of results. And from a user's perspective, relevant results; a simple, clear interface; and the possibility to expand or restrict the search if needed (Gil 2018). All these attributes in their own way emphasize efficiency.

Search engine specialists agree that Google's entrance into the search engine market in 1997 was a watershed event in the history of Internet search engines, due to its PageRank algorithm, whose efficiency, especially in terms of accuracy and relevance, was substantially better than that of previous search engines (Vaughan-Nichols 2017; Search Engine Land 2017). As a result, over the next decade Google's share of the global search market grew so rapidly that by 2008 it has reached just over 90%, a highly monopolistic market share that it has maintained ever since (StatCounter 2018). As technology specialist Steven Vaugh-Nichols explains:

> none of the other BG (before Google) web [search] engines had what it took to compete with Google's genius. [Hence, today] we really do live in AG (after Google) days. I doubt that will change anytime soon. Google certainly continues to work hard to make sure that will be the case (2017).

Google's PageRank Algorithm

At the core of Google's search engine technology is the PageRank algorithm. The founders of Google, Larry Page and Sergey Brin, published an explanation of PageRank in an academic paper the same year Google was founded (Brin and Page 1998). Before Google, existing search engines depended on keywords as the main factor in ranking their search results. In contrast, the PageRank algorithm ranks websites by 'votes': if two sites have a link to a third site, and one of them is heavily 'voted' on its own, then the 'vote' of that popular site has a higher statistical weight in determining the ranking of the third site they both link to. Hence, a page's rank is actually determined by the ranking of the pages that link to it. It sounds simple but, in fact, the algorithm conducts complicated calculations performed thousands of times over. However, this algorithm is only a small part of Google's ranking algorithms; since its founding Google has been constantly tweaking and adding to its algorithms, generally in secrecy, to the point that it is impossible to know the set of statistical weights and considerations (referred to in the business as 'signals') that determine a pages's rank fate: whether a page will benefit from the spotlight of the first Google result page or disappear into the oblivion of subsequent, rarely visited pages.

As soon as search engines appeared, it became clear that the search result ranking of websites had a major economic, cultural, and political significance. This led to the rise of the Search Engine Optimization (SEO) industry, whose sole purpose is to raise the ranking of their customers' sites in the search results (Search Engine Journal 2017). Some in the SEO sector estimate that Google's expanded PageRank algorithm utilizes around 200 signals: some are obvious, such as user location and language; since 2007 it has also incorporated the search history of individual users, their Gmail correspondence, and myriad other user data (Moz 2015).

The 'Single Perfect Answer'

Google Search's obsession with efficiency—taken to its terminal logic—was revealed in an interview with Eric Schmidt, who was Google's CEO between 2001 and 2011, and the executive chairman of the company between 2011 and 2017. In the interview Schmidt said:

> When you use Google, do you get more than one answer? Of course, you do. Well, that's a bug. We should be able to give you the right answer just once (Ferenstein 2015).

Such an extreme goal of providing a 'single perfect answer' for everyone for every query epitomizes the value of efficiency: Google wishes to spare users the time it takes to click through a few suggestions before hitting upon the one that purportedly answers their query. In fact, for certain kinds of knowledge, Google Search (with other search engines) has been able to provide this kind of information for some time now, by providing the 'perfect' result at the top of its first result page. Yet, such answers are only possible within a narrow range of topics such as factual and scientific knowledge (e.g., 'what is the address of this restaurant'; 'what is the atomic number of Helium'). Regarding most social, cultural historical, and political issues (e.g., questions on racism, gender, attitudes towards refugees and migrants, etc.) a supposed 'single' answer is untenable or, worse, suspect and dangerous.

Google's Cultural and Ideological Dominance: Has it Become our New Reality?

The immense power, prestige, and often quasi-religious adulation for Google's search engine has been on display for a long time. For example, a *New York Times* column by Thomas L. Friedman, from 2003, was titled, "Is Google God?"[2] Recently, *Time* Magazine selected Google's search engine site as the 'most influen[tial]' site in the Internet's history (Fitzpatrick, Eadicicco, and Peckham 2017). A truly powerful indication of the massive, global cultural impact of Google was the selection of the neologism 'google'—often appearing as a gerund: 'googling'—meaning to search the web, as the 'most useful word' of 2002 by the American Dialect Society, and then again as the 'word of the decade, 2000–2009.' Since 2006, the term has entered most dictionaries as a legitimate verb (Axon 2010; Wikipedia 2018).

Moreover, there are deeper and more telling assumptions of Google's cultural power that are divulged by the frequent use of this verb. Google's own long-time corporate mission statement is 'to organize the world's information and make it universally accessible and useful' (Google 2018). Yet, as Virginia Heffernan points out, '"to Google," of course, does not mean "to organize"; it means to submit to Google's organization while pretending you're blazing' through all of the information online. She continues:

> Part of the shared pretending done by Google and anyone who uses it is to act as if 'information' on the web already existed in some kind of natural state prior to Google's mission to organize it. Instead, [online] information (…) would be nothing without the company's decision to recognize it as Googleable, rank it in the algorithm's esoteric hierarchy, and incentivize its renovation so it might make itself prettier to Google. 'To Google' something, therefore, is to accept the fiction that Google is both the whole information world--and the only path through it (2017).

The cultural and ideological power of Google is even more pronounced, not only because of Google's dominance but also because of the secrecy of its ranking algorithms. In practice, Google's PageRank algorithm has created its own hierarchy of online information which, due to its massively monopolistic edge, has become for most Internet users the *de facto* access point to the entire online world.

Nonetheless, users have definitely embraced Google's search engine in terms of satisfaction, and especially in terms of relevance and efficiency, in a way that conforms to Schmidt's ideal of the right answer just once (Ferenstein 2015). Over the past two decades, many studies conducted by academia and by the SEO sector have focused on the importance of search result ranking in Google. Their findings suggest that over the past decade or so nearly all users have reduced the number of result links they click and look at (users' clicks on links are termed 'CTR,' an abbreviation of 'Clicks-Through Rates'). For some time now, more than 90% of CTR occur only on the first page of Google Search results. Moreover, the first top three results on the first result page receive around 60% or more of all CTR (Jansen & Spink 2006; Lorigo et al. 2008; Chitika 2013; Lincoln 2017; Smart Insights 2018; Zero Limit 2018). Thus, we see how users have internalized Google's order of results, which in turn signifies users' ideological acceptance of Google's monopolistic domination of the selection, curation, display, and provision of online information. To put it bluntly, what we see is what we Google.

Culture Jamming

At first glance, *Shmoogle* definitely falls into the artistic and critical tradition of culture jamming, whose goal is to critically play with 'heavily circulated images [and texts] that are already infused into the public's general consensus' (Chaudhary 2010, 1). Culture jamming, as an artistic and critical practice, came into prominence in the 1980s–1990s. During the 1990s, the practice was often identified with *Adbusters*, a collective of activists from Vancouver who at the time manipulated advertisements and brands, critiquing corporations and consumerism. They were not the only ones to do so. One of the most accomplished practitioners of culture jamming as a critical artistic practice are *The Yes Men,* Andy Bichlbaum and Mike Bonanno. For example, they marked the twentieth anniversary of a highly lethal gas leak in Bhopal, India, from a Union Carbide Corporation factory owned by Dow Chemical Corporation. Masquerading as a Dow Chemical spokesperson, Andy Bichlbaum was unwittingly interviewed by BBC World News, claiming that Dow planned to liquidate Union Carbide for 12 billion dollars and utilize the money for medical care for the Bhopal victims, for site clean-up, and to fund research into the dangers of other Dow products. Though the escapade was quickly exposed, it nonetheless brought Western media's attention back to the continuing suffering of the survivors and to the petty amount of compensation that a powerful Western corporation could get away with in a developing country (The Yes Men 2012).

Critics position the practice of cultural jamming as an ideological descendant of the culture industry critique of the Frankfurt School, and the critical-artistic practices of the Situationist International movement, in particular with the media theorization of Guy Debord and his anti-capitalist practice of *détournement* (Carducci 2006, 118). The principle of *détournement,* (French, meaning: diversion, misappropriation, misuse, hijacking) is an aesthetic act which mixes pre-existing elements of mainstream art and media, recombining them in new ways. In an (anonymous) article on *detournement,* the Situationists identified 'the two fundamental operations of *détournement*' as:

> the loss of importance of each *détourned* autonomous element (...) and at the same time the organization of another meaningful ensemble that confers on each element its new scope and effect (Situationiste International [1959] 2006).

This astutely describes the operation of *détournement* that *Shmoogle* performs on Google's PageRank-ed results. In this respect, *Shmoogle* is a work of algorithmic culture jamming in which *shmoogle's* algorithm creates a different 'meaningful ensemble' from the *detournement* of PageRank's algorithm; additionally, through this operation it also attempts to expose the ideological dimensions of this secret code at the heart of Google's search engine. Which is to say that, in addition to culture jamming, *Shmoogle* offers an institutional critique of Google Search.

Shmoogle and Institutional Critique

In 1989, Andrea Fraser gave four performances, titled *Museum Highlights,* at the Philadelphia Museum of Art, posing as a fictive tour guide named 'Jane Castleton.' Fraser writes:

> While Jane is a fictional docent, I would like to consider her less as an individual 'character' with autonomous traits than as a site of speech constructed within various relations constitutive of the museum (1991, 107, note 4).

This point is critical for understanding the piece as a form of institutional critique: it exposes and highlights various relations that are unacknowledged, unspoken, or unseen, yet nonetheless constitute relations that define the museum as an institution. For example, in her explanations as a guide, Fraser introduces into her talk elements that seem intentionally absurd and humorous, yet possess a biting critical sting. For example, addressing an exit sign above a door in a gallery, she describes it as: 'Firm in painting, delicate in color and texture, this picture is a brilliant example of a brilliant form' (ibid. 115).[3] Such archly ironic interjections question normative hierarchies of attention and spectatorship in art exhibition spaces. They introduce into the guided tour items that are conceived as extra-artistic items and, therefore, excluded from attention or discussion. However, these may impact and influence the visitor's experience, as well as put into question various ideological norms. Giving lavish attention in an old-fashioned art jargon to the banal exit sign interrogates what constitutes art in the museum's visual sphere and the authority behind such decisions. These questions are especially fitting for the Philadelphia Museum of Art, which houses the largest and most important collection of Marcel Duchamp's work, including many of his Readymades.

In her *Museum Highlights* performance, Fraser quotes a varied array of textual sources—often in incongruent or dissonant contexts. This heteroglot performance piece utilizes a *détournement* of the museum guide's normative role to express institutional critique. Similarly, *Shmoogle* uses a *détournement* of Google's PageRank hierarchy as a means to expose and highlight its unspoken ideology. Indeed, *Shmoogle* goes beyond culture jamming by putting into question the norms, assumptions, and expectations of Google Search users, based on results that are in many ways no less curated than art exhibits. A further semblance between Fraser's *Museum Highlights* and *Shmoogle* is the performative aspect. *Museum Highlights* is a performance piece proper; in *Shmoogle*, due to its interactive character, there is also an integral performative element. In this sense, one can say that both works share the tactic of *performative détournement*.

Although most of Fraser's performance of institutional critiques focuses on art world ideological norms, it is also aimed at and critiques class and gender norms (Fraser 2005, 108–9, 113–14, 117). In this regard, Fraser represents what is termed the 'second wave' of institutional critique, from the 1980s to 1990s, involving such artists as Louise Lawler, Fred Wilson, Mark Dion, and artist collectives such as Group Material, Critical Art Ensemble, and the Guerrilla Girls (Björk and Kokkonen 2014, 3–4; Fraser 2005, 100). This 'second wave' was preceded by an earlier generation of artists working in the late 1960s and 1970s, including: Hans Haacke, Daniel Buren, Marcel Broodthaers, Michael Asher, and Martha Rosler (Fraser 2005, 100; Alberro and Stimson 2009, 3–7; Björk and Kokkonen 2014, 5–9). A significant distinction between these two phases is that the artists of the first, with the exception of Rosler, criticized solely art world institutions and practices. In contrast, artists from the second phase focused also on ideological norms and topics, not necessarily associated with the art world. Genealogically, critiquing the dominant ideologies of art can be delineated back to Marcel Duchamp and his readymades and to the Dada movement (Alberro 2009, 3, 5; Buchloh 1990, 107). As Peter Burger in *The Theory of the Avant-Garde* writes:

> [W]ith the historical avant-garde movements, the social subsystem that is art enters the stage of self-criticism. Dadaism (...) no longer criticizes schools that preceded it, but criticizes art as an institution, and the course its development took in bourgeois society (1984, 279).

From Mimetic Critique to the Hypericon

Benjamin Buchloh's essay, 'Conceptual Art 1962–1969: From the Aesthetic of Administration to the Critique of Institutions,' provides an astute and intricate analytic historization of 1960s American conceptualism. He traces the origins of institutional critique back to the later phase of conceptual art, with artists such as Haacke and Buren. Additionally, in this context it is germane to note the position of conceptual art in the genealogy of new media art. Early texts on new media consider digital art as a novel form of conceptual art (Cramer 2002, 1; Manovich 2001, 235; Gere 2004, 15; Greene 2004, 157, 163).[4]

Returning to Buchloh essay, he writes that minimalism and conceptual art constituted:

> the postwar period's most rigorous investigation of the conventions of pictorial and sculptural representation and a critique of the traditional paradigms of visuality (1990, 107).

In the wake of these radical critiques, he locates toward the end of the 1960s the rise of 'a critique that operates at the level of the aesthetic "institution"' (ibid. 136). It arose out of a recognition by some conceptualist artists that their critique needed to go beyond purely aesthetic strategies, 'since these practices are always already inscribed within the conventions of language and thereby within institutional power and ideological investment' (ibid.). In the essay Buchloh dialectically deploys the term 'administrative aesthetic' (106, 140, 141). He characterizes the style of conceptual and minimalist art as 'bureaucratic' or 'administrative,' since it employs such forms as texts, lists, tables, seriality, etc., which at the time were perceived as nonaesthetic (106, 118–19, 128, 140). Dialectically, Buchloh argues that this administrative aesthetic 'mimed the operating logic of late capitalism and its positive instrumentality' (143). Buchloh considers Haacke and Buren's move from conceptual art to institutional critique as one that began by mimicking the administrative aesthetic, but utilizing it to critique the very ideology of capitalist instrumentality (ibid).

Almost thirty years later, Buchloh's dialectic principles and his emphasis on a mimetic criticism are more relevant than ever. In today's nonmaterial information society, many artistic endeavors mimetically appropriate code and online aesthetics in order to transform and critique them in countless ways. *Shmoogle* appropriates the design of Google's search and its result pages that embody the operational logic of contemporary information capitalism. Thus *Shmoogle's* mimetic doubling of Google

Search's design opens up a space for a critical investigation of this complex institutional apparatus.

In order to fully appreciate the critique performed by *Shmoogle* we refer to W.J.T. Mitchell's concept of the hypericon. Mitchell characterizes the hypericon in the context of mimetic doubling and self-referentiality: they 'depict the act of picturing, imagine the activity of imagination, figure the practice of figuration.' Thereby they 'provide models for thinking about images'; emphasizing that hypericons serve as 'concrete vehicles in a metaphoric treatment of abstractions' (1986, 5, 162). In a later text, Mitchell describes the hypericon as a special mimetic image 'that incapsulates an entire episteme, a theory of knowledge' (1994, 49).

Shmoogle constitutes a hypericon, in all these senses. As an interactive work, it constantly generates new images of Google's search results in random order. More importantly, it provides concrete images that mimetically represent the abstract, invisible, and secret algorithm of PageRank. Shmoogle performs a *détournement* upon the original search results, yet it simultaneously provides—in parenthesis—for each result its original position in Google's hierarchy. Shmoogle thus offers a double object/image to the mind's eye, which begins to captures Google's epistemic and ideological biases in the very difference between the two competing hierarchies. It opens a gap between Google's dominating hierarchy and Shmoogle's anarchic disorganizing intervention, exposing Google's overarching algorithmic control of online information.

Notes

1. We view ideology from a Gramscian perspective (Gramsci 2011).
2. Scott Galloway also depicts the public perception of Google as godlike, but vigorously criticizes such impressions; in his book, the chapter on Google is titled: "Google: Our Modern-Day God" (2017, 5, 124–28).
3. This is a quote of a one-sentence description of a painting from the *Pennsylvania Museum of Art Handbook,* from 1931 (Fraser 1991, 115). Most of Fraser's text is culled from myriad quotes from early 20th-century texts about the Philadelphia Museum of Art and the city of Philadelphia in general.
4. Many of the authors cited above build on Sol LeWitt's essay, 'Paragraphs on Conceptual Art' (1967) in which he states: 'In conceptual art the idea or concept is the most important aspect of the work. When an artist uses a conceptual form of art, it means that all of the planning and decisions are made beforehand and the execution is a perfunctory affair. The idea becomes a machine that makes the art' (166).

References

Alberro, Alexander. 2009. "Institutions, Critique and Institutional Critique." In *Institutional Critique: An Anthology of Artists' Writings*, edited by Alexander Alberro, and Blake Stimson, 3–19. Cambridge MA: MIT Press.

Alberro, Alexander, and Blake Stimson (eds.). 2009. Cambridge MA: MIT Press.

Axon, Samuel. 2010. "'Google' Is Word of the Decade." *Mashable*. https://mashable.com/2010/01/11/google-tweet-word/#9RL.XMhF6ZqC.

Bey, Hakim. 1985. *The Temporary Autonomous Zone, Ontological Anarchy, Poetic Terrorism*. Brooklyn, NY: Autonomedia.

Björk, Helena, and Laura Kokkonen. 2014. "Introduction: Claiming Actuality." in *Never Underestimate the Institution*, edited by Nora Sternfeld, and Henna Harri, 3–6. Helsinki: Department of Art Aalto University.

Brin, Sergey, and L. Lawrence Page. 1998. "The Anatomy of a Large-Scale Hypertextual Web Search Engine." *Seventh International World-Wide Web Conference*, Brisbane, Australia. Doi: 10.1016/S0169-7552(98)00110-X.

Buchloh, Benjamin H.D. 1990. "Conceptual Art 1962–1969: From the Aesthetic of Administration to the Critique of Institutions." *October* 55:105–43. Doi: 10.2307/778941.

Burger, Peter. 1984. *The Theory of the Avant-Garde*. Minneapolis MN: Minnesota University Press.

Carducci, Vince. 2006. "Culture Jamming." *Journal of Consumer Culture* 6(1): 116–38. Doi: 10.1177/1469540506062722.

Castells, Manuel. 1996–2010. *The Information Age*, 3 Vols. Oxford: Blackwell Publishers.

Chaudhary, Shikher. 2010. "Culture Jamming," *Helterskelter*. http://helterskelter.in/2010/08/culture-jamming/.

Chitika. 2013. "The Value of Google Result Positioning." *Chitika Online Advertizing Network*. https://chitika.com/2013/06/07/the-value-of-google-result-positioning-2/.

Cramer, Florian. 2002. "Concepts, Notations, Software, Art." *Netzliteratur.Net*. https://www.netzliteratur.net/cramer/concepts_notations_software_art.html.

Ferenstein, Greg. 2015. "Google's End Game is a Single Perfect Search Result." *Ferenstein Wire*. https://medium.com/the-ferenstein-wire/google-s-end-game-is-a-single-perfect-search-result-as-eric-schmidt-explains-in-one-quote-ab9d75309243.

Fitzpatrick Alex, Lisa Eadicicco, and Matt Peckham. 2017. "The 15 Most Influential Websites of All Time," *Time*, October 20, Tech. 2018http://time.com/4960202/most-influential-websites/.

Fraser, Andrea. 1991. "Museum Highlights: A Gallery Talk." *October* 57 (Summer):104–22. http://www.jstor.org/stable/778874.

Fraser, Andrea 2005. "From the Critique of Institutions to an Institution of Critique." *Artforum* 44:100–106. https://monoskop.org/File:Fraser_Andrea_2005_From_the_Critique_of_Institutions_to_an_Institution_of_Critique.pdf.

Friedman, Tom L. 2003. "Is Google God?" *New York Times*, June 29. https://www.nytimes.com/2003/06/29/opinion/is-google-god.html.

Galloway, Scott. 2017. *The Four: The Hidden DNA of Amazon, Apple, Facebook, and Google*. New York: Penguin Random House.

Gere, Charlie. 2004. "New Media Art and the Gallery in the Digital Age." *Tate Papers* 2 (Autumn). https://www.tate.org.uk/research/publications/tate-papers/02/new-media-art-and-the-gallery-in-the-digital-age.

Gil, Paul, 2018. "The Best Search Engines of 2018". *Lifewire*. https://www.lifewire.com/best-search-engines-2483352.

Google. 2018. "Mission Statement." *Google.com*. https://www.google.com/about/.

Gramsci, Antonio, Quintin Hoare, and Geoffrey Nowell-Smith. 2011. *Selections from the Prison Notebooks*. New York: Columbia University Press.

Greene, Rachel. 2004. *Internet Art*, London: Thames and Hudson.

Harvey, David. 2014. *A Brief History of Neoliberalism*. Oxford: Oxford University Press.

Hassin, Tsila. 2005-. *Shmoogle*. http://shmooglekoogle.org/.

Heffernan, Virginia. 2017. "Just Google It: A Short History of a Newfound Verb." *Wired*. https://www.wired.com/story/just-google-it-a-short-history-of-a-newfound-verb/.

Jansen, Bernard J., and Amanda Spink. 2006. "How Are We Searching the World Wide Web? A Comparison". *Information Processing and Management* 42:1, 248–63. Doi: 10.1016/j.ipm.2004.10.007.

Lewitt, Sol. 1967. "Paragraphs on Conceptual Art." *Artforum* 5(10): 79–83. https://www.corner-college.com/udb/cproVozeFxParagraphs_on_Conceptual_Art._Sol_leWitt.pdf

Lincoln, John E. 2017. "Google Click-Through Rates in 2017 by Ranking Position." *Ignite Visibility*. https://ignitevisibility.com/ctr-google-2017.

Lorigo, Lori, et al. 2008. "Eye Tracking and Online Search: Lessons Learned and Challenges Ahead." *Journal of the American Society for Information Science and Technology*, 59(7): 1041–52. https://doi.org/10.1002/asi.20794.

Manovich, Lev. 2001. *The Language of New Media*. Cambridge MA: MIT Press.

Mitchell, W.J.T. 1986. *Iconology: Image, Text, Ideology*. Chicago IL: University of Chicago Press.

Mitchell, W.J.T. 1994. *Picture Theory: Essays on Verbal and Visual Representation*. Chicago IL: University of Chicago Press.

Moz. 2015. "Google Algorithm Change History." https://moz.com/google-algorithm-change.

Pariser, Eli. 2011. *The Filter Bubble: What the Internet Is Hiding from You*. London: Penguin Press.

Poster, Mark. 1990. *The Mode of Information*, Chicago IL: University of Chicago Press.

Rogers, Richard. 2018. "Aestheticizing Google Critique: A 20-Year Retrospective", *Big Data & Society* 1(13), Doi I: 10.1177/2053951718768626.

Search Engine Journal. 2017. "20 Years of SEO: A Brief History of Search Engine Optimization." https://www.searchenginejournal.com/seo-101/seo-history/.

Search Engine Land. 2017. "History of Google Algorithm Updates." https://searchengineland.com/library/google/google-algorithm-updates.

Search Engine Watch. 2005. "*Shmoogle* : Randomizing Google Results to Show Good Stuff Beyond Top 10." https://searchenginewatch.com/sew/news/2061942/*Shmoogle* -randomizing-google-results-to-show-good-stuff-beyond-top.

Search Engine Watch. 2018. "Top Google Result Gets 36.4% of Clicks." https://searchenginewatch.com/sew/news/2049695/top-google-result-gets-364-clicks-study.

Situationniste Internationale, [1959] 2006. "Détournement as Negation and Prelude." *Internationale Situationniste* 3. topologicalmedialab.net/xinwei/classes/readings/.../DebordTheoryOfTheDerive.pdf.

Smart Insights. 2018. "Search Engine Statistics." https://www.smartinsights.com/search-engine-marketing/search-engine-statistics/.

StatCounter. 2018. "Search Engine Market Share Worldwide." http://gs.statcounter.com/search-engine-market-share.

Vaughan-Nichols, Steven. 2017. "Before Google: A history of search." *Hewlett Packard Enterprise.* https://www.hpe.com/us/en/insights/articles/how-search-worked-before-google-1703.html.

Wikipedia, (2018). "Google (verb)." https://en.wikipedia.org/wiki/Google_(verb).

Yes Men (The), 2012. "Flashback: The Yes Men Explains Dow Chemical/Bhopal Disaster Prank," YouTube video, 7:35, *LeakSourceArchive*." https://www.youtube.com/watch?v=a8V-QIGaq7m4.

Zero Limit. 2018. "Part 1: Organic vs PPC in 2018: The CTR Results." *Zero Limit Web Digital Marketing.* https://www.zerolimitweb.com/organic-vs-ppc-2018-ctr-results-best-practices/.

Zuboff, Shoshana. 2019. *The Age of Surveillance Capitalism: The Fight for a Human Future at the New Frontier of Power*. New York: Hachette.

Sculpting Time
The Art of Collective Memory

Ryszard W. Kluszczyński

Introduction: Memory, Archive, Identity

In the last two decades of the twentieth century the theme of memory became one of the focal concerns of cultural studies. In that time, memory research developed vigorously, as evinced by numerous studies produced in a variety of languages by many authors, and opened huge conceptual and interpretive vistas for further explorations. Individual and collective memory, private and social memory, institutional, cultural, and communicational memory, the relations between memory and history or the role of technology and media in exteriorising, archiving, and distributing memory are but a handful of memory-related subjects that proved irresistibly attractive to the humanities and social sciences towards the end of the twentieth century. In this way, cultural studies research was augmented with new issues to explore and new notions to hone, which inspired new scholarly ventures. All these processes and developments ultimately contributed to a paradigmatic shift and the rise of a new perspective as the 'memory turn' was proclaimed in cultural studies (Kitzmann, Mithander, and Sundholm 2005).

The themes investigated in this comprehensive field have not lost any of their relevance or sway today. On the contrary, robust advancements in digital and Internet technologies have generated new subjects to examine and describe. In terms of my argument, virtual environments and digital archives are the most topical issues of this kind, as their development involves new forms of memory work and is essentially based on participatory methods of creation and utilization.

The further development, transformations, and reorganization of research into memory—in particular, the central focus on archives, their forms, and the ways they are produced and deployed—propelled the emergence of another subfield of cultural studies which complements memory research. The inclusive span, the dynamics, and, primarily, the significance of these pursuits resulted in announcing another cultural shift, one that has come to be referred to as the archival turn (Rosengarten 2013). Rosengarten observes that the archival turn is intimately connected with the previous pivotal transformations, such as the ethnographic turn and the curatorial turn. This shows how closely all the four turns are interrelated as they comprehensively affect culture in the broad sense of the term and, specifically, art. As a matter of fact, in their different ways they make the products of the others into objects of attention. However, this principle is undermined in the world of new media, which is by the way becoming a general rule today.

Of course, the interest in memory, which is so pronounced in contemporary reflection on culture, is not limited to memory as such and its mechanics (commemoration, preservation, and sharing). Importantly, it also investigates the mediational functions of memory, that is, its relations to the experience it shapes and the object it concerns. Submerged in the field of temporality, the complex nexus of experience, memory, and identity appears in the modern humanities research as a unique cognitive pattern or as a script, i.e. as a structure which both ascribes to these three components their respective places and roles within a superior order as well as determining the relevant research procedures. The meaningful relations among the elements of this structure certainly count among the most important reasons behind the upsurge in the preoccupation with memory-related themes we have long observed. Memory research is, in my view, significantly linked to interest in the confluence of these themes, which demarcates a field of exploration where the concepts of time, representation, trace, and presence play a seminally important role alongside their impact on the formation of experience, memory, and identity. This thematic node has inspired several theories which are instrumental in interpreting current social processes, such as the widely discussed concept of postmemory developed by Marianne Hirsch (1997) to explain current attitudes to the Holocaust and genocide; the framework of prosthetic memory proposed by Celia Lury (1998), who enquires into the functions of photographic seeing; or the slightly less known notion of modular memory put forward by Barry King, furthering our understanding of the modern formation processes of liquid, unstable subjectivities (King 2000, 9–12).

Art and Memory: Archival Art

Widespread as it is, the fascination with memory and archives is by no means limited to the institutionally circumscribed research field of the humanities and the social sciences. It is also rife in art, which is the primary context for my argument in this paper. The direct intertwining of the interest in memory and the strategy of framing artworks as collections or databases has engendered the archival art movement. Drawing on the insights of Hal Foster (2004, 4–5), Ruth Rosengarten (2013) states that this approach is most frequently manifested in installation art 'in which the artists present found material (image, object, text), with sources at times familiar and drawn from mass culture, and at other times arcane.'[1] Christian Boltanski, Ilya Kabakov, and Anette Messager are among the best known and most appreciated artists engaged in such pursuits.

The ideas and practices of archival art could not but be fruitfully carried on within and by new media art, particularly digital art. The specific properties of computer technology (a meta-medium for all forms of digital art) inevitably channel it towards the concept of the archive. An important influence in this respect was exerted by Lev Manovich (2001), one of the first theorists of new media art, who insisted that the database was its basic structure. Manovich's source for this thesis was not the experiences of interactive art, which I believe to be the foundation of the archive experience in digital art; instead, he derived it directly from a comprehensive view of new media as a totality. The qualities of new media he registered comprise a network of connections from which the database emerges as their logical extension.

Nevertheless, I will not treat each work of digital art as a form of archival art by default due to its medium. If I did, this gesture would empty the notion of archival art of all the cognitive value it has vis-à-vis new media art. An artwork can only be classified in this category on the basis of how its structure uses the medium, what concept and content it conveys, and in what way it organizes the behaviors of the audience. Such works-archives are plentiful in new media art. They are projects which invite the viewers to explore collections put together for them and, in this way, to discover or to produce the meaning and the value of their experience together with its individual and social references.

An essential form of the archival movement within new media art is showcased in the tendency which I called elsewhere the interactive strategy of archives (Kluszczyński 2010, 1–27), whose first important examples can be found in George Legrady's *An Anecdotic Archive from Cold War* (1993) and Agnes Hegedüs's *Things Spoken* (1998).

In his work, Legrady adapted the methods typically employed by Boltanski and Kabakov to the environment of hypermedia art. Legrady is both an organiser and a protagonist of interactive discourse in which he shares with his audience objects gathered from the public space and multiple private sources. This discourse can be said to work towards objectivising the collection and turning it into an intersubjective representation of the theme it addresses. With a view to ensuring the objective tenor of *An Anecdotic Archive from Cold War*, Legrady compiled numerous digitalized objects epitomizing the Stalinist era in East European countries, especially Hungary, such as photos, posters, banknotes, recordings, etc. Nevertheless, the artist also inscribed his own, personal narrative in this objective, representative body of materials—a story of how he had left Hungary with his family in 1956 to return to it years later and confront his memories with the country's present-day realities. This individual narrative neither divests the amassed resources of objectivity nor dismantles their fundamental historical framework of reference, yet at the same time it transfigures them into the artist's personal expression, makes the work represent him as well, and turns it into a part of his life history.

In *Things Spoken*, Ágnes Hegedüs puts together an ensemble of objects which belong to her personal world and encourages the audience to take a journey which unveils her multifarious identity founded in individual memory. The digitalized objects represent various periods and events in her life and her relationships with other people. The interactive user can explore the visual space, browse through the collection of objects, and listen to the stories of the artist and various people who are associated with her in one way or another. While these narratives directly pertain to the collection and its respective items, they also use these objects to reveal the artist herself, serving in this way as her unique, interactive, and virtual self-portrait.

Arguably, Hegedüs constructs her *Things Spoken* in conjunction with the radically subjective attitude which is embodied in archival art in such works as Mary Kelly's *Post-Partum Document* (1973–79), Tracey Emin's *Everyone I Have Ever Slept With, 1963–1995* (1995), and Sophie Calle's (slightly later and slightly differently conceived) *Take Care of Yourself* (2007).[2] All these works revolve around individual narratives which are used by the artists to inspire their audiences to ponder on intersubjective problems.

The two installations evoked above and their representational or interactive likes espouse a perspective in which the archives (the fundamental part of the projects, as it were)—whether emerging from the private sphere or from the public space—are constructed by the artists themselves and, as such, directly represent only them, personally, politically, or ecologically.[3] Other works in this category, ones which relinquish the subjective enrootment and pivot on objective or intersubjective issues, such as numerous database installations by Geert Mul, represent the artists aesthetically. Common to them all is, emphatically, the fact that the collections that make up their texture are fully created or composed by individual artists.

This is not, however, the only possible format for a collection artwork, especially in new media art. For example, in his later works Legrady opens the archives in his installations to the audience, who are prompted to participate in building them, thus becoming their co-creators. In *Pockets Full of Memories* (2001) and *Cell Tango* (2006–8), interactive users produce virtual archives which incorporate their own engagement and existences (Gagnon 2008, 23–27). In this way, Legrady uses his art to reflect on collective memory, while exploring the links between memory and archive in the context of hypermedia art (Legrady and Honkela 2002, 163–69).

Towards Artworks-Memorials

I was directly inspired to study transformations in archival art within the field of new media by my work on the Japanese artist Masaki Fujihata, in particular his project *Voices of Aliveness* (2012). However, I find the practices of Sanja Iveković, Luz María Sánchez, and Krzysztof Wodiczko equally important and meaningful. In the following, I will thus scrutinize selected works of all the four artists within the theoretical framework outlined above in order to explore how their projects are embedded in and contribute to the themes of memory and commemoration. Besides the subjects and approaches encapsulated in the respective works, I will also examine such issues as the conflict between memory and history, the strategy of bottom-up narrative as a tool for invalidating and abolishing this opposition, the incompatibility of institutional commemoration and the collective nature of social memory, and the role participation plays in overcoming this incompatibility. By focusing on the selected works by Fujihata, Iveković, Sánchez, and Wodiczko, I will offer an account of the totality of memorial art as a significant and diversified practice spreading across media art today, affecting the work of several artists and nurtured by a varied array of factors.

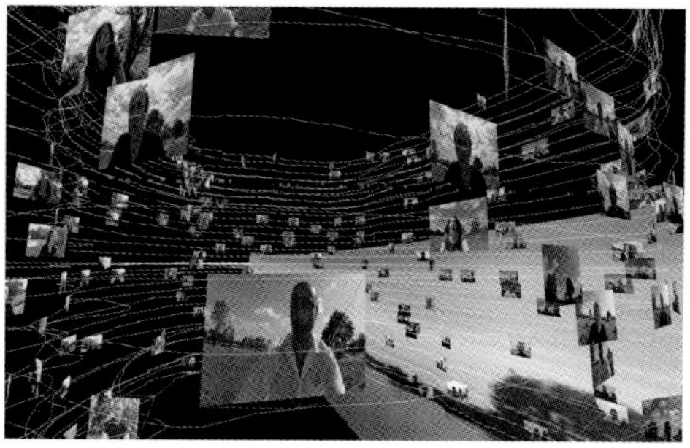

Fig. 1. Masaki Fujihata, *Voices of Aliveness.* © Masaki Fujihata, 2012, produced at The Art College of Nantes-Metropole, France. (Plate 19, p. 315)

In Fujihata's practice, the interweaving of interactivity and digital databases brings forth artworks as collections of individual behaviors, which at the same time become sources and foundations of both social archives and new visions of subjectivized history. In this context, I am particularly interested in Fujihata's *Voices of Aliveness* (Fig. 1), an installation which not only ushered new foci into his art, but also shed a different light on his prior projects.

> Voices of Aliveness invites people to ride a bicycle on a special path prepared exclusively for this project, called the 'shouting circuit.' The bicycles are equipped with a GPS recorder and a video camera. As participants ride the bicycle, traces of their path and shouts will be transformed into cyberspace in the shape of a ring. The collected rings from each participant will be compiled to form a tower-like 'time tunnel'—a cyber tunnel of collected shouting (…).[4]

Fujihata referred to *Voices of Aliveness* as a 'meta-monument of collective memory' (2013). This meta-monument is basically a collection of screams, a storage of voices, and a unique vocal database in one—briefly, a meta-monument as an archive.[5] Discussing his work, Fujihata called it a symbolic image which is simultaneously a site and a means of generating memory. He also emphatically insisted that it was not a conventional monument preserving collective memory. Instead, the project was supposed to produce memory and transform it into a novel kind of monument. Briefly, memory-as-monument (Fujihata 2013).

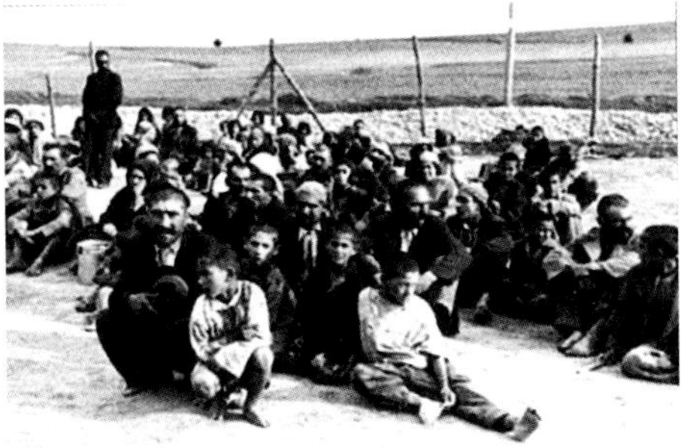

Fig. 2. A group of Roma prisoners sitting on a field inside the fence of the Belzec concentration camp awaiting instructions from their German warders. Photo taken in Belzec, Poland, 1940.
Credit: USHMM, courtesy of Archiwum Dokumentacji Mechanicznej.

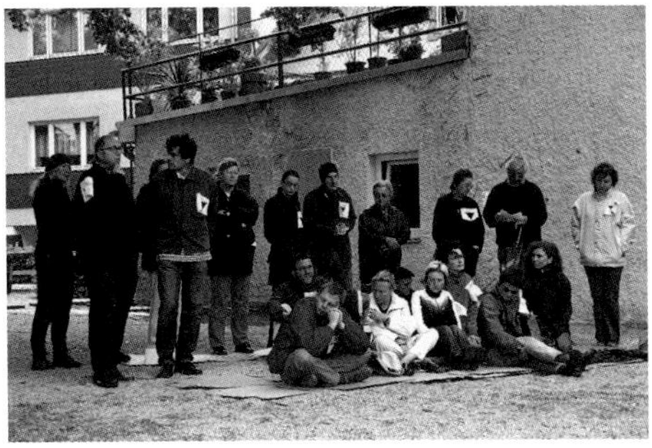

Fig. 3. Sanja Iveković, *Rohrbach Living Memorial*, 2005.
Photo courtesy of the artist.

This engenders a multi-layered identification of the archive, memory, and the monument. As memory itself is recognized as a monument, the latter is envisioned as a mental product and an intentional object of consciousness. This has further implications. By being submerged in mental space, the work partly residing in the memories of participants, Fujihata's artwork-monument becomes a living memorial. This concept was earlier proposed by the Croatian artist Sanja Iveković to label her projects: the *Rohrbach Living Memorial* (2005) and *On the Barricades* (2010). In coining this notion, Iveković was inspired by the idea that in the wake of today's crisis of the memorial, conceived as less a form of commemoration and more an ideological instrument for the exercise of power, it could only reclaim its function and social value if it was transferred into the depths of human consciousness. For this reason, Iveković staged her memorials as collective performances enacted by the audience, who in this way transformed into participants in the event. In the *Rohrbach Living Memorial*, the participants performatively enacted a photograph presenting the Roma who had been murdered in a concentration camp during the Second World War (Figs. 2–3). In *On the Barricades*, the participants imaginatively and symbolically identified with the students killed during a peaceable demonstration.[6] Participation in the performance was supposed to transmute the mental representation of the commemorated event (knowledge) into its sustained presence in the personality of each performer (part of their attitude to the world). Consequently, the memorial turned into an aspect of the identities of the participants and, personalized and internalized, it started to live their lives.[7]

In ways slightly different from Iveković's works, Fujihata's installation has this aliveness arise and unfold on a multiplicity of levels. While Iveković's projects, albeit communally, socially, and politically embedded, call for individual, meditative responses from each viewer-participant, Fujihata's work embodies a more complex, differentiated and multidimensional structure of experiences.

At one level, the monument emerges from individual performances carried out by the event participants within a script of actions prepared for them. Consequently, the living monument is fashioned out of mental traces linked to this event, of each participant's individual memory of his/her own, personal participation, which is distinct and distinctive in each case. At another level, the monument is built on communal relationships within the collaborative creation of the holistically captured event. In this sense, the living monument is crafted of the memory of the event as a collective action, the memory founded on social experience. At the third level, the monument is erected as visualized memory—as a virtual monument—in the virtual environment which can be accessed through a range of interfaces. At this level, the artwork-as-memory shuffles off any direct dependence on the subjectivities of the participants and transcends the boundaries of the mental space, becoming its collective extension and a virtual archive.

The visual iteration of Fujihata's virtual monument contains references to the remaining two dimensions. The grid visualizes the geolocation data generated during the event and, as such, organizes the space and aesthetics of access to video images. It is usually first experienced in contact with the artwork and turns one's thoughts to its collective dimension, making it into the basic, all-encompassing structure. Out of this structure individual audio-visual sequences and autonomous video recordings emerge to represent respective performances. Immersed in the virtual, both construct the work as a collective memory of the initial collective bike-riding performance and its virtual commemoration. What comes to pass in the context of the virtual environment is the abolition of the individual, establishing the communal as an overall perspective built for communalized individuality—for collective memory which respects individual memories and is capable of coalescing them into a coherent, nonantagonistic, diversity-embracing ensemble.

Fujihata's artwork-memorial thus embodies augmented memory, which develops (like the work itself) in a hybrid environment and, as such, offers various access pathways and ushers in several commemoration modes. This leads us back to reflections on the archive, for this augmentation of memory is nothing other than a virtual archive which intertwines audio-visual video data with geolocation data. In his art, Fujihata draws implications from the processes that transform our reality and undermine the distinction between the real and the virtual. The development of the concepts and practices of augmented reality brings forth a hybrid world—a real-cum-virtual space of experience into which Fujihata inscribes the augmented memory of *Voices of Aliveness*.

As already mentioned, the concept of an artwork as a structure of memory, which couples the monument with the archive, is not a distinct invention of Fujihata alone. It has also been evoked and used by Sanja Iveković, by Luz María Sánchez, and, in a different way, by Krzysztof Wodiczko. In his case, projects were preceded by extensive research on the issues they thematized in their dedicated data archives, and involved public, anti-monument presentations in which temporary living archives were staged by and out of people who (sometimes live) talked about and shared their experiences with the audience, affording them an insight into their memory spaces.

Most of Wodiczko's works, like those of other artists I discuss in this paper, are community art. There are several reasons why I find such collectively made works extremely valuable and engaging. First, they aptly delineate the power relations established by the institution of art in relation to the community, positioning the artist, his team, and the community members as partners. Second, this practice helps make the artwork yield an actual insight into the collectively and individually relevant problems of the community. In this way, the artwork addresses and tackles the real challenges the community faces. Third, within this approach the developing artwork can become a valuable anti- or meta-monument, for it is only in a collective form that it can indeed represent the entire community involved. Fourth and last, this practice turns the artwork into a database—an archive that prevents it from ultimate petrification and promotes its enduring presence also in the mental spaces of the community of its participants (Kluszczynski 2017, 162–69).

In Wodiczko's rendering, living archives approximate the living library concept, which has been internationally practiced by activist communities for nearly twenty years now: 'The Human Library is designed to build a positive framework for conversations that can challenge stereotypes and prejudices through dialogue. The Human Library is a place where real people are on loan to readers' (Human Library n.d.).[8] The descendants of the atomic bomb victims, whose voices are heard in Wodiczko's Hiroshima projection, or the female victims of abuse and violence, who are scattered amidst the crowd and simultaneously shown on the screen during the Tijuana projection, exemplify the proximity—nearly sameness—of the living archive and the living library. In Fujihata, such an approximation does not occur because his archive, immersed in the virtual environment as it is, takes the form of an augmented archive.

By conceptualizing *Voices of Aliveness* as a meta-monument—a form of commemoration—we can revise the idea of Fujihata's earlier works, in particular the Field Works series which he has been developing since the early 1990s.

From Time-Structures to Augmented Memory-Spaces: Towards the Participatory Memorial

Impressing Velocity (1992–94) marks a breakthrough in Fujihata's creative practice, as it brings together and elaborates on his previous focal concerns while also propelling them in new directions. The work grew out of an action-event in which a group of students led by Fujihata climbed Mount Fuji. Their climb was documented through video images and, more importantly, by means of a GPS system, which rendered a wealth of data about the time and speed of the climb, the distances covered, the altitudes reached, and the stop-overs made on the way. The multimedia artwork capping the project combined several components, the major three being the event as a performance, the images of a dual kind (video recordings and visualizations of the GPS data), and the installation presenting these images. All three elements were equally important and formative of the structure of the work, while their relationships jointly determined its character, aesthetics, and the experience offered to the audience.

In subsequent projects within the series, these relationships changed. The installation ultimately came to be identified with the interface, whereas the visual environment—an archive of compiled and arranged images—became the artwork proper. At the same time, the original action—a specific performance that initiated each subsequent work—served in hindsight as the framework of reference for the collective memory inscribed in it, the commemorated center of the artwork. The tripartite structure was preserved, yet the interrelations among its components were altered.

Impressing Velocity was followed by a series of projects which Fujihata called by the collective name of *Field Works*. Each project in the series is underpinned by the same artistic procedure and boasts the same final structure: an action undertaken in the real space leads to the construction in the collective cyberspace of a virtual memory of this action, in the form of a video archive annotated with the GPS location data and displayed as an installation (Fujihata 2013a). Already applied in *Impressing Velocity*, the same strategy is observable in the consecutive *Field Works* projects, including *Tsumari* (2000), *Hayama* (2001), *Alsace* (2002), *Lake_Shinji* (2002), *Mersea Circle* (2003–5), *Talking Tree* (2005), *Landing Home in Geneva* (2005), *Simultaneous Echoes* (2009), and, finally, *Voices of Aliveness*, the central axis of this argument. These projects are usually presented to the audience in an interactive form. They always arise as a result of field activities, hence the title of the series. Produced in a collaborative undertaking, each work-archive is a representation of the community that contributed to its making. At the same time, it functions as a meeting place for the individuals that make up such a community. It is a site of individual and simultaneously collective creation. Such creative practices are invariably determined by the relationships among the participants which are organized around the particular thematic concern designated by the artist, such as crossing borders (*Alsace*), language-territory relations (*Landing Home in Geneva*), and human relationships (*Talking Tree*).

In each of the *Field Works* projects, virtual collective memory concerns straightforwardly defined subjects. It is the memory of the community whose actions produced the work, a community that arose and formed around this work.

As already mentioned, *Voices of Aliveness* is, basically speaking, a collection of screams. Commenting on his work, Fujihata pointed out that people produce their first cry the moment they are born. Consequently, the project presupposes affording its participants an opportunity to be born again, whereby they are their own agents and their own source. This agency ensues from the fact that this time the birth is a spiritual, psychological, mental, and crucially symbolic event. Participation in the project becomes a specific rite of passage and the rebirth becomes a gesture of self-constitution which enables the individual to re-define or re-assert his/her status: the world of his/her endorsed values and the ways of life that emerge from them together with goals to pursue. This may be why *Voices of Aliveness* is defined as a monument. This monument commemorates the radical shift in the lives of the participants in the event—the moment they were born again.

At the same time, *Voices of Aliveness* is more than a gesture of commemorating the event from which it emerges (as a memorial). It is also a reflection on the meaning and form of commemoration processes launched in the public space. This is why Fujihata has referred to *Voices of Aliveness* as a meta-monument. The project embodies discourse on memorials and monumentalism in which the causes, aims, functions, and implications of the presence of memorials in our lives and in our world are deconstructed. In an age in which states and communities use monuments to battle with each other and wage wars, such a discourse acquires an inherently political dimension.

Voices of Aliveness is a very peculiar memorial. Because if it is a monument to an individual's (and perhaps community's) rebirth, it was put up by—or at least with a very significant contribution from—those whom it commemorates. It is a bottom-up, communal, and participatory memorial which neither represents power nor imposes any dominant ideology, nor glorifies violence, nor marginalizes anybody. It is a monument as an eruption of joy, a monument and a nonmonument at the same time.

Soundscapes of Violence

In her art, Luz María Sánchez proposes a slightly different approach to the idea of an artwork as a memory archive and to participatory engagement. Sánchez's multi-channel installations employ sound archives she accumulates around events involving violence and death. Sometimes, the components of her archives are produced by the artist herself. This is the case in her eight-channel sound installation entitled *2487* (2006). The title refers to the number of people who lost their lives while trying to cross the US-Mexico border between 1993 and 2006. Sánchez pronounced and recorded the first names and surnames of 2,487 victims of political violence (what name other than political violence can we give to state borders in today's world, in which identification with the place is systematically and profoundly undermined?) about whom she had found out during her research (the total estimated number is indeed considerably greater). The articulated names are recorded in separate three-second-long files, which the installation plays in a random sequence, whereby its space is filled with the sound structure whose dynamic aesthetics reproduces disorderly processes of social migrations, as a result becoming an original form of migratory aesthetics. Expressed in the vocal rendition of the name list, the artist's action is both discreet and poignant, minimalist and expressive, evoking very strong emotional responses in the audience. The installation also includes a book which lists the same names along with home town, age, date and cause of death, if known. A pertinent visual context of the installation is provided by the score which illustrates how the voice-generating system functions.

The installation can be exhibited in a variety of forms and settings. At the San Antonio Artpace (2014–15) it was put on display in a gallery environment. A white bench for the audience was surrounded by sixteen low-mounted speakers. Such a spatial arrangement favored a contemplative approach to the work, intensifying its commemorative character (as a border death memorial). In turn, at Fort Winfield Scott at Langdon Court in San Francisco (2016), the installation was set up in a bunker, with the speakers placed along a corridor. As the visitors walked in the installation space, their movement became an important factor in experiencing the work, which acquired the symbolic form of a quest for a new life, one which actually ends in death.

The installation also has another version in which the participatory component is foregrounded, as the names of the victims are delivered by various voices—of women, men, and children. This version accrues additional meanings since the diversity of the uttering voices corresponds to the multiplicity and analogous diversity of the victims. Whereas in the former version the tribute to the victims is paid by the artist herself, in the latter they are celebrated by the community of the living. As such, the latter version more emphatically highlights the communal character of the individual experience of death, fostering the realization that individual death ruptures social bonds, produces a hiatus, and triggers the feelings of loss in those who remain.

Like Iveković and Wodiczko, by relying on different strategies Sánchez gave her work the character and the function of a memorial, whereby she either lent the dead her own voice or called forth a community of voices to commemorate them.

Focused on the dead, the *2487* installation forms one part of the installation project entitled *Diaspora I/II* (2006). Its second part—the *riverbank*—is devoted to survivors. This is an arrangement of objects (a sculpture of found clothing and personal belongings) gathered by Sánchez on her several trips to the border near Laredo in 2006 (Agresta 2015). The items were abandoned by the immigrants who had managed to make it from Mexico to the US across the Rio Grande. Some of the objects are pieces of garments which the immigrants discarded after swimming across the border river, putting on new garments in order to blend in and start a new life at the new place.

Sánchez employs yet another strategy in her multi-channel installation, *Vis. [Un]necessary Force_1.01* (2017), in which the sound sources (loudspeakers) are installed inside purpose-made gun-shaped (Caracal F 9x19 mm) sculptural forms produced by means of the additive method (additive sculpture) or 3D printing.[9] The sounds collected in the archive compiled by the artist, which forms the body of the work, are the noises of shootings that people accidentally chanced upon, recorded with their smartphones, and posted the recordings on YouTube.

Vis. [Un]necessary Force_1.01 is a participatory installation in two essential dimensions. It is participatory in terms of production because the sound data were generated by multiple individuals, who in this way contributed to the work. It is also participatory in terms of audience experience, as the viewers decide how the installation is used: by using a special button they can choose to play (or, for that matter, not to play) the sounds from the gun-sculptures and listen (or not) to the shooting. The viewers-users of the installation can also avail themselves of the laser placed in each of the objects in order to turn other people around them into potential victims and identify themselves with the perpetrators. The installation is also complete with information about the sources of which the audience hear, providing maps and detailed descriptions of the incidents (Fig. 4).

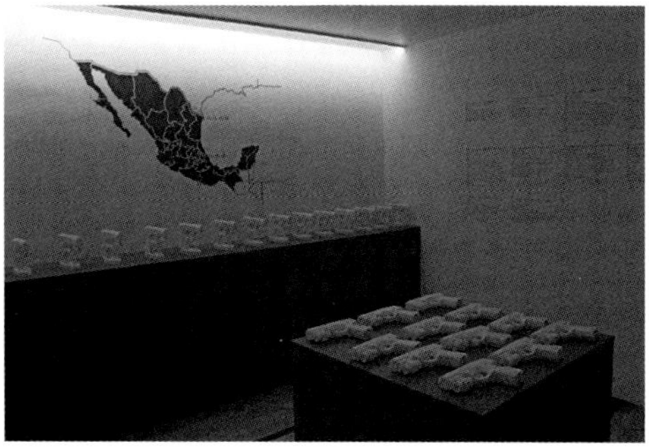

Fig. 4. Luz María Sánchez, V.U(n)f_1.01 (2017). Multi-channel asynchronous sound installation: 40 portable digital speakers* in the shape of a Caracal 9mm pistol [3D prints], 40 Micro-SD cards, 40 mp3 sounds, modular structure. Dimensions: variable. Duration: undetermined. Installation view: Modos de oír. Prácticas de arte y sonido en México. 28 November 2018 – 31 March 2019. Laboratorio de Arte Alameda, Mexico City. Mexico.
Photo credit: Marcos Ysair Pérez Botello. (Plate 20, p. 315)

The artist leaves it up to the audience to manage all the processes involved in the installation, which not only emphasizes its participatory character but also weaves an interactive thread into it: the way in which the work is experienced depends on the interactive behaviors of particular viewers. Because of the rich texture of connections that underpins *Vis. [Un]necessary Force_1.01*, the work, crucially material as its experience is, can also be regarded as network-based, by which I mean that it relies on a system of mutual links which bind the installation's varied components, inducing audience behaviors.[10]

In both installations by Sánchez sound is used asynchronically. But while in the former the asynchrony is an outcome of random sound generation (the technical aspect), in the latter it results from the irregular, free behaviors of the audience (the human aspect).

The forms of socialization and participation, as well as the forms and strategies of the art of collective memory, that underpin Sánchez's art differ from those used by Fujihata, Iveković, and Wodiczko. When developing her *2487* project, Sánchez drew on the work of NGOs which collected information about people attempting to get across the Mexican-American border. In *Vis. [Un]necessary Force_1.01*, she built on the pursuits of people involved in civic journalism who collected and published violence data. In Mexico, where the authorities seek to curtail the information flow concerning events and problems addressed in Sánchez's projects, civic activity is of primary importance—as is Sánchez's art as such—and carries a crucial political investment. Her works perfectly exemplify the interpenetration of the artist's individual perspective and the social perspective, while creatively employing the collectively amassed resources. This interpenetration marks both the structure of the works and the methods of acquiring, processing, and utilizing information.

Sánchez's art can easily be recognized as a perfect example of artistic research. Her projects grow out of explorations of the social situation, field research (which is in fact very risky, given the issues she tackles), and the comprehensive study of multifarious materials. The databases she creates as a result of her research and analyses also feature in her art as outcomes of her own enquiries unrelated directly to the endeavors of civic organizations. For example, *Detritus* (2011–13) is an installation which emerged from the artist's examination of the media representations of violence and the war on drugs. Her basic sources were articles published in 2006–12 in two national Mexican newspapers. This database includes 15,585 complex data entries. In exhibitions, the images from the databases are sometimes accompanied by violent audio clips from police frequencies in Mexico, or by deliberately selected music. Yet even in such works, the outcomes of efforts undertaken by a community (in this particular case, by the journalist community) are observably and meaningfully bound up with the perspective of the artist, whose analytical endeavors highlight changes in the media policy adopted by the Mexican authorities vis-à-vis violence, particularly drugs- and kidnapping-related violence.

Conclusion

The works of Masaki Fujihata, Sanja Iveković, Luz María Sánchez, and Krzysztof Wodiczko, discussed in this paper, reveal a fundamental disparity between the social collectiveness of memory and the institutional forms of information and commemoration. When memory is institutionalized, commemoration all too often mutates into a political and propagandist action, which is patently evinced in the current thriving of agendas rooted in historical politics and a striving to replace history and transform memory into the site where their own preferred vision of the past is forged. The living, collective, and social memory around which communities are formed is instrumentalized by states and other institutions of commemoration, which harness it to achieve their own goals and satisfy their own needs. This entails not only a profound institutional re-programming of the image of the past, but also—as shown by Sánchez's art—filtering information about current events in order to fix the future image of the present even as it is becoming.

Every present is a locus of differentiated processes, discrepant axiological systems, and contradictory needs. There are no commemoration forms which could accommodate them all. But there is an urgent need to preserve in memory the totality of their diversity. Bottom up, participatory commemoration forms and micro-communal memorials are a response to this exigency.

Bottom-up participation is the most important challenge of today's globalizing world. It is also this world's greatest need. Among the responses to this need, there are artistic participatory strategies which result in instituting new commemoration forms, ranging from creative collective performances organized as the axial components of artworks to the inclusion of the work of civic movements within artworks. These participation strategies aim to transform the homogeneous, universally shared memory and history into an intricate tangle of history in which every community can find a place for itself. Memorials as databases, as archives of memory, seem to be fitting and viable tools for meeting this need. They enable communities to regain their collective memories—memory-archive or memory-rhizome.

The bottom-up participatory commemoration forms also offer opportunities for resolving the conflict between memory and history. Pierre Nora has argued that:

> Memory and history, far from being synonymous, appear now to be in fundamental opposition. Memory is life, borne by living societies founded in its name. It remains in permanent evolution, open to the dialectic of remembering and forgetting, unconscious of its successive deformations, vulnerable to manipulation and appropriation, susceptible to being long dormant and periodically revived. History, on the other hand, is the reconstruction, always problematic and incomplete, of what is no longer. Memory is a

perpetually actual phenomenon, a bond tying us to the eternal present; history is a representation of the past (1989, 8).

Works of art in the form of databases or memorials spin bottom-up narratives which are capable of overcoming the memory-vs.-history opposition. Even if they do not bring the two concepts into synonymy, they can make them encapsulate different aspects of the same process, in which living history unites with regulatory memory in order to jointly connect the past and the future.

As already emphasized, Fujihata's meta-monument is a manifestation of joy. Of subversive joy, to boot. The living memorials by Sanja Iveković, the anti-monuments by Krzysztof Wodiczko, and the sound-archival memorials by Luz María Sánchez are anything but joyful. Nevertheless, the projects of all four artists share the same political status and the grid of social references they weave. For all of them are expressive of a rebellion and side with the citizens. These projects focus on various species of historical politics with their social ramifications: reminding us about their victims (Iveković); attending to violence and its institutionalized forms (Sánchez); exploring the control of social space in conjunction with the marginalization and rejection of the Other (Wodiczko); or investigating cyberspace and encouraging bottom-up, grassroots commemoration practices (Fujihata). They have in common that they reject all kinds of violence, negating the practices that honor and commemorate it.

Notes

1. Of course, it does not mean that such themes are not explored in other forms of art. Such a notion is easily belied by, for example, numerous found-footage films, such as Péter Forgács's video series *Private Hungary*.
2. Calle's work is an assemblage of the artist-inspired responses of various people to her personal experience.
3. Unlike the female artists referred to above, Boltanski, Kabakov, and Legrady (as mentioned) develop their projects employing public documents rather than private resources only, and thus propose a specifically objective view of the themes they survey. Their objective investment notwithstanding, they nevertheless also refer to personal attitudes, memories, and interests. Consequently, their works are individualized as well.
4. http://voicesofaliveness.net/concept/.
5. Ibid.
6. For a more detailed discussion of Iveković's two works, see (Kluszczyński 2017, 155–62).

7. Sanja Iveković prepared a letter-invitation to the performance:
   'I invite you to come to Rohrbach and take part in the street performance *The Rohrbach Living Memorial* which is dedicated to the hundreds of thousands of Roma, victims of Nazi persecution genocide. This invitation is open to men, women, and children of non-Roma and Roma origin. Although there is still no national memorial which would embody the official memory of Roma victims of the Holocaust in Austria, the photograph which shows the Gypsies waiting to be transported to a concentration camp is slowly becoming a part of our collective memory.
   I invite you to join us and become that image.
   Together, we will demonstrate that we are not willing to forget the crime, so that it may not be repeated in the future. By participating in this action we will also show that we are able to challenge not only the «invisibility» of the Roma people in the troubled past, but also the widespread public prejudice and official discrimination which they are facing at present.
   Traditionally, memorials are constructed from a material that presumes durability and permanence. Here, the opposite is the case: this «memorial» will last only a few hours and will consist of us – living creatures. However, I believe that in such a way we will take the role of active witnesses who will become responsible for the memory's preservation after the image fades from sight and will thus enable our fight against the passivity of stone, metal or concrete.
   The extreme fragility and the temporariness of «The Rohrbach Living Memorial» should also be understood as a good metaphor for the extreme precariousness of a long and difficult process which is in front of us: the recognition of the Roma and their culture within today's Europe.'
8. http://humanlibrary.org. See also Carageragea n.d.
9. *Vis. [Un]necessary Force* is the title of the whole project and series of artworks which compose it.
10. There is an Internet version of the work as well.

References

Agresta, Michael. 2015. "Narco Violence, as Seen by a Journalist-Turned-Artist." *Texas Observer*, November 5.

Carageragea, Violeta Maria. *Human Library Facilitation Guide*. https://learningforchange.net/knowledge-base/human-library-facilitation-guide/.

Fujihata, Masaki. 2013. "Voices Of Aliveness–Masaki Fujihata, [talk at] Prix Forum 2013, TOTAL RECALL–The Evolution Of Memory". October 10, 2013. www.youtube.com/watch?v=hXSEJ5USSlw.

Fujihata, Masaki. 2013a. "Field-works." September 2013. http://www.field-works.net.

Foster, Hal. 2004. "An Archival Impulse." *October*, 110 (Autumn): 3–22.

Gagnon, Jean. 2008. "George Legrady–Cell Tango et Pockets Full of Memories: Structure de l'archive virtuelle." *Ciel variable*, no. 80 (Automne): 23–27.

https://id.erudit.org/iderudit/13205ac.

Hirsch, Marianne. 1997. *Family Frames: Photography, Narrative, and Postmemory*. Cambridge MA: Harvard University Press.

King, Barry. 2000. "Being Virtual: Modularity As a Cultural Condition." *Afterimage* 28(2) (September): 9–12.

Kitzman, Andreas, Conny Mithander, and John Sundholm (eds.). 2005. *Memory Work: The Theory and Practice of Memory*. Frankfurt am Main: Peter Lang GmbH.

Kluszczyński, Ryszard W. 2010. "Strategies of Interactive Art." *Journal of Aesthetics and Culture* 2: 1–27. DOI: 10.3402/jac.v2iv0.5525.

Kluszczyński, Ryszard W. 2017. "Monument-as-Archive. Artistic Strategies from Anti- to Meta-Memorial." In *Augmenting the World: Masaki Fujihata and Hybrid Space-Time Art*, edited by Ryszard W. Kluszczyński, 150–75. Gdansk: Center for Contemporary Art.

Kusahara, Machiko. 2017. "The Early Years of Fujihata's Art." In *Augmenting the World: Masaki Fujihata and Hybrid Space-Time Art*, edited by Ryszard W. Kluszczyński, 46–67. Gdansk: Center for Contemporary Art.

Legrady, George, and Timo Honkela. 2002. "Pockets Full of Memories: An Interactive Museum Installation." *Visual Communication* 1(2) (June): 163–69. DOI: 10.1177/147035720200100202.

Lury, Celia. 1998. *Prosthetic Culture: Photography, Memory and Identity*. London: Routledge.

Manovich, Lev. 2001. *The Language of New Media*. Cambridge MA: MIT Press.

Nora, Pierre. 1989. "Between Memory and History: Les Lieux de Memoires." *Representations* 26, Special Issue: *Memory and Counter-Memory* (Spring): 7–24.

Rosengarten, Ruth. 2013. *Between Memory and Document: The Archival Turn in Contemporary Art*. Lisbon: Museo Coleção Berardo. Kindle.

The Logic of Participation in Digital Art and Network Culture

Manuela Naveau

Introduction

Why are unconscious or involuntary forms of participation overlooked today? Why do we not take seriously the fact that participation models are infiltrated by profiteers, that the systems of digital participation are deliberately rendered opaque or presented as mere decorative accessories?

There is no doubt that our world is undergoing radical change. Digitalization is the topic of the moment. Political representatives' clumsy efforts to address this issue, whether at the national or EU level, reveal that they have little clue about its dynamics. Attempts to explain digitization often compare our era to the industrial revolution. At the same time, however, we are informed that linear thinking similar to rack-wheels, where cause and effect are mutually dependent, is no longer valid in our networked world, where everything occurs simultaneously.[1]

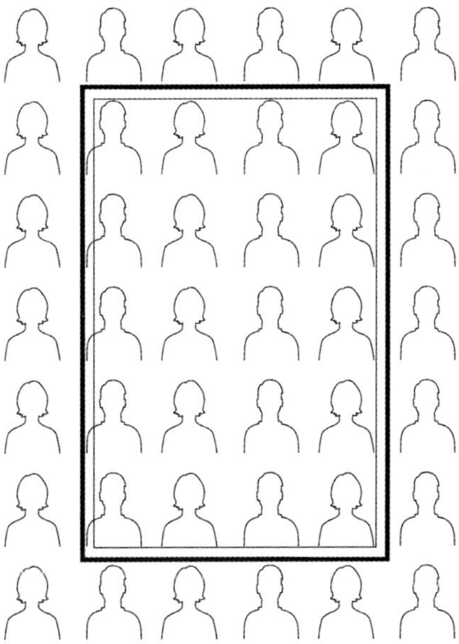

Fig. 1. Manuela Naveau, *CROWD & ART – Kunst und Partizipation im Internet*, 2017. Image: Nicolas Naveau.

Politics in Europe is struggling to keep pace with technological advances and, although we are embroiled in discussions of digitalization, we seem not to understand what it means. Gerfried Stocker writes that:

> 'Digital' is after all not merely a technical category but rather an ontological state of information, an aggregate state, a distinctive form of how information exists and appears, which is created when information is detached from the conditions of material data carriers and as code, as a series of zeros and ones, takes on the ephemeral form in which it can be converted and passed on into all conceivable forms of expression and copied and shared—in other words communicated—without limits. The computer has long completed its transformation from literal 'computer' to universal communication and culture device; as its users, we are no longer operating a machine—instead we are nodes in the network (quoted in Naveau 2017, 7).

Here Stocker gets to the heart of an issue for which, as yet, we have no suitable conceptual model.[2] As a result, our networked reality forces us to reflect on terms such as 'digital' and 'physically real,' 'mass' and 'individual,' 'public' and 'private.' Moreover, we must also become critically conscious of forms of network participation that are otherwise overlooked because they occur unconsciously or involuntarily. Grasping ourselves as nodes operating in a network will constitute a fundamentally new understanding of social reality. Each individual is a part of an incomprehensibly enormous public, which encompasses us whether we like it or not. Ultimately, this leads to the question of how we handle our bodies in different public spaces, and the different forms of participation and involvement that our bodies can perform or support.[3]

Observing Forms of Participation

In 2010 I began a series of theoretical engagements that inspired my book, *Crowd and Art*. I started with the following observations: for one thing, I saw that specialist literature on artistic participation did not attend to current forms of online participation models; indeed, I discerned a widespread obfuscation of digital participation. I also recognized that forms of participation in art have changed due to computer technology, the Internet, and smart phones—in other words, the development of digital hardware and software that we have in our pockets. Initial research revealed a plethora of publications, conferences, and papers on participation not only in art, architecture, theater, and performance, but also in political and social domains, particularly education (Naveau 2017, 21). At that point, digital participation was discussed largely by media critics (Naveau 2017, 32). This was also a time of revolt and political upheaval: in particular Twitter (which is based on the TXTMob program used by democratic activists in the US) enabled protestors to coordinate their actions and communicate up-to-date information about police activity. Besides their use of Twitter, protest movements combined social networks in a clever way, which expedited further uprisings beginning in 2011. Motivated to create new forms of political organization, protesters in Moldavia (2009), Iran (2009), Egypt (2011), Spain (2011/12), and Turkey (2013) placed great hopes in social networks.

Which forms of participation did these networks foster? What is the significance of appropriation, deception, instrumentalization, or outsourcing in the digitalized world? What were the participants' goals or agendas, and what does participation have to do with knowing or not knowing? As a curator and researcher at Ars Electronica Linz, I have sought to address such questions by investigating today's networked reality by means of selected artistic works, adopting neither a romantic nor a dystopian attitude towards the digital. I attend to artistic practices that tried out forms of online networking and participation early in their development. These projects are mostly subversive, in that they reveal mechanisms of participation and demystify discourses of participation. They show that participation as an end in itself—'participation for participation's sake,' as it were—makes little sense, but that participation, knowingly or not, has to do with the generation of knowledge. Four case studies and accompanying interviews are set out below.

The World in 24 Hours by Robert Adrian X[4]

In the 1970s and early 80s, the Canadian artist Robert Adrian X, who was then living in Vienna, began to focus on the new field of 'the electronic,' in collaboration with artist colleagues around the world. Fascinated by how digitalization simplified the process of exchanging pictures, he conceived and initiated an art project which was commissioned by Ars Electronica in 1982. Based on telecommunication technologies, the project connected 15 cities in different parts of the world. It was shown at the Ars Electronica Festival (September 27–28, 1982).[5] His goal was to foster a collective investigation of electronic space using telecommunications media accessible at that time. Three telephone lines were simultaneously used for phone calls, faxes, and artistic exchanges via Slow-Scan Television (SSTV) and the international computer mailbox and conference system produced by the company I.P. Sharp (IPSA).[6] All these were available throughout the 24-hour period. The technologies were chosen in large part because they were relatively inexpensive, easy to use, and accessible in Linz in a form authorized by the Austrian Postal and Telegraph Administration (ÖPTV).

The beginning of the Internet era saw widespread fascination with opportunities for global communication through new networks and platforms; exchanging images; and

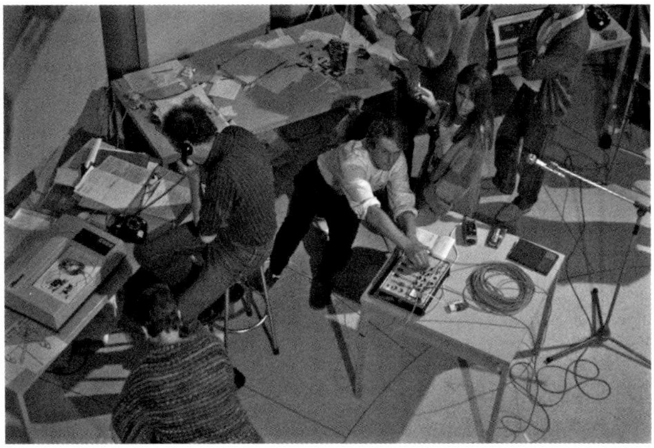

Fig. 2. Robert Adrian X, *Die Welt in 24 Stunden*, Ars Electronica Festival, Linz, 1982. Photo: Sepp Schaffler. (Plate 21, p. 316)

meeting other people in virtual space. This fascination drove extremely expensive telematics projects, building on previous experiments conducted by satellite in the 1970s. *The World in 24 Hours* was special, however. It heralded a new era in telecommunication art because it used relatively convenient communication channels, which could address much wider audiences than the satellite projects. Crucially, it saw the creation of the online artists' platform *Artbox/Artex*. Although its essential concept closely resembles later chat rooms, and indeed general communication in Web 2.0 and contemporary social media platforms, *Artbox/Artex* laid the groundwork necessary for audiences from all around the world to actively participate in artistic processes.

According to Adrian X, *The World in 24 Hours* was a 'communication sculpture'—a communicative experiment performed with and by artists. It was never about creating a 'work of art' (Adrian X sees the concept of 'work' as a nineteenth-century relic); the goal instead was to understand emerging media through the application of artistic methods. Only when the possibilities opened up by a medium are understood can we critically intervene in how it is disseminated and used. As communicators, Robert Adrian X and many of his artist colleagues felt that they had a duty to prevent these media from being left to profit-oriented companies. They wanted to ensure that all individuals (including artists) could not only access the media, but also have a say in how they develop and are used.

Lizvlx and Hans Bernhard on [V]ote-auction[7]

This project started when a US student named James Baumgartner had the idea of installing a platform on the Internet that could buy and sell votes, as part of his final master thesis. Immediately after the website was published, he received an injunction from the New York

Election Committee. He contacted the artistic activist group The Yes Men, one member of which (then known as RTMark) was his professor. He knew that this kind of website could only be tested outside the jurisdiction of the United States, and contacted Hans Bernhard and Lizvlx from *ubermorgen.com*, who then developed the work on this basic idea.[8]

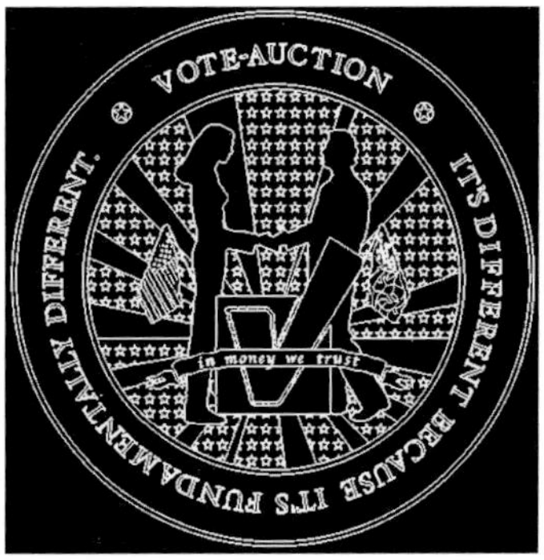

Fig. 3. ubermorgen.com, *[V]ote-auction.*

The concept behind the platform *[V]ote-auction*, subtitled *Bringing democracy and capitalism closer together*, was as follows. During the 2000 American presidential election (Al Gore vs. George W. Bush), it offered voters the chance to sell their votes to the platform, which would then auction them off to the highest bidder on election day (November 7, 2000). The project, of course, was a hoax, as there were neither purchases nor sales, only offers. The project received enormous news coverage, which peaked in a 27-minute segment on CNN. The FBI, NSA, and CIA prosecuted the artist duo; Attorney General Janet Reno even got involved. It is estimated that 450 million people learned about the project through media reports.

The various parties involved—US citizens, Internet engineers, courts, lawyers, and journalists—performed different forms of participation. On the one hand, there were the millions of Americans who could be not exactly sure how they were participating in *[V]ote-auction* in selling their votes. This, I claim, constitutes uninformed (or, as I go on to call it, unwitting) but voluntary participation. Even those who, despite not offering their votes for purchase, sent fan mail or hate mail were voluntary but unwitting participants. I regard courts and lawyers as unwitting and involuntary participants, in that *[V]ote-auction* exploited their professional obligations to further the project. As representatives of the law they had to react, which meant they were deceived by Lizvlx and Hans Bernhard. Inflexibly fulfilling their offices without critically examining the situation, these legal professionals became involuntary, unwitting participants. Journalists, however, voluntarily exploited the situation for their own ends, even if they reported on the project facts without knowing its background.

Art and participation always seem to follow an agenda. The difference between a 'cause' and an 'agenda' is interesting. A cause denotes highly subjective wishes, hopes, and commitments, and can therefore only apply to conscious, voluntary participation. An agenda is a more abstract term, oriented towards structures and functionality in pursuit of optimal results. The Latin term simply refers to 'what must be done' (Wikipedia n.d.a), and does not pertain to individual desires.

In light of this, the artist group *ubermorgen.com* holds that we can speak of 'participation' even in projects such as *[V]ote-auction*. In this case study, therefore, I intend not only to introduce *ubermorgen.com*'s artistic practice, but also highlight its communicative achievements. *ubermorgen.com*'s work shows how participation is an underestimated part of communication. When considering their practice, one is struck by how participation affects the body and how individuals need mental strength to withstand situations of constant confrontation.

Aaron Koblin's The Sheep Market[9]

Koblin—who found himself in a constant 'schizophrenic battle between [his] own luddite philosophical tendencies and an overwhelmingly nerdy curiosity in all technological tools' (Koblin 2006b)—sought to combine his artistic aspirations with the newest technological developments. He was inspired by the fact that people can still accomplish some tasks more easily than computers or gadgets can. In addition, he was interested in contemporary art and artists' working practices. These elements combined to produce *The Sheep Market*. Over a period of 40 days, Koblin hired workers on Amazon's Mechanical Turk (MTurk) to 'draw a sheep facing to the left.'[10] He set up a website which, aside from providing background on the project, presented a mosaic of 10,000 miniature drawings of sheep. If the user clicks on an image of a sheep, an application opens to show how the sheep was drawn in 'playback mode'. It gives users the impression that they can sense the person who drew the selected sheep. As if drawn by an invisible hand, the picture completes itself, sometimes fast, sometimes slowly. Sometimes nothing happens for several seconds.

Amazon's Mechanical Turk is an online marketplace where clients divide jobs (called HITs 'Human intelligence tasks') into the smallest possible pieces ('microjobs') before outsourcing them to mostly anonymous workers around the world. The marketplace's name refers to the famous chess-playing automaton constructed in 1769 by Wolfgang von Kempelen, the Austro-Hungarian court official and inventor. The automaton was not actually a robot—instead, a person sat inside taking the controls (Wikipedia n.d.b). Accordingly, Mturk is about human intelligence and, more specifically, the accumulation of individual solutions. Since the Turkers'[11] solutions are not coordinated or complementary, concepts of 'swarm solutions' or 'swarm intelligence' do not apply to Mturk (see Kwastek 2013).

Each online worker participating in *The Sheep Market* was paid two cents for drawing a sheep. Koblin collected 10,000 sheep, which works out at about eleven sheep per hour.[12] Capitalizing on this minimally paid labor, Koblin sold the sheep drawings 'en bloc' as stamps accompanied by a certificate of authenticity, which raises issues of copyright and authorship.

Thanks to the Internet, the outsourcing of production has taken a new turn: Whereas once it was successful artists who were in a position to afford a team of specially selected

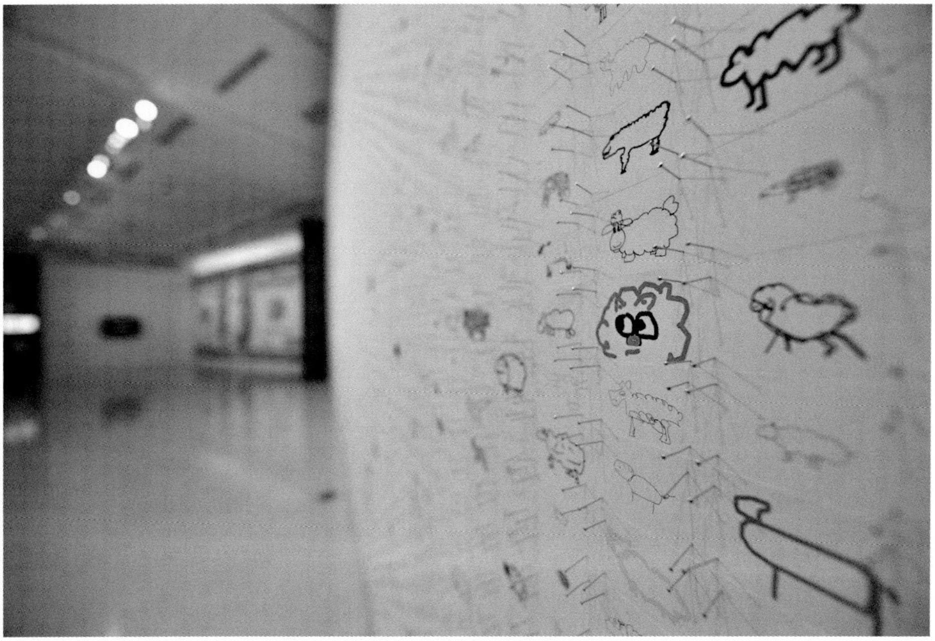

Fig. 4. Aron Koblin, *The Sheep Market,* Ars Electronica Export in Taichung 2012.
Photo: David Sun / Chi River Art International.

assistants, now a range of tasks can be done quickly and very cheaply by anonymous people over the Internet. In contrast to traditional participatory art, which outsources artistic practice to audiences that have at least a minimal relationship with art,[13] Aaron Koblin had the work outsourced to workers who were not necessarily related to art at all.

While the Warhols, Hirsts, and Eliassons of our time seek out specialized workers, Koblin had no influence over the selection of Turkers involved in *The Sheep Market*. Whereas Hirst is said to work with 180 assistants who all have to deny their individual style, Koblin did not necessarily want his sheep perfectly replicated on canvas, as if executed by a computer. Indeed, Koblin looked for the personal angle on every sheep. The interface he set up allows the individuality of each draftsperson to come across. To gather as diverse a collection of sheep drawings as possible, the artist gave participating workers free rein in how they represented the sheep. He intentionally avoided crediting those paid to draw the sheep. *The Sheep Market* not only highlights important issues such as ever-changing work environments; labor processes; the value of work in the virtual economy; the attention economy; and digital reproduction. It also explores questions of authorship and copyright, as well as the status of human creativity in our networked society. In this work the concept of participation is decisively expanded, investigated, and removed from its familiar context. In the process, it throws the characteristics of the social-media age into sharp relief. It foregrounds immateriality, simultaneity, processuality, and the delegation of tasks to—or appropriation of information from—unwitting participants.

Case study 4: Paolo Cirio and Alessandro Ludovico's *Face to Facebook*

Face to Facebook is a Facebook hack, multimedia installation, and—this is the artists' definition—'a social experiment.' The artists stole data and profile pictures from one million Facebook users and analyzed them in a database. They took information such as name, country, and the Facebook groups joined. Facebook's friend system made it possible for the artists to cull a relatively global sample of data. They wrote a software program which extracted 250,000 (mostly smiling) faces from the data group according to six types: 'climber,' 'easy going,' 'funny,' 'mild,' 'sly,' and 'smug.' The artists then put them on a dating website—or rather online partner search platform—they had set up: *http://www.lovely-faces.com*. Although the website had to be taken down due to a complaint by Facebook, the artists were not prosecuted. However, the artists did show the dating website in kiosk mode, and presented prints of 250,000 small-format versions of the faces at exhibitions. They also showed a diagram representing the project's software development process on an old overhead projector. Cirio and Ludovico were not primarily concerned with demonstrating the system failures of Internet giants, and they did not steal private data to generate personal or monetary profit. On the one hand, they appropriated the data to challenge tech-companies' omnipotence and make people think—or perhaps even laugh. On the other hand, the artists wanted to impart knowledge through artistic intervention and start a discussion. They were largely successful in all this, as is demonstrated in many different responses to the project, not least the legal correspondence with Facebook's law departments, which is available, along with the artists' defense, on their website.[14]

Discussion and confrontation were central to the project, which sought to spark critical thought about profit-oriented online platforms (see Cirio 2012). This is evident in the project's communication machinery, (presumably) planned press releases (see Paolo Cirio and Alessandro Ludovico n.d.c), and the artists' communication with the persons who participated or were affected, including lawyers, Facebook users, journalists, and festival goers. The primary goal was to encourage people to examine the agenda behind participation models such as Facebook, which use 'participation' and 'sharing' as marketing buzzwords. The project also implied that we must relinquish the false belief that the social media cosmos could establish a participation ethic. Given their obvious agenda, such platforms are the least appropriate possible vehicle for such aspirations.

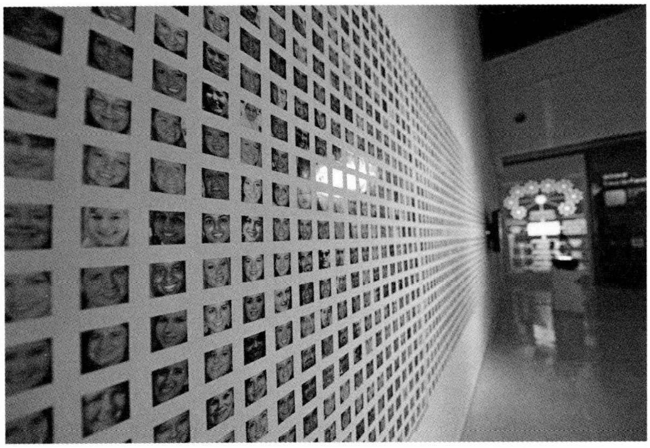

Fig. 5. Paolo Cirio and Alessandro Ludovico, *Face to Facebook,* Ars Electronica Export in Taichung 2012. Photo: David Sun / Chi River Art International. (Plate 22, p. 316)

The Logic of Participation: A Taxonomy

The concept of participation—which is composed of the Latin words *pars* (part, share) and *capere* (seize, appropriate, catch, grasp, capture)—implies both an offer or invitation (by someone who wants to initiate participation), and a reception performance of the part-taking person. The literature on participation in art therefore deals with both invitations to participate from artists and forms and degrees of involvement from audiences. To question the concept of audience, I will call audiences the 'others.' The following analysis distinguishes, therefore, between forms and offers of participation on the one hand, and audience experiences on the other.

My analysis of participation models differentiates among different invitations to participate (these include, for example, traditional participation, or participation as appropriation, deception, or instrumentalization). In addition, it also takes a closer look at the different ways in which recipients can participate, effectively or otherwise (these include, for example, conscious, unwitting, voluntary, and involuntary participation). Four forms of inviting and receiving participation can be extrapolated from the case studies (Fig. 6).

The traditional principle of participation implies a clear offer and a deliberate and voluntary action performed with others. With the principle of appropriation there is no offer at all. The initiator of the participation process takes information from unwitting, involuntary involved persons that are mutating into participants even without ever being asked. The principle of deception comes into play when participants are entrapped and deceived about the ends to which their participation is directed. The way in which they think they are participating differs from the in way in which they actually are participating (the offer to take part pretends something else). The last category refers to the principle of instru-

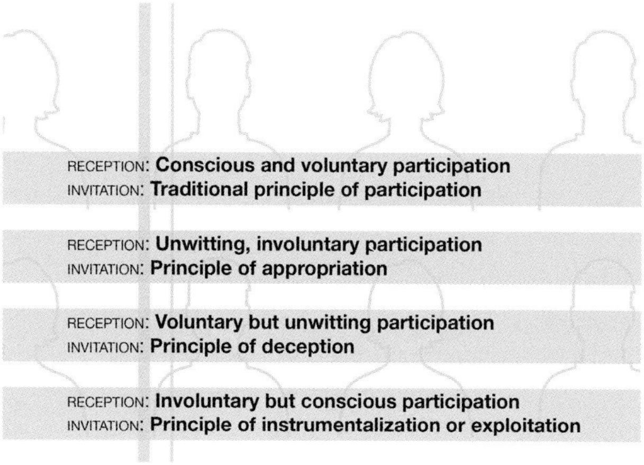

Fig. 6. Forms of inviting and receiving participation.

mentalization, which is much in evidence on the social media platforms described in this chapter: Facebook users (should) know that their participation is actually being instrumentalized for other purposes. The offer is clearly formulated, and we know where we are participating. However, our participation performance is in addition used actually to feed something else and we consciously know this, even when we are not welcoming this policy.

As the case studies indicate, participation projects can be assigned to several of these categories, since the forms of participation they involve can change. Participation in the *Face to Facebook* project, for example, is unconscious and involuntary in the initial phases, but not later on. However, the involuntary participants contacted through the dating site could decide themselves whether they wanted to remain on the database, or be removed. Involuntary participation, then, can sometimes turn into voluntary participation, and unwitting participation can sometimes become conscious participation.

The Logic of Network Cultures: Are We Merely Taking Part or Do We Have a Share?

The Internet is intrinsically participatory.[15] As such, it brings to light forms of participation that have been obscured by overly celebratory attitudes towards participation. Artists encourage participation in several ways, whether actively soliciting it in their work, passively allowing it to occur, or appropriating from unwitting audiences. Regardless of the means, however, these artists pursue participation as a way to explore sociopolitical issues, interfere with existing systems, or investigate how technology might develop in the future (as in *The World in 24 Hours*). Such artists must confront the fact that serious participation requires management and leadership and comes with very high communication demands (as in *[V]ote-auction*). Their projects often look into the intentions of online platforms and examine concepts of authorship and copyright in light of networked reality (as in Aaron Koblin's crowdsourced art). Their goal is not to present a dystopian picture of the online world, but rather to demythologize mechanisms of participation and inspire critical reflection (as in *Face to Facebook*).

All of the participatory artworks and processes presented here are connected by one aim: that of exploding discrete, dedicated art spaces so as to seek out conflict with people outside the art world. Such works are less concerned with creating utopias than with scrutinizing and distilling reality. Their makers are more concerned with making processes visible than generating objects or relics for exhibition. It is clear (at least in retrospect) that the rules of a particular project's participation are laid down by its initiators, even if this is not recognized during the participation process (as in *[V]ote-auction, Face to Facebook*). Even though participatory art clearly devalues notions of singular authorship, these have not been eliminated. Therefore, we cannot speak of flat hierarchies, equally distributed responsibilities, or collective control.[16] As the artistic projects presented here attest, the current critical discourse on participation has been shortsighted in neglecting the art of media artists or other artists who initiate open production, presentation, or distribution processes via the Internet and networked devices.

In his writings of the late 1950s, Umberto Eco formulated notions of the '*open work of art*' and the '*work of art in motion.*'[17] These anticipated the later idea that, in participatory art, the number of participants cannot serve as a yardstick of success. Instead, participation is above all a method of generating knowledge.The critical discussion of participation is not confined to aesthetics. Participation (whether conscious or unconscious, voluntary or involuntary) is based on how participants generate observations, experience, and knowledge.

As the case studies have shown, participation leaves dedicated (art) spaces in order to enter into communication with 'others,' to whom traditional understandings of 'audience,' 'recipient,' or 'observer' no longer apply. The 'others' are invisible in digital networks; indeed, they are mostly unknown. What does the 'others' mean here? And what does 'audience' mean there? My consideration of these questions is based on two important theoretical insights: that participation in art is more than a community project (see Kravagna 1998), and that artistic participation does not necessarily require an audience's consent, as Claire Bishop (2010, p. 264) has pointed out.

The case studies have shown that participatory artists place little value in their works being attributed to them. Subversively, it is often important that they are not recognized as artists. In addition, artists often collaborate with people from different disciplines, with whom they share credit, from programmers, creative technologists, to 'Internet engineers' (Naveau 2017). Internet phenomena such as 'Karen Eliot' (also spelled 'Karen Elliot')—a pseudonym for various artists working under the same name—underscore the need to rethink how we conceive the value of authors, copyright, and owners in participatory art projects. Yet even if there is broad agreement that artistic participation is more about process than producing physical objects, my discussion had emphasized how authorship has not been eliminated, only reassessed.

'Art with an agenda' seems a fitting slogan to describe different approaches to participatory processes. I use it to suggest that 'official' participatory processes—those initiated by social institutions and political representatives—often see participation as ornamental. Rather than generating value through knowledge, they are often interested simply in engaging large numbers of people. In this context, 'artistic or creative participation' is understood as a kind of playful cooperation. Unfortunately, it is often used and propagated by political or economic bodies to legitimize themselves and their agendas, while often having no genuine interest in participation as such. In opposition to this, it is tempting to imagine a form of participation that is oriented towards self-sufficiency. Participation for its own sake, however, is a contradictory idea.

Although at first glance it might seem that participation has to do with communication, exchange, and mediation, my case studies confirm that getting participation to be taken seriously requires an enormous effort. This applies above all to works dealing with unwitting and involuntary forms of online participation, because the artists behind these projects often simply irritate participants (although this sentiment frequently dispels as the artistic process goes on). As the case studies show, critical engagement with the Internet is required because it is the medium of numerous governmental and corporate agendas. For all the collaborative and collectively produced online content, the forces shaping and using online participation demand skeptical examination.

The examples I have explored make it clear that a too rigid concept of participation limits epistemological possibilities. Much of the value of participation has to do instead with play, chance, coincidence, and open-ended practices. In a digital world, 'coincidences are subject to rules' (Kim 2015, 10). Coincidence has a hard time in the digital world, where it is either banished or preprogrammed. Truly coincidental occurrences, however, entail disruption. The 'others,' that is participants, can be a disruptive factor in that their reactions (such as cheating) and activities (such as unpredictable actions) can be neither preprogrammed nor predicted. Human beings guarantee the coincidental in digital networks. As such, they not only design and keep the machines running, but also sometimes stop them from working as planned.

The role of what I have called the 'others' in online networking requires critical attention. Often, they are described as a 'mass,' 'mob,' 'crowd,' or 'collection of individuals' online. On the one hand, we lack terms to describe networked multitudes; and on the other, we must not fall prey to the temptation of believing that ethical and moral correctness are primary concerns for online agents. People behave online more or less as they do in the physical world.[18] Drawing on Jaron Lanier and Robert Adrian X, I would suggest that the key issue here is how we deal with invisibility in the Internet and how we render visible the processes that generate knowledge.

Final Comments

The information age began with figures such as Paul Otlet and H.G. Wells, who wished to democratize knowledge by making it more accessible. However, now we find ourselves in the age of participation. Taking a cue from the German distinction between *teilnehmen* (taking part in something) and *teilhaben* (having a share in something), I would suggest that users can influence whether the Internet is populated almost exclusively by commercial platforms or it is steered towards decentralized connectedness and a self-organized society. This might take the form of network and 'platform cooperativism' (Scholz n.d.), and the digital commons.[19] Users also have influence over the data fed into Internet servers and how they are communicated—a right and privilege of our time, which it is crucial to exercise. Although most users will not own servers and cannot determine their use, we can still understand how algorithms in digital systems are shaping our data and participation.

I close with this question: to what extent does network-based participatory art enable a critical model of our world, compensating for the lack of a corresponding theoretical model of a network culture?

> Nobody knows what it's all about. You can't analyze the past and extrapolate the trajectory into the future. Hegel has nothing to say about this, since you can't trust people who've never used a telephone. This simultaneity we have now, this is totally new. And the devices are getting increasingly invisible, just like the internet itself is invisible. And it's simply difficult for us to theoretically deal with this situation (Adrian X, 2013).

Notes

1. See also Robert Adrian X (2015)
2. The complexity of contemporary society has been examined by philosophers and theorists such as Bruno Latour (2013). Judith Butler (2016), for example, investigates the presence of both physical and virtual spaces in our bodies, and political scientist Jodi Dean (2016) explores how the appearance of individuals and multitudes might provide new approaches to politics. Despite this, we still lack a fundamental understanding of network culture based on digitalization of our world.
3. My discussion of bodies is influenced by Butler (2016) and her notion of bodily presence without speech. Bodily presence is manifest not only in direct linguistic utterance, but also representative signs or images. As such, it can appear also online.

4. For an online transcript of the interview, see Adrian X (n.d.)
5. The work put Linz in contact with numerous cities according to the following schedule: September 27: Vienna, Bath and Amsterdam at 12:00; Frankfurt at 13:00; Pittsburgh at 18:00; Toronto at 19:00; Wellfleet at 21:00; San Francisco at 22:00; Vancouver at 23:00. September 28: Sydney at 03:00; Tokyo at 04:00; Honolulu at 05:00; Florence at 10:00; Istanbul at 11:00; and Athens at 12:00.
6. In an interview (Adrian X n.d.) notes that although five phone lines were planned, only three worked. Fax machines or telecopiers were not common yet in 1982. That would come later in the 80s, as the thermal paper they initially required was replaced by standard paper. SSTV is an analog transmission method used by amateur radio operators to transmit static pictures The *Artex*-System was a network tool, but can also be described as one of the first communities of digital artists. It was an important precondition for projects such as *The World in 24 Hours*, Roy Ascott's *La Plissure du Texte* (1983), and *Planetary Network*. The enthusiasm with which the roughly 35 *Artbox* or *Artex* collaborators discussed their ideas and projects is reminiscent of early social media platforms.
7. For this project's webpage, see [V]ote-auction (n.d.).
8. The artists behind *RTMark* already knew Hans Bernhard, because they worked with the artist group *etoy*, of which he was a founding member. According to Baumgärtel (2001), the *RTMark* artists saw themselves more as political net activists than net artists. They supported *etoy* in its legal dispute with the US toy distributor eToys in the same way that they supported *ubermorgen.com*.
9. This project's website can be found through Koblin (2006a).
10. Since each participant was allowed to draw up to five sheep, the precise number of people involved was under 10,000. It is remarkable how many Mturk workers accepted the job considering the extremely low renumeration of just 10 cents for five sheep,
11. At Amazon Mechanical Turk, the online workers are often briefly referred to as Turkers.
12. If one person were to draw the 10,000 sheep in the same period of time (40 days), they would draw about 11 sheep per hour and about 250 sheep per day.
13. Many artists can afford permanent assistants. The delegation of artistic work did not begin in the Italian Renaissance, when artists such as Tintoretto and Veronese had work done by assistants. Artists have always been managers and entrepreneurs: think of Andy Warhol´s *The Factory* or the artistic and scientific assistants of Olafur Eliasson or Damien Hirst.
14. See Cirio, Paolo, and Alessandro Ludovico (n.d.a) for reactions from participating persons; Cirio, Paolo, and Alessandro Ludovico (n.d.b) for Facebook's legal correspondence; and Cirio, Paolo, and Alessandro Ludovico (n.d.c) for the artists' defense.
15. Tim O'Reilly (2005), who coined the term 'Web 2.0,' describes webservices' software design as 'Architecture of Participation.'
16. This experience is consistent with my observations on projects from the *Digital Communities* category of the Prix Ars Electronica. Even if a community keeps its project going through its members' constant contributions, dedicated 'caretakers' who feel responsible for it are still needed.

'Caretaking,' however, also implies management, and control—terms usually considered undesirable in connection with communities, collaboration, and participation.

17. In *The Open Work*, Umberto Eco (1989) explains how artworks—now that the traditional criteria of aesthetic evaluation no longer apply—can function as 'epistemological metaphors.' He argues that:
    The discontinuity of phenomena has called into question the possibility of a unified, definitive image of our universe; art suggests a way for us to see the world in which we live, and, by seeing it, to accept it and integrate it into our sensibility. The open work assumes the task of giving us an image of discontinuity. It does not narrate it; it is it. It takes on a mediating role between the abstract categories of science and the living matter of our sensibility; it almost becomes a sort of transcendental scheme that allows us to comprehend new aspects of the world (90).
18. See in particular Jaron Lanier's remarks on the 'pack switch' in the speech he gave in acceptance of the 2014 Peace Prize of the German Book Trade: 'People are like wolves, according to this theory; we are members of a species that can function either as individuals or in packs. There is a switch inside us. We are prone to suddenly fall into pack thinking without even realizing it. If there is one thing that terrifies me about the Internet, this is it. Here we have a medium which can elicit 'flash mobs' and routinely creates sudden 'viral' popularities. So far, these effects have not been evil on an epochal level, but what is there to prevent that?' (quoted in Kim 2015, 22).
19. The work of the P2P Foundation is particularly relevant in this regard. See P2P Foundation (n.d.).

References

Adrian X, Robert. n.d. "THE WORLD IN 24 HOURS." http://alien.mur.at/rax/24_HOURS/24-catalog-e.html.

Adrian X, Robert. 2015. "Robert Adrian X turns 80 – In any case, it´s all about the telephone." https://ars.electronica.art/aeblog/en/2015/02/21/robert-adrian-x-turns-80/.

Adrian X, Robert, and Grundmann, 2013. Interview by Manuela Naveau, Wien/Linz) Audio recording in the author's archive.

Adrian X, Robert. 1989. "Elektronischer Raum." *Kunstforum International. Im Netz der Systeme* 103: 142–47.

Bishop, Claire. 2010. "Antagonism and Relational Aesthetics." In *The 'do-it-yourself' artwork. Participation from Fluxus to new media*, edited by Anna Dezeuze, 257–80. Manchester and New York: Manchester University Press.

Bishop, Claire. 2006. *Participation. Documents of Contemporary Art*. London/Cambridge MA: MIT Press.

Bishop, Claire. 2009. "Claire Bishop and Boris Groys on participatory art (Tate Modern 2009)." Transcription of discussion held at the *Futurism* exhibition, London, June 12 – September 20, 2009. http://pric.wordpress.com/3-theory/claire-bishop-and-boris-groys-on-participatory-art-tate-modern-2009/.

Butler, Judith. 2016. *Anmerkungen zu einer performativen Theorie der Versammlung*. Translated by Frank Born. Berlin: Suhrkamp.

Cirio, Paolo, and Alessandro Ludovico. n.d.a. "The Social Experiment." http://www.face-to-facebook.net/social_experiment.php.

Cirio, Paolo, and Alessandro Ludovico. n.d.b. "Legal." http://www.face-to-facebook.net/legal.php.

Cirio, Paolo, and Alessandro Ludovico. n.d.c. "Press Folder." http://www.face-to-facebook.net/press.php.

Cirio, Paolo. 2012. "Face to Facebook 2011." In *TEA BREAK: Selected Writings of Collective Wisdom / 2012 International Techno Art Exhibition*. Taichung: National Taiwan Museum of Fine Arts.

Dean, Jodi. 2016. *Crowds and Party*. London and Brooklyn NY: Verso.

Eco, Umberto. 1989. *The Open Work*. Translated by Anna Cancogni. Cambridge MA: Harvard University Press.

Kim, Anna. 2015. *Der Sichtbare Feind. Die Gewalt des Öffentlichen und das Recht auf Privatheit*. St. Pölten–Salzburg–Vienna: Residenz.

Koblin, Aaron. 2006a. "The Sheep Market." http://www.thesheepmarket.com/.

Koblin, Aaron. 2006b. "The Sheep Market: Two Cents Worth." PhD diss., University of California Los Angeles.

Kravagna, Christian. 2010. "Working on the community: models of participatory practice." In *The 'do-it-yourself' artwork. Participation from Fluxus to new media*, edited by Anna Dezeuze, 240–56. Manchester and New York: Manchester University Press.

Kwastek, Katja. 2013. "Crowdsourced Art – Die crowdbasierten Kunstprojekte von Aaron Koblin." Unpublished paper in author's archive, Munich.

Latour, Bruno. 2013. *An Inquiry into Modes of Existence: An Anthropology of the Moderns*. Cambridge MA: Harvard University Press.

Miessen, Markus. 2012. *Albtraum Partizipation*. Berlin: Merve.

Naveau, Manuela. 2017. *Crowd and Art – Kunst und Partizipation im Internet*. Bielefeld: Transcript Verlag.

O'Reilly, Tim. 2005. "What Is Web 2.0." http://oreilly.com/web2/archive/what-is-web-20.html.

P2P Foundation. n.d. "Home."—https://p2pfoundation.net/.

Pfaller, Robert. 2008. *Ästhetik der Interpassivität*. Hamburg: Philo Fine Arts.

Schäfer, Mirko Tobias. 2011. *Bastard Culture! How User Participation Transforms Cultural Production*. Amsterdam: Amsterdam University Press.

Scholz, Trebor. n.d. "Platform Cooperativism". http://platformcoop.net/.

[V]ote-acuction. n.d. "[V]ote-auction—Bringing democracy and capitalism closer together—UBERMORGEN.COM, 2000–2006." http://www.vote-auction.net/.

Wikipedia. n.d.a "Agenda." http://de.wikipedia.org/wiki/Agenda.

Wikipedia n.d.b "Schachtürke." http://de.wikipedia.org/wiki/Schacht%C3%BCrke.

Aesthetic Strategies in the Wasteocene

Yvonne Volkart

> Intimacy will be the new code word. There will be no 'away' towards which we can throw anything. … It's like the story of Anansi and the Tar Baby: the harder he tries to pull away, the more he finds himself stuck (Timothy Morton 2011, 82–83).

> The indeterminacy of waste draws our attention to the *imprescriptibility* of ethical responsibility to future generations and environmental sustainability. … To paraphrase Barad, it matters how waste comes to matter (Myra Hird 2016).

You access a website that displays one sentence against a white background: 'GOOGLE.COM EMITTED 1020.89 KG OF CO2 SINCE YOU OPENED THIS PAGE.' While everything else on the screen stands still, the number rapidly increases: 3244.35, 3766.07, 26035.67. You can hardly follow, let alone pronounce the numbers. The clock—your clock—is ticking.

I have been describing Joana Moll's real-time, Internet-based installation *CO2GLE* (2014). One click on the webpage sets off a CO_2 counter, indicating that Google emits about 500 kilograms of CO_2 every second due to users accessing the company's website from around the globe. *CO2GLE*'s 'informational' aesthetic gives the impression that everything is under control: technologies count, measure, and track fluid atmospheric molecules, translating them into scalable and tradable digits. But you know better: there is no control, and you do not understand what these numbers really mean. Nonetheless, you are part of this multiplying force, this indeterminacy. You are part of this normalized, machinic CO_2 output. You are a co-emitter of enormous abstractions, gigantic amounts of waste, because you are online, and use Google. You are entangled in this activity, participating in it, along with unfamiliar machines and people. And you understand one shocking thing: that uncanny things are going on behind the reassuring screen, with its minimal, black-and-white display. There is a multiplicity of beings out there which you never, ever, fully understood. All the while, your clock is ticking.

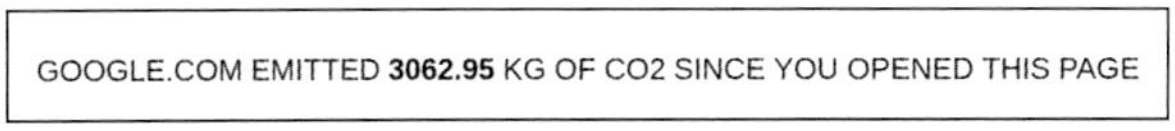

Fig. 1. Joana Moll: *CO2GLE*, 2014. Courtesy of Joana Moll.

CO2GLE raises my paper's basic question: What is the material impact of digital technology? Although my primary concern is with problems of waste posed by digital technology, particularly by the smartphone, I begin with Moll's insistence on the more general materialities of communications through the Internet:

Almost nobody recalls that the Internet is made up of interconnected physical infrastructures which consume natural resources. How can such an evident fact become so blurred in the social imagination?[1]

Both Moll's (rhetorical) question and *CO2GLE,* her ostensibly immaterial installation, lead to the heart of a crucial problem with which we are confronted today: that matter doesn't matter in technocapitalist times. Raw materials, human and animal bodies, plants, and labor are constantly depleted, devalued, and turned into waste. This process of wasting has a long history: colonialism, industrialization, and digitalization are its different faces (Kaitatzi-Whitlocks 2015). They are each rooted in a belief in the abundance of resources and the assumption that increased technological skill will master the naturecultures. Subsequently, our cybernetic conceptions of information have come to disavow the material impact of technology, and are based instead on the idea of immateriality (Hayles 1999). But times change, and while waste, people, and mobility ramify, raw materials and fertile soils decrease. This is how 'waste' became material 'information' (Offenhuber 2016): Waste tells us that information is (corpo)real. And it tells us not only that the endless devaluation of life continues in times of 'immateriality,' but that it is precisely digitalization and its exploding need for resources that have accelerated this devaluation. Scarcely anybody thinks about the fact that an 'immaterial' digital package—a simple message spoken into my smartphone—must circle around the planet before reaching my child chatting downstairs (Crawford and Joler 2018). Put another way, CO_2 emissions, electronic waste, the loss of biodiversity, and other forms of waste have all drastically increased over the last 25 years, corresponding exactly with the advent of the network society. Adding to *CO2GLE,* we can point to how language-based AI networks and Amazon have become monstrous emitters of CO_2. Welcome to the Wasteocene.[2]

This paper begins from the premise that we live in times of waste. Waste haunts us: we settle in waste, we have to live with waste, not against it. In confronting the displaced and working through the abject, eco-aesthetic practices may provide a productive basis for thinking through the problematic of contemporary waste. Diverse practices and progressive strategies have been developed, especially in transmedia- and research-based arts. These span the tracking trash, including waste from the extraction of raw materials; circuit bending; hardware hacking; practices of accumulating or co-composing electronic waste; computer gaming; resurfacing 'deleted' data; as well as work with dismantlers in their workshops.

Drawing on my collaborative research on the routes and transformations of the smartphone and its wastes,[3] as well as on theories of new materialism (Barad 2014; Haraway 2016) and media ecology (Maxwell 2014; Parikka 2015), this paper analyzes three contemporary new media projects, each of which provides a different perspective on electronic waste. It asks where and how their aesthetic media strategies open up transversal potential. Art is approached as a field of both symbolic expression *and* material articulation. It is an 'interface' that can 'shift' the normalized view of things, introduce difference, and thus open up other possibilities for action. *Shifting Interfaces,* the title of this collection, might well describe the role of art, thus conceived. In approaching art in this way, I do not mean to overestimate its function, and recognize how aesthetic processes and relations have to be continuously restaged. Still, this notion of art allows us to grasp waste as a paradoxical figure, a boundary shifter, a disruptive force, and alien matter. Waste 'is' something, or rather performs something, that might become a promise: a promise of monsters (Haraway 1992). As such, waste is always already alien and alienated. I claim that the potential of waste aesthetics lies in the calculated way it stages this foreignness of things, in all their incalculable foreignness. And it is this foreignness of the aesthetic—this unfamiliar perception—that gives rise to a surplus, an aesthetic excess, a transversal becoming.

Waste Always Wastes Anew

In German, garbage is called *Abfall,* meaning matter that has fallen away: it is the 'fallen-away,' the fallen-down, the ab-ject. In life cycles, something always falls away, becomes useless, haunts order and culture. Mary Douglas has defined dirt as 'matter out of place' (Douglas 2002); Bertolt Brecht spoke of dirt as 'matter at the wrong place.' These definitions of waste as abject, displaced matter—the down- or up-cycled plane in which the ontology of garbage is always situated—both depend on perspective.

In a workshop I co-organized, the German historian Bernd-Stefan Grewe said that 'Waste is matter that is too much mixed.'[4] It is matter that you can no longer separate into its valuable and valueless components, at least not in a way that provides new economic resources. Mixing entails depreciation: what is mixed is unspecific, therefore tends towards worthlessness. With waste, everything becomes similar. Gray-brown indifference. Waste, therefore, is also monstrous, alien, feminine, queer (Grosz 1994; Volkart 2006; Davies 2015).

In the end, waste is matter that does not go away; there is always something left over, a remainder, based on the physical law that something cannot turn into nothing. Waste disposal or cleaning generally means transforming waste into smaller pieces and transferring it to another place. By definition, garbage cannot be dispelled without leaving traces, despite our hopes when we invent clinical terms like 'recyclable' and 'waste management,' or technical 'solutions' like geoengineering and the like. On the contrary, waste, garbage, rubbish, or 'the fallen-away' does not resolve itself, but only wastes anew: Every 'fallen-away' generates a new 'fallen-away,' another useless leftover that might be harmful. By definition, waste is endless transformation: it is in movement, process, an endless state of becoming (something else).

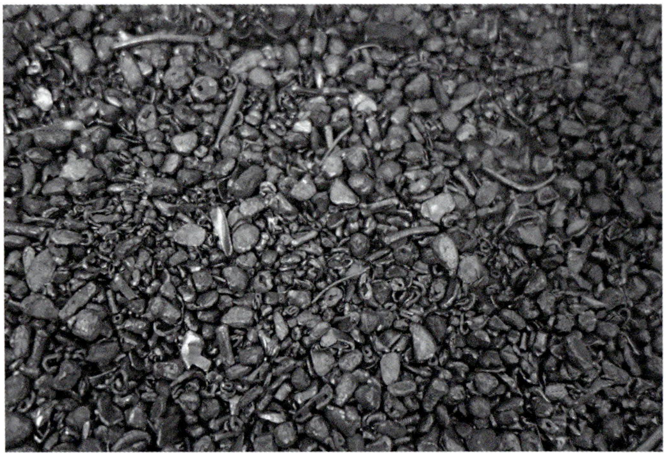

Fig. 2. Granules resulting from the recycling process of e-waste at Immark AG Regensdorf/ Switzerland, 2015. Courtesy of Team *Times of Waste.* (Plate 23, p. 317)

It is, as Armiero has said, not an 'object,' but a relation: a 'wasting' 'based on consuming and "othering,"' in his geopolitical gloss.[5] Waste, Myra Hird suggests, 'is anything but static and submissive: waste flows and mobilizes relations' (2016, 4). When a small device like a smartphone has been recycled, even with state-of-the-art-technology, there is always something left: after the device has been dismantled and shredded, various granulates remain; slag and ashes containing toxic particles survive the smelting of its gold, silver, and palladium, while the slag left by battery recycling contains other particles. They all are 'sleeping' and might be mobilized and become toxic at any time through water or acid rain.[6] Endless transformations, endless ways. If you recover the gold in a mobile phone, then you lose the elusive rare earth element neodymium. This is a law of thermodynamics which recycling cannot overcome. Switzerland, for example, both has one of the best functioning e-waste-recycling systems in the world, and ranks among the highest producers of e-waste (around 22 kilograms of electronic waste per person) and waste in general. While humans scour the earth's crust for raw materials, they not only eradicate environments, but fill them with new wasting materials (Hird 2016).

Stories from the Long Life of a Smartphone

The point of departure for our collaborative research on the aesthetic and transmedial dimensions of 'Times of Waste' was the following question: Where do the leftovers from discarded smartphones go? What kinds of processes do these smartphones undergo, and what discourses and practices are involved in them? As we were attending to issues that had not yet been researched in detail in the field of arts practice or had become topical only recently, our simple questions soon became very complicated. The smartphone contains so

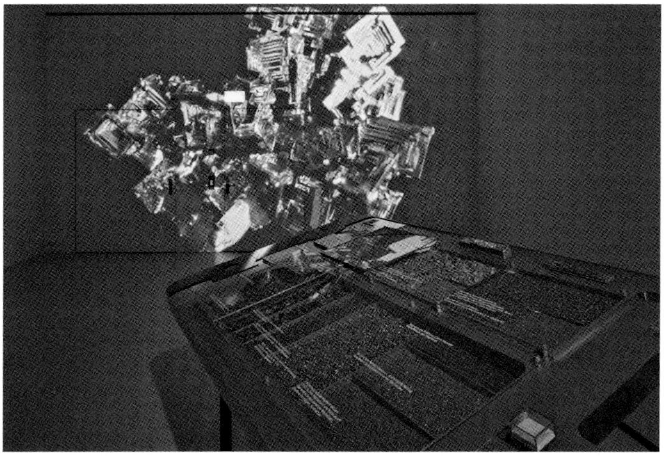

Fig. 3. Research team 'Times of Waste': Exhibition Times of Waste—The Leftover, Museum der Kulturen, Basel 2017.
Courtesy of Team *Times of Waste*, © Omar Lemke. (Plate 24, p. 317)

many elements. There are manifold possible ways of either disposing of or furthering the life cycle of every section and/or component, and many gaps in those routes. Often, the companies are unwilling to disclose the details of these breaks. In addition, smartphones do not always follow the 'conventional routes' of e-waste management because they contain large amounts of reusable metals. They are normally not mixed and shredded like 'conventional' e-waste, but taken directly to the smelter. Although there are a large number of gold smelters in the world, only a few smelt precious metals from scrap. Most of these are in Europe, and they even smelt e-waste from Africa that has been delivered through legal routes. One such route, for example, has been built between a state-of the-art new dismantling factory in Uganda and a Belgian gold and scrap smelter that was founded during colonialism. Hence, although there are many routes through which e-waste is removed (in both legal and illegal ways) to the Global South, there are also some paths leading (back) to the Global North.

Approaching the smartphone as an assemblage of various components and matters continuously wasting anew, it was clear that we could not track routes taken by recycled smartphones by means of attached sensors. Instead, we relied on the situated knowledge of people working with waste, closely cooperating with waste disposal institutions, scientists, NGOs, and artists-researchers. Beyond e-waste's various routes to national dismantlers, shredders, and landfills, we explored numerous areas of recycling and waste disposal: the growing international market for second-hand mobile phones; laboratory processes using microorganisms to extract rare earth elements from household slag; the company Schweizer Metallhandels AG Deutschland, which trades in so-called strategic metals (like rare earth elements); political initiatives; and cobalt and copper mining in the Democratic Republic of Congo.

Our research team assembled an extensive object biography of the smartphone, which we presented as an online archive and exhibition, including maps, sound, photographs, videos, and text.[7] An audio-essay, for example, relates the long life of the smartphone from the perspective of the element of neodymium, which narrates its origin as follows:

> I came into the world as dust, as the remains of an exploded star. I embedded in the steppes of Mongolia and became a rare earth metal. I have been called Neodymium for 150 years. I am a chemical element and have many siblings. We are scattered all over the earth's crust; yet even in faraway wastelands it is worthwhile to extract us. Like all earth elements, we live in groups. I do so along with praseodymium, thorium, lanthanum, uranium. Humans wrest us apart, for we are needed in pure form. So uranium and thorium migrate into the tailings—waste lakes. Most of us are garnered in China. Cheaply produced, we are sold at a price. Our defense is our radioactivity, lung cancer for the miners. It is only with me that high-performance magnets consisting of iron, neodymium, and boron can be manufactured, those that are needed for the microphones, loudspeakers, and vibrators in the smartphone. I am multifarious. I've come a long way. And I am small. …[8]

We adopted this autobiographical micro-perspective to create a strange sense of intimacy with a non-human entity, and to defamiliarize our normalized (often hierarchical) attitudes towards our consumer devices. We wanted to challenge the unspoken 'law' in science and critical art that you should not assume the speaking position of the fantasmatic (subaltern) Other. Interestingly, it was one of our research partners Patrick Wäger at the Empa—among others responsible for international audits of e-waste recycling—who fostered this idea. He supplied us with a conference paper in which he tried to overcome conventional scientific language by having the rare earth element neodymium speak.

Media theorist Jussi Parikka claims that we should conceptualize media according to a geological framework, as extensions of the earth. Quoting conceptual artist Robert Smithson, Parikka says, 'Technology is made of the raw materials of the earth' (Parikka 2015a, 5). Unexpectedly, our research was propelled in a similar direction, in that it came to emphasize how computers are a part of the cycle of geology, of sediments and rocks. Their main components both derive from the earth's crust and return to it. Computers are zombies, write Parikka and Heitz (Parikka 2015b, 141–53). They do not decay, as I have explained above, but only waste anew, for metals never die. Metals, in other words, not only live on and on after a smartphone's death. They also already have long histories, and long journeys behind them, before they are incorporated into devices. As one of our interview partners, Rainer Bunge from the HSR Hochschule für Technik Rapperswil, put it, 'It is quite likely that a modern smartphone comprises at least a few atoms of copper originally mined during the Bronze Age' (Times of Waste Research Team 2016).

We also learned that the greatest waste occurs not after but before smartphones are produced and used, in the process of mining raw materials. This was unexpected, considering that smartphones' first-use life cycle is only short (between 12 and 24 months), and 3.8 billion smartphones were in use in 2018, which might have suggested a preponderance of wastage post-use.[9] Still, although the amount of smartphone e-waste after use is small compared to that of a printer or car, it is likely to increase, given that the global smartphone market has almost doubled in the four years from 2014 to 2018. Ever-new mobile phone gadgets, which require almost 50 pure elements, appear. Moreover, recycling rarely delivers purity, while mined elements are cheaper than recycled equivalents. As always, in the end it is a question of economy—an economy of wasting.

As part of the research project, we also produced a comprehensive sustainability scale with which to assess the production and use of materials. The first variables indicated on the scale are more positive from the point of view of sustainability than those towards the bottom of the scale:

1. Sufficiency and sharing models
2. Longevity of products
3. Use of second-hand products
4. Repair of defective equipment (difficult with smartphones)
5. Re-use of individual parts/component recycling
6. Various levels of material utilization (e.g., smelting metals) = classic recycling
7. Energy utilization (plastic, etc.)
8. Regulated landfilling
9. Illegal landfill

Remarkably, material recycling ranks sixth on this scale. This is due to the fact that even the best recycling involves major losses. However, recycling still produces less waste than excavating the earth's crust. The most environmentally friendly product is one that does not have to be produced—an unpopular message, which smacks of unsexy ascetics. But is it really more appealing to continue consuming into the distant future—indeed endlessly—through perfect recycling?

The e-waste situation can be summarized as follows: Although certain countries have promised innovations and adopted reasonably good practices, there remains no overarching solution to the ongoing accumulation of an enormous volume of permanent e-waste. In fact, the picture is even worse. The United States exports 85% of its e-waste, having never ratified the Basel Convention, which regulates the transportation of hazardous waste and bans waste exports to the Global South. And the European Union, for its part, does not properly recycle 65% of its e-waste (Interpol/CWIT 2015). Furthermore, rapid changes in the elemental composition of high-performance electronics (and their wastes) mean that there is no certainty about the long-term behavior of slags. There is no ultimate solution, only indeterminacy, and wasting anew.

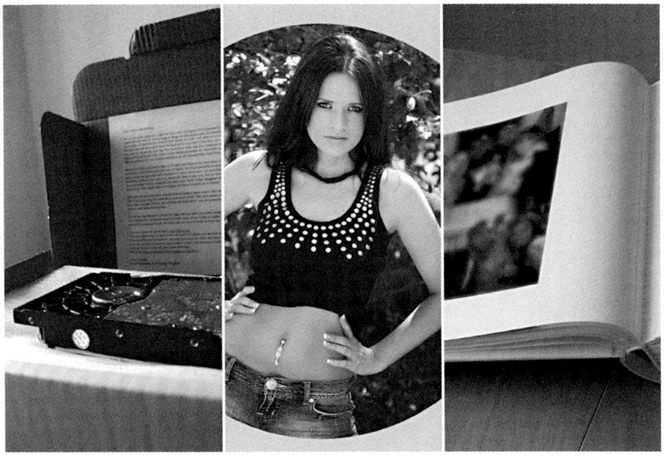

Fig. 4. KairUs: *Forensic Fantasies*, 2016. Courtesy of KairUs. (Plate 25, p. 318)

Discarded Data Contacts Us

The practice-based research project *Behind the Smart World* (2010–16) reveals strange entanglements among immaterial data, its material storage, and databased control of people. A community-oriented laboratory project, it was led by KairUs (Linda Kronmann and Andreas Zingerle) in close cooperation with the Linz network cooperative 'servus.at' (Uschi Reiter) and Linz's biannual festival *Art Meets Radical Openness* (*AMRO*).[10] *Behind the Smart World* began by looking at 22 discarded hard disks that KairUs had bought in Agbogbloshie, Ghana, the world's largest e-waste dump. Using open-source data recovery software in cooperation with forensic experts, they managed to make the contents of six hard disks readable, although their former owners had deleted most of the data they contained. KairUs passed these data, and the hard disks themselves, to eleven artists (including themselves) for further processing.

KairUs found 'sensitive' material on two hard disks, which they processed for the trilogy *Forensic Fantasies* (2016). In one case, the disk's former owner in England could be tracked down via social media, complete with his current address and social circles. This resulted in part one of the trilogy, *#1: Not a Blackmail*, which consisted of a letter and a parcel containing the recovered data ready to be posted to the previous owner. In the other case, which resulted in part two of the trilogy, *#2: Identity Theft*, the disk contained recurring photographs of young white women from different dating portals, and other evidence of organized romance scams. This hard drive, it seems, contained material from African romance swindlers. These people falsely pose online as the women in the images, feigning love with interested men. Romance scamming is a recognized problem in West Africa, and is addressed in the so-called 'Nollywood' films. KairUs printed and hung the found images in an exhibition, combining them with found film footage so as to make the women unidentifiable. The third part of the trilogy, *#3: Found Footage Stalkers*, mixes and shows found private photos in the form of a family album. The trilogy is presented in analog media—postal parcel, letter, picture on the wall, family album—as if the artists wanted to trigger a transmedial shift in which the found digital 'story' is told from another material perspective. Furthermore, the artists found it important that they did not profit personally out of the wasted material, or cross ethical thresholds of privacy and surveillance. By choosing a conceptual, 'dry' framing, they resisted the seductions of spectacle and exhibitionism. Refusing to act as wasters, they instead attempted to build a community of 'recyclers-artists' starting from waste. Both *Forensic Phantasies* and *Behind the Smart World* demonstrate that today—in the age of Big Data and dataleaks—data are a resource, a new form of capital. Those who own others' data have power and make profit. Accordingly, KairUs do not approach data as they are economically valorized, but work through their material presence as both wasted matter and resource. They zone in on data's ability to generate haunting and unpredictable effects. In short, they reveal their (im)materiality as a material force that always wastes anew.

Lithium: The Fallen-away of the Mother Mountain

Returning to the perspective of the 'fallen-away,' we might say that most resources—whether oil, gas, carbon, (precious) metals, or rare earth elements—were once waste, the 'fallen-away.'[11] Oil and gas are remains of plants and animals that have sedimented over millions of years. Metals and rare earth elements are the remnants of stars that exploded in the wider galaxy before falling to earth and becoming embedded in its crust. From the perspective of geological deep time, there even seems to be a certain similarity between the waste before and after its technological use. At both stages waste settles over inhuman timescales, accumulating, sedimenting, and mobilizing particles and entities on a molecular level. Indeed, small gray slags are reminiscent of mountain rubble. From the perspective of deep time, all these materialities come to share an identical character as fallen-away, transformative, and living. Still, although it may not always be productive to distinguish waste and nonwaste from an inhuman aesthetic perspective, we still have to reconsider that in smaller, more human time scales there are qualitative and molecular differences, e.g. between fallen-away pebbles and the fallen-away of depleted soil. And these differences have consequences for all organisms that live with, in, and through these wastes.

The video *We Power Our Future with the Breastmilk of Volcanoes/Lithium Dreams* (2015/18), produced by English artists Unknown Fields Division, presents the raw material of lithium as both an astronomical and corporeal waste: an indigenous myth tells of a volcano's female maternal fluids that have crystallized as a barren desert.[12] In footage of a drone flight over high Bolivian salt pans, recently the site of lithium production, we wit-

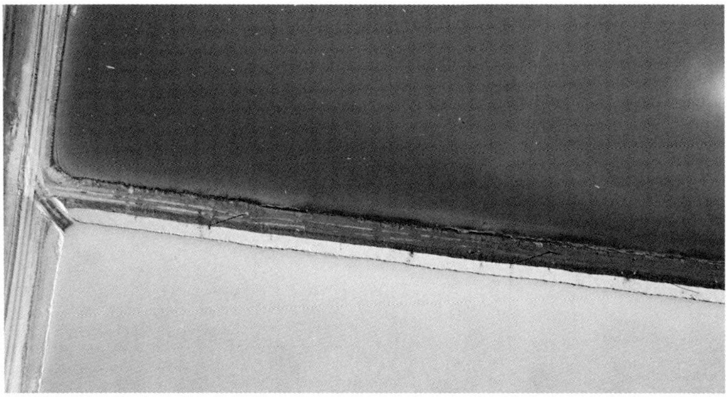

Fig. 5. Unknown Fields Division: *We Power Our Future with the Breastmilk of Volcanoes / Lithium Dreams*, 2015/18. Courtesy of Unknown Fields Division. (Plate 26, p. 318)

ness abstract images, patterns, and pools, whose colors change from poisonous greens to azure blues, such that this footage comes to resemble an abstract painting. A woman's voice speaks broken English. She tells of the Big Bang, which released elements that are sought after today; of the lithium that is increasingly needed in immense quantities in lithium-ion batteries for electric cars and similar 'green' technologies; of the salt desert in the Andes that contains the coveted lithium; and of mountains that are sacred to her people. She relates the mythic origins of this plateau, explaining how it had been formed by the tears cried by the volcano mother for her lost lover, which mixed with the mother's flowing milk.

This giant's white waste is transformed into colored fluidity, volatility, and dirt by a transnational factory. The video juxtaposes astronomical, geological, mythical, and capitalist dimensions of time and space. The beauty of the images and the voice's strange tranquility are set in harsh contrast to the forceful colonial excavation that imprinted the land with strange patterns. This aesthetic collision foregrounds the disproportionate temporality of raw material extraction: Although mythical and deep times have reigned until now, geological materials are destroyed in no time at all. The verbal conjunction of myth—in the story of breast milk and tears—and progress—in the form of lithium evaporation—in the work's title generates a 'desert-like' mood of acute pain. It repeats the mother mountain's ancient mourning, with a difference: The loss of the mountain's lover becomes a theft suffered by the indigenous population, who have different values and thus fewer rights. By connecting geological deep time with both mythology and accelerated machinic time, Unknown Fields Division reveals how various time scales obtain simultaneously in the present tense of the film. This strategy stresses how manifold perspectives and actors exist in the world, and that other temporalities, beyond the capitalist time of excavation, have value. Thus, the juxtaposition of the abject (in the form of dirt), wounds, and waste against beautiful visual abstractions allows us to experience non-linear progress and loss of control at both temporal and aesthetic levels. A transversal flux of cyclic events and affective subjectivizations is introduced: we see and hear cracks and fractures, we compassionately feel that something wrong is going on, and we sense that 'green' technology is not the next step in the historical lineage of technological innovation. The film aesthetically conveys a sense that it is just a material variation, another stage of loss and depletion in times of waste. The video involves us in an excess of more-than-human mourning, combined with a paradoxical situation of simultaneous proximity and distance generated by the drone's bird's-eye view. The affective aesthetic that emerges is strange and obscure, beyond conventional human perception. It offers an opening onto the previously unthought, the alien, and perhaps the transversal.

A project like *We Power Our Future with the Breastmilk of Volcanoes/Lithium Dreams* demonstrates that raw material production is waste production, and that a smartphone or green Tesla is always already waste. It reveals how the wasting of land amounts to what Rob Nixon (2011) terms 'slow violence,' a continuous but often unrecognized destruction of the planet. Geologic sediments are literally evaporating; solid matter is dissipating into the air; lithium salt feeds the speed of the machine; and geological matter becomes part of atmosphere, dissolves into almost nothing, but still something. Two further elements of this complex transmedial project address this condition. One is a crafted glass battery[13] containing a mass of alternating aluminum and graphite anode and cathode, submerged in a lithium brine electrolyte collected from Bolivia's electric Salar de Uyuni. The battery creates a slow reaction, like the drip charge of a crying mountain. The other is *Uyuni Super Mine: History's Largest Multinational Mine*, a trailer by Jon Skerrit for something approaching a Japanese racer game. In it we see the landscape flashing by become fuel for 'our' racing car. All three works narrate variations of old myths of depletion and loss—yet another form of waste that always wastes anew.

Transmedia-Waste-Aesthetics

Matter plays a crucial role in the current geopolitical discourse surrounding e-waste. It is restless, generative, and active, but also strange, opaque, and indeterminate. Scientists and technicians try to calculate degrees of uncertainty of waste materials' possible behavior and develop strategies for dealing with this uncertainty. Institutional regulations assure us that everything is under control, citing regulations, laws, and other systems. Only when a catastrophe occurs and becomes public is the indeterminacy and dangerousness of living material assemblages acknowledged, often in a hysterical way, before they fall into oblivion once again. Modes of dealing with the indefinite and precarious against the background of our mortality, which constitutes the essence of our material existence, have been reduced to exclusive expert knowledge in the techno-scientific sphere. The aesthetic practices presented here intervene in this containment, closure, and repression of environmental knowledge. Although the artists seek contact with so-called experts, they use defamiliarizing techniques to publicize waste's uncanny liveliness and materiality. All the projects use waste as an artistic material. Exposing the dump and working through its foreignness, they show how it is a figure of both thought and body of matter, which mobilizes relations and queerness. Ironically, most toxic wastes endanger fertility and masculinity. All the projects link 'e-waste' with infowaste and 'human waste' (Kaitatzi-Whitlocks 2014), revealing its geopolitical, postcolonial, and capitalist dimensions. All the projects are mediatized condensations of reality which link theory and practice. Lastly, all the projects show waste as versatile matter *and* medium that affects humans, landscapes, matter, data, and capital.

The ontologies of waste and transmedia aesthetics are profoundly similar. Both have the tendency to waste, to continually develop variations on the same story, and to preserve unexpected traces of original media and matter. Both render familiar things strange, disrupt smooth flows, and create a network of mobile relationships and meanings that transgress the boundaries of artworks' supposed autonomy. The transmedia-waste aesthetic inspires a strategy of mixing, blending, and commoning that goes beyond the dead ends of the Anthropocene. Knowing that there is no ultimate solution to the Wasteocene, these modes of living and acting reject fantasies of finding a single global resolution. Where there is the problem of borderless wasting, there is no ultimate elimination, no technical solution possible. There are only actions that relate to others in ways that vary dominant modes of wasting which, as Armiero says, are based on 'consuming and othering.' If there is hope, it is in small-scale relations that grow, distribute, and infect others, actions that waste, always anew, in a paradoxical way. Unsuspected resistances might be at work in such dynamic relations, deadly forces that might have transversal potential—or might not.

Notes

1. See Moll n.d. for the artist's explanation of the project.
2. 'Wastocene' plays not only on the term Anthropocene, but all the other geological '-cenes.' We adopted this wordplay in our research project 'Times of Waste' (see endnote 3). Environmental historian Marco Armiero coined the term Wasteocene for similar reasons. In his lecture 'Welcome to the Wasteocene: Sabotaging the Anthropocene through Guerilla Narrative,' Armiero writes: 'Waste can be considered the essence of the Anthropocene, embodying humans' ability to affect the environment to the point of transforming it into a gigantic dump. For this reason I have argued that this new epoch might be called the Wasteocene (Armiero and De Angelis 2017). However, the Wasteocene frames waste not as an object—'waste'—but as a relation—'wasting'. My hypothesis is that while wasting relationships are based on consuming and 'othering,' that is, on sorting out what and who is waste, commoning practices are based on reproducing resources and communities.' https://www.space4235.com/archives/fumogeni-2-marco-armiero (accessed January 28, 2019).
3. 'Times of Waste' research project, 2015–18 at the Academy of Art and Design FHNW Basel, funded by the Swiss National Science Foundation. Team: Flavia Caviezel (lead), Mirjam Bürgin,

Anselm Caminada, Adrian Demleitner, Marion Mertens, Yvonne Volkart, and Sonia Malpeso. See Times of Waste Research Team 2015–18a.

4. *Times of Waste* workshop, Critical Media Lab/Institute of Experimental Design and Media Cultures, Academy of Art and Design FHNW Basel, October 15, 2015.
5. See endnote 2.
6. See Times of Waste Research Team 2018.
7. The object biography is a method adopted among others in sociology, anthropology, and history (Appadurai 1986; Kopytoff 1986). It investigates and constructs the life and career of a thing like the life and career of a human being. Doing so, it processes generic lifespans of exemplarily chosen things, in order to grasp their global and historical entanglements.
8. See Times of Waste Research Team 2015-18b.
9. newzoo.com and www.statista.com.
10. http://kairus.org.
11. This insight draws on Yvonne Volkart's conversations with Thomas Lichtensteiger, geologist at Eawag aquatic research Switzerland (November 2018), and Michael Doser, physicist at CERN (December 2016).
12. http://www.unknownfieldsdivision.com/summer2015bolivia+atacama-lithiumdreams.html (accessed January 28, 2019).
13. Fabricated with assistance from Eduardo Andreu Gonzalez and Aimer Ltd. Battery science consultant Donal Finegan.

References

Appadurai, Arujun (ed.). 1986. *The Social Life of Things: Commodities in a Cultural Perspective*. Cambridge: Cambridge University Press.

Armiero, Marco, and Massimo De Angelis. 2017. "Anthropocene: victims, narrators, and revolutionaries." http://roar.uel.ac.uk/5355/1/Armiero%20and%20De%20Angelis_Anthropocene_final%20clean.pdf.

Barad, Karen. 2014. "On Touching—The Inhuman That Therefore I Am (v1.1)." In *The Politics of Materiality*, edited by Susanne Witzgall and Kerstin Stakemeier, 163–70. Zurich and Berlin: Diaphanes. English version: http://planetarities.web.unc.edu/files/2015/01/barad-on-touching.pdf.

Crawford, Kate, and Vladan Joler. 2018. "Anatomy of an AI System: The Amazon Echo as an Anatomical Map of Human Labor, Data and Planetary Resources." https://ainowinstitute.org.

Davies, Heather. 2015. "Toxic Progeny: The Plastisphere and Other Queer Futures." *philoSOPHIA: A Journal of Continental Feminism* 5(2) (Summer): 231–50.

Douglas, Mary. 2002. *Purity and Danger: An Analysis of Concepts of Pollution and Taboo*. London and New York: Routledge.

Greenpeace/Elisabeth Jardim. 2017. "10 Years of Smartphone." http://www.greenpeace.org/usa/wp-content/uploads/2017/03/FINAL-10YearsSmartphones-Report-Design-230217-Digital.pdf.

Grosz, Elisabeth. 1994. *Volatile Bodies: Toward a Corporeal Feminism*. Bloomington and Indianapolis IN: Indiana University Press.

Haraway, Donna. 1992. "The Promises of Monsters: A Regenerative Politics for Inappropriate/d Others." In *Cultural Studies*, edited by Lawrence Grossberg, Cary Nelson, and Paula A. Treichler, 295–337. New York: Routledge.

Haraway, Donna. (2016). *Staying with the Trouble: Making Kin in the Chthulucene*. Durham NC: Duke University Press.

Hayles, N. Katherine. 1999. *How We Became Posthuman: Virtual Bodies in Cybernetics, Literature, and Informatics*. Chicago IL and London: The University of Chicago Press.

Hird, Myra J. 2016. "The Phenomenon of Waste World Making." *Rhizomes: Cultural Studies in Emerging Knowledge* 30. https://doi.org/10.20415/rhiz/030.

Interpol/CWIT Project. 2015. "Countering WEE Illegal Trade: Report Summary 2015." https://www.cwitproject.eu.

Kaitatzi-Whitlocks, Sophia. 2014. "E-Waste, Human Waste, Infoflation." In *Media and the Ecological Crisis*, edited by Richard Maxwell, Jon Raundalen, and Nina Lager Vestberg, 69–84. New York and London: Routledge.

Kopytoff, Igor. 1986. "The Cultural Biography of Things: Commoditizaion as Process." In *The Social Life of Things: Commodities in a Cultural Perspective*, edited by Arjun Appadurai, 64–91. Cambridge: Cambridge University Press. http://ls-tlss.ucl.ac.uk/course-materials/ANTH7021_68798.pdf.

Maxwell, Richard, Jon Raundalen, and Nina Lager Vestberg (eds.). 2014. *Media and the Ecological Crisis*. New York and London: Routledge.

Moll, Joana. n.d. "CO2OGLE". http://www.janavirgin.com/CO2/CO2GLE_about.html.

Morton, Timothy. 2010. *The Ecological Thought*. Cambridge MA and London: Harvard University Press.

Morton, Timothy. 2011. "Zero Landscapes in the Time of Hyperobjects." *Graz Architectural Magazine* 7: 78–87.

Nixon, Rob. 2011. *Slow Violence and the Environmentalism of the Poor*. Cambridge MA: Harvard University Press.

Offenhuber, Dietmar. 2017. *Waste Is Information: Infrastructure Legibility and Governance*. Boston MA: MIT Press.

Parikka, Jussi. 2015a. *The Anthrobscene*. Minneapolis MN and London: University of Minnesota Press.

Parikka, Jussi. 2015b. *A Geology of Media*. Minneapolis MN: University of Minnesota Press.

Times of Waste Research Team. 2015. "Times of Waste." https://www.times-of-waste.ch.

Times of Waste Research Team. 2016. "Metals never die." *Lost & Found, continent* 5(1): 11–13. http://continentcontinent.cc/.

Times of Waste Research Team. 2018. "Smartphone Objectbiography". https://objektbiografie.times-of-waste.ch.

Volkart, Yvonne. 2006. *Fluide Subjekte: Anpassung und Widerspenstigkeit in der Medienkunst*. Bielefeld: transcript.

III

Bacteria and Smart Objects

....

Presences Beyond the Human

Rehabilitating Bacteria
An Epistemological Art/Science Interface

Jens Hauser

Bacteria are getting big.[1] They are increasingly considered in art, philosophy, the natural and the health sciences, becoming omnipresent in art-based research and research-based art. Consequently, in the contemporary field of hands-on biomedia art practices, bacteria as the oldest, smallest, structurally simplest but ubiquitous organisms vital for all other life forms are being rediscovered and given a crucial role in both artistic discourse and aesthetic settings. While in earlier biomedia art bacteria remained ontological blind spots—cells, tissues, and genetic sequences were considered more suitable biological entities to microscopically 'embody' or 'encode' individual organisms for which they stood in *pars pro toto*—bacteria and their related metaphors currently serve as epistemological indicators. In parallel to the shift in scientific focus areas from the Human Genome Project (1990–2003) to the Human Microbiome Project (2007–16) research-based art has contributed to a change in cultural perception, via its selective emphasis on which biological agent, scale, function, and agency to either foreground or background. Incrementally taking prokaryotic organisms as agents, material medium, motives, metaphors, and models of knowledge production, art has contributed to changing guiding tropes from individual codes to cellular cities and societies, from programmable workhorses to complex ecologies where bacterial forms of organization serve as role models in a larger bio-semiotic web of relationships. As such, bacteria offer an exceptional scope of potentials for artistic practice, since they cover the whole range of modes of biomediality (Hauser 2014 and 2016)—as *milieus* for other organisms, technologized *means* of transformation and production, and analytic tools for *measuring* such as bio-indicators in environmental control—and become at the same time subjects, objects, and tools in artistic displays: from simulated intestinal flora and ecologies in Winogradsky columns, to photo- or magneto-tactic bacteria as photographic media or mirror, to extremophile bacteria as cleaning agents, or programmed bacteria as chassis in synthetic biology.

Artists working with these 'significant others'[2] today often combine profound biological knowledge, biotechnological skills, philosophical, science and culture historical awareness, post-anthropocentric attitude, and a lot of humor. For example, in Adam Brown's installation *The Great Work of the Metal Lover* (2012; plates 31-32, pp. 321-322). extremophile bacteria are the protagonists employed for their innate capacity to metabolize toxic metals: In a bioreactor simulating a reduced atmosphere without oxygen, *Cupriavidus metallidurans* bacteria produce particles of 24 karat gold that the artist can then harvest to create little nuggets, hence, seemingly solving an ancient alchemist riddle. The title is a reference to mediaeval attempts to transmute base metals into gold by synthesizing the alchemical substance known as the philosopher's stone, *Lapis Philosophorum*. Here, 'transmutation' is achieved biologically by *Cupriavidus metallidurans* exposed to a concentrated toxic gold chloride solution. Such extremophile bacteria play an important role in our understanding of the origins of life. They serve the development of geo-biological exploration tools, e.g. gold bio-indicators and biosensors, and are used to filter toxic metals out of industrially polluted soils or waste water; in short: they clean up humankind's mess and can make our 'shit' into gold. When visitors perplexedly scrutinize the hardly visible gold particles expressed in the bacterial broth at the bottom of the installation's flask a typical question audiences ask is 'how do you do this?' Brown's equally typical answer is 'I do not do anything, the extremophiles do.' The dialogue may continue as follows: 'What are extremophiles?'—'Extremophiles are microorganisms supposed to love extreme conditions. Do you love candy?'—'Yeah.'—'Do you like spinach?'—'Not really, I eat it when there's nothing else before I starve'—'The *Cupriavidus metallidurans* feel the same: I get them started with sugar; then I stop and feed them only a gold chloride solution until they metabolize—they don't like it any more than you love spinach.' Here, the precise strain of bacteria not only is given its name, credits for its work and agency, it is also assigned differentiated patterns of attitude that make it appear as even more reasonable than a stubborn child refusing to eat its greens. While humans may look down upon these bacteria as primitive enemies, or mere workhorses in biotechnology, Brown stages the continuity of life's fundamental principles by utilizing single-cell organisms to produce that most coveted of all metals associated with everlasting longevity.

Historically, bacteria and the visual arts entertain an ambivalent relationship. Single-cell organisms remained invisible until Antonie Van Leeuwenhoek's first microscopic observations, in 1676, of what he called 'animalcules'; and such 'tiny animals' growing in visible colonies would rather be held responsible for contamination and bio-deterioration of human-made art and cultural heritage than considered for sparking aesthetic fascination. Traditionally described from Koch to Pasteur as 'invading animals' and notorious pathological agents in the age of science's discovery of the causality between disease and bacteria, in the nineteenth century, the epoch of hygiene's vocabulary and cultural metaphors led to biased preconceptions of bacterial presence. Addressed as 'colonization,' 'immigration,' 'infiltration,' 'intrusion,' these threatening 'enemies' were opposed to primary human body cells labeled, in turn, as 'good republicans, peaceful, sedentary, with legitimate kinship and safe loyalties within the body' (Sarasin 2007, 445, 450, and 454). When optical media became increasingly available, early twentieth-century micro-cinematography for example by French pioneer Jean Comandon represented bacteria such as syphilis spirochete by embedding them in martial, sensationalist narratives, like an anthropomorphized 'great combat carried out in the organism' with 'one of the most dreadful microbes' (Landecker 2006, 128). Biomorphic modern painters, for example Wassily Kandinsky, indeed searched for natural harmony by including depictions of microbes and microtubules forming flagella that help single celled organisms move. But it is only with the general shift from object-hood to processuality linked to the cybernetic paradigm in the second half of the twentieth century that artists have been pulled into the force field between their motifs and the related technical media actually producing new scientific insights. Right during the period that Lucy Lippard diagnoses as that of the 'dematerialization of the art object' (Lippard 1968) with a greater focus on ideas and conceptual artistic thought, Jack Burnham's *Beyond Modern Sculpture* from 1968 provides a biologistic retrospective over 2,500 years of sculpture, and declares art's survival dependent on its transition from material objects to permanently changing systems: from idealistic-vitalistic imaginary shapes to organicistic processes, a transition 'from a psychically-impregnated totemic object toward a more literal adaptation of scientific reality via the model or technologically inspired artefact' towards 'life-simulating systems through the use of technology' (Burnham 1968, 7 and 76). Influenced by cybernetics, environmental concerns, and Ludwig von Bertalanffy's early systems biology, Burnham hopes that such art may encourage spectators to adopt planetary-holistic thought and develop environmental consciousness—not contra but qua technology. He puts the typical anthropocentric and mesoscopic compression into perspective and outguesses that the arts need to generate both system awareness for the invisibility of the microscopic and for the incomprehensibility of the macroscopic, requiring a departure from form and a reorientation toward functionality 'away from biotic appearances toward biotic functioning via the machine' (Burnham 1968, 76). This paradigm has fostered experimental artistic movements that in actual practice use the most diverse technological and biological agents, including bacteria, since the 1990s. It also anticipates current developments of 'microperformativity' (Hauser 2014 and 2015), denoting the shift from *actions* of mesoscopic bodies to *functions* of microscopic bodies, when gene or cell fragments, proteins, enzymes, or bacteria etc. are employed to act as identity proxies, or to stage nonhuman agencies in relation to performative techno-scientific systems. In this context, bacteria are today centrally addressed in terms both of instances of the machinic and of the organic.

Bacterial agency has become a vector in artistic practice that encompasses both ontological role models of bacteria as subjects or objects of inquiry and their epistemic roles as instances of *biomediality*: On the one hand, bacteria can be staged and addressed in their own right as acknowledged agents; on the other hand, as media. Biomediality can be understood as a *special form of loose coupling*[3] of de- and re-organized biological entities (e.g. at lower levels, nucleotides, cells etc., or at higher levels, laboratory organisms) that, despite their own vital potential, are conceived as indifferent vis-à-vis what they mediate. Such atomism differentiates itself qualitatively, however, from that of media understood merely physically, in that organisms as a central point of reference are or were in the first instance organized. The term biomediality defines that ensemble of all topical and functional *enabling factors* that arises as a result of the organization of living organisms or biological processes, and can be functionally divided into *biological media, biomedia,* and *media of biology*. Because of their ubiquity, adaptability, and functional ambivalence, prokaryotic organisms are particularly suitable to cover all three aspects:

- First, *biological media* in the sense of a *milieu*, comprise existential enabling conditions that turn biological systems into something, surround a body and enable its internal functions, or solicit changes in living beings or organic entities. This definition of media is relevant e.g. in art practices with tissue engineering where incubators, culture vessels, and growth media to activate cell and tissue growth are required. But it can also relate to simplified bacterial cells used as *chassis* in synthetic biology, the *bacterial flora* establishing a vital condition for their host, or subsequent organ systems such as the digestive tract. The mutualistic relationship of bacterial communities that form an organism's microbiome can be considered to provide a *milieu*.
- Second, media in the sense of *means* of transformation or generation: Biological systems are considered as processing bodies, which can be technically programmed to carry out functions even beyond the inherent parameters of their organic purpose. *Biomedia* as a 'particular instance in which the "bio" is transformatively mediated by the "tech"' (Thacker 2004, 6) shift the ability to transmit, store, and process into the biological realm, making use of living systems' internal mechanisms. This encompasses spliced or synthesized gene sequences, the use of viral promoters to transfer DNA sequences, but also the use of microorganisms to produce substances or pigments, or to be used at an ecological scale for the purpose of bioremediation.
- Third, *media of biology* are those employed to *measure, analyse,* and *observe* biological systems: one organic system reveals something about another. Positioned in the tradition of microscopy and other optical or physical instruments of perception and analysis, biological systems are themselves technologically turned into representational units and serve knowledge production. At the smallest level, these can be, for

example, fluorescent biomarkers, DNA chips, but also bacteria either used because of their innate technical capacity or technically designed as bio-, geo-, or eco-sensors. At a macroscopic level, organisms themselves can serve as ecological indicators, e.g. like sensitive amphibians.

These three categories of biomediality imply examining the transformation of the concept of media as such through the influence of biology and technologies applied to living systems in order to move beyond the currently dominant understanding of media to merely serve communication. This requires the inclusion of previous natural-scientific conceptions of media. As Erik Porath reminds us, still in the seventeenth century, the loan word *medium* in its translation from Latin into modern European languages was used more frequently in the context of the natural than of the human sciences, 'before it was superseded in the second half of the eighteenth century by a broader understanding of "medium" as "intermediary element" and, more generally, as "means" or "tool",' while today's focus on functions of communication causes us 'to lose sight of the natural-scientific relevance of mediality,' so that 'cultural studies and traditional media studies generally lack consideration of natural science and its history' (Porath 2008, 254, and 256). Likewise, in order to consider medial functions of biological agents such as bacteria, it is indispensable to historically consider the interplay and entanglement between their epistemic and aesthetic dimensions. For example, the still valid method of carrying out bacteria taxonomy called Gram-staining, originally developed to help contrast bacteria against host cells in tissue, is based on the visualization of physical differences of cell walls via color difference that allows for quickly determining bacterial infection while waiting for culture results. Developed by Danish physician Hans Christian Gram in 1884, this efficient staining technique permits the distinguishing of Gram-positive bacteria with thick cell membranes that appear as purple from Gram-negative bacteria with thin membranes that turn reddish pink, so as to help doctors decide rapidly which antimicrobial agents to prescribe. Conversely, aesthetical practice with bacteria has contributed to major scientific findings, such as Scottish microbi ologist Alexander Fleming's seminal discovery of the antibiotic properties of the penicillin fungus thanks to his ephemeral hobbyist microbe paintings in Petri dishes. He took advantage of different natural pigments expressed by *Bacillus prodigiosus*, the chromo-bacterium *violaceum*, *Sarcina* and *Staphylococci* bacteria to depict motifs such as ballerinas, soldiers, or mothers feeding children, but also, in 1923, a plate with shapes resembling planets and stars in a night sky which came under attack of a 'rising sun'—the penicillin fungus whose enzyme lysozyme killed his inoculated bacterial motifs (Fleming 1937; Dunn 2010).

Within the framework of a multifold approach involving aesthetics, epistemology, as well as the history of art and science, a large and continuously proliferating cluster emerges of how bacteria are being materially used and culturally addressed in artistic practice (including, but not limited to): unspecified bacterial flora or vital milieu, symbolic proxies of aliveness and processuality, microperformative movie actors, living pigments to produce colors and forms, contaminating pathogenic invaders and aggressors, cleaners or healers, biodeteriorators as well as preservers and bioremediators of cultural heritage, patina, or microbial dust as time capsules in suspended animation, astro-biological messengers, evolutionary indicators, quickly learning and adapting biosemiotic organisms, bio-, geo-, or eco-sensors, supra-individual inhabitants of microbiomes shifting role models from individual to collective identity, naturally technical power plants, gold miners, metabolic producers, composters, factories of olfactive compounds, architectural constructors, breeding containers, or standardized chassis in synthetic or DIY biology with implanted genetic circuits, bio-controllers, or computers, or bacterial plasmids as malleable workhorses in molecular analysis.

A short and potted history of process-based art since the 1960s shows that bacteria have first and foremost been employed as a natural, pervasive, and easy to obtain ingredient to fight against object-hood, and to campaign for processuality within the spectrum of ecology and systems aware practices. For example, in Michael Badura's closed biosphere *Die Eingeweckte Welt* (the bottled world), consisting of a multitude of interconnected glass jars, decomposition processes and metabolic exchange take place involving divers plant and food waste materials, water, and gases: 'Irrevocably gone are the days of statuary (classical) sculpture! Its place is taken by the "model" which is assembled from diverse, subjectively "microscoped" (acquired) items' (Badura 1967). This is also true for HA Schult's large museum-scale installation *Bio-kinetic Situations* involving large cultures of differently colored bacteria and fungi, employed and controlled to stage the idea of loss of control in an industrialized society. Conceived as action art beyond the human-centered happening then in vogue, Schult's work aims at 'establishing actions and changes as legitimate artistic refinements to generally bring to mind phenomena of reality' (Schult 1969). Such strategies further evolve in the 1980s to target more specifically art historical references concerned with 'aliveness,' such as represented by Courbet, Kandinsky, or Mondrian, prone to conceptual deconstruction and institutional critique. In 1987, Peter Gerwin Hoffmann samples the diverse microorganisms populating Wassily Kandinsky's painting *Parties Diverses* (1940), and cultures them in a rectangular array of Petri dishes, shifting the focus from representation to manipulation of the organic. Choosing Kandinsky as a protagonist of modern, supposedly organic art concerned with questions of balance and proportions, Hoffmann claims to materially fulfil promises that the classical avant-garde, still trapped in the image and in metaphor, has finally not delivered, and updates them within the epoch of genetic engineering:

> The living beings (plants and animals) that surround us are artificial, which means that they acquire significance immanent in art and that their remaining form of existence must be understood and interpreted as artwork. […] The work of the artist obtains a

> new status, because it is of highest social priority for the future that artworks (such as the cow or the manipulated bacterium) are being analysed and reflected based upon the criteria if art, and not upon the criteria of economics, politics and science (Hoffmann 1987, 142–43).

Hoffmann brings the cow and the genetically engineered bacterium into play to 'dissolve the polarity between art and nature' in favor of a 'polarity between real art and art art,' (Hoffmann 1987) thus echoing Fluxus artist Allan Kaprow's earlier critique of representational-symbolic, self-referential 'art art' (Kaprow 1971), and situating artistic practice within the biotechnological realm. Hoffmann continues that 'the way microbes live on the image of Kandinsky, as a work and world of art, equals the way we live.' (Hoffmann 1987) There is no coincidence when he chooses as historical reference point a modernist painter who has himself moved from geometric abstraction to biomorphic forms inspired by microbiology, while producing as his own conceptual result a highly abstracted grid of geometrically arranged Petri dishes with bacterial cultures in growth that lack any aesthetic appeal or purposeful color control, in order to favor conceptual process over formal effect.

Deconstructing the abstract, modernist agenda remains a powerful strategy for artists staging bacteria even two decades later. In her live biological installation mimicking a Mondrian painting, *Decon* (2007), Marta de Menezes replicates the Dutch painter's taste for perfectly geometric arrangements of primary colors, shapes, and textures. But the colored compartments contain not only azoic pigments recently banned in the European Union because of their environmental toxicity, but also *Pseudomonas putida* bacteria used in bioremediation of polluted soils and waters. Over the time of the exhibition, they digest the industrial colors, and symbolically deconstruct, decontaminate, and decompose Mondrian's obsession to control natural forces by human organization. While the bacteria, on one side, make the artwork into something literally alive, on the other, they refer to the painter's own writings on art as a substitute for life: 'Art is a substitute for as long as life lacks beauty, it will disappear as life becomes more balanced.'[4] However, the prokaryotic aliveness poses challenges to safely maintaining the piece after the opening, and prompts the question about the gallery's responsibility where bio-remediating *Pseudomonas putida* bacteria act as pathogens that can cause life-threatening infections to immune-compromised humans, since the color degradation was accompanied by the release of a strong odor which made the space uninhabitable, fluids of nutrient media 'poured down the white walls,' provoking unexpected growth of 'cultures of fungi and bacteria with varied colors and textures'—the gallery concludes: '*Decon* holds the risk of its own threat. It does not show the "biohazard" sign. And yet it is potentially a contaminant capable of strong violence.'[5]

Although biomedia art works since the year 2000 have advocated post-anthropocentric worldviews by staging cultured animal or plant cells, bacteria were rarely seen as covetable collaborators. Alongside fungal infestation, they appear mainly as a threat to destroy the actual living artworks through contamination. While transgenic, bioluminescent moss in Jun Takita's *Light, only light* (2004) and co-cultured animal cells in *Victimless Leather* (2004), a biotechnological installation by the Tissue Culture and Art Project, are foregrounded, the bacterial background appears as undesired deterioration. Both art works demand a great deal of care and maintenance, and centrally emphasize their very aliveness via decay and the threat of death. However, contamination by unwanted agents is avoided at the price of constant replacement of the desired living forms exhibited.[6] Takita's ephemeral transgenic artwork *Light, only light* takes the form of a light-emitting moss sculpture in the shape of the artist's own brain obtained via a magnetic resonance scan printed three-dimensionally and covered with bioluminescent *Physcomitrella patens* moss to which the genetic sequence coding for the enzyme luciferase of a firefly has been transferred. Presenting us with a plant that emits light, and given the historical association of light with life, Takita presents the transgenic as an ambiguous cognitive achievement of the human brain. Meanwhile the brain shape of the work is strongly reminiscent of the skull as a *Vanitas* motif and allegorical *Memento mori* (Hauser 2013, 281). On the level of material semantics, lastly, flies in art history have often been read as symbols of life's impermanence, or as companions of the devil. Similarly, *Victimless Leather* consists of a series of fragile miniature leather-like jackets grown in incubators out of immortalized animal and human cell lines that form a living layer of tissue supported by biodegradable polymer matrices—something usually dead is staged in its becoming animated. This mix of 'semi-living' human and mouse cells evokes the utopian ideal of producing leather garments without having to kill animals, despite the need for animal-derived products, such as DMEM medium with added foetal calf serum. Aesthetically, the garments appear as suspended, head- and member-less bodies, falling down as seasonal garments like ripe fruit, staged to die, while all is done to prevent bacterial agency taking over.

Around the turn of the millennium, in the shadow of the Human Genome Project, micro-performative agents associated with human bodies have been getting the credit, while bacterial activity was mostly addressed as undifferentiated bacterial flora only, as *biological media* in the sense of the enabling condition of a *milieu.* A striking example is Wim Delvoye's series of bio-cybernetic automata, defecating digestion machines branded *Cloaca* (since 2000). They constitute a case example of art involving functional metabolic processes: food, enzymes, mechanical simulation of a digestive tract, control of pH level and temperature, and *E. coli* bacteria—but the latter remaining totally blind spots. While the artist programs the complex fecal language by precisely addressing enzymes such as pancreatin, amylase, maltase, lipase, lactase, protease, and the compounds alkaline oxen-bile, bilirubin, sodium bicarbonate, sodium hydroxide, and hydrochloric acid, he does not 'give the bacteria a shit' when producing artistic excrement (Hauser 2007). Although the

E. coli are well responsible for the olfactive contamination generally invading the gallery space, the emphasis put on the smallest gastronomic detail and the clockwork mechanism of the digestive simulator and its scatological output favour an anthropomorphic reading of *Cloaca*—even though it actually functions with enzymes sourced from pigs and cattle. The scatological vector serves as a swift link to art history—from Jacques de Vaucanson's eighteenth-century mechanical duck, purportedly digesting grain, via realist Gustave Courbet and symbolist James Ensor to Piero Manzoni's *Artist's Shit* (1961)—and equally to contemporary art in the age of synthetic biology. Interestingly, these practices connect a large variety of biological entities when appropriating so-called *biobricks*, standardized DNA sequences to implement new functions into model organisms, especially bacteria and yeast. Tuur van Balen employs both biobricks and bacteria to address the ecological, ethical, and societal consequences of the release of genetically engineered organisms into the environment. In his project *Pigeon d'Or* (2011) he attempts to make pigeons defecate soap, so as to turn animals commonly seen as 'flying rats' into swarming urban disinfection machines. In a first step, in collaboration with biochemist James Chappell, he modifies *in vitro* the metabolism of gut bacteria with the help of two customized biobricks—one lowers the pH level in the *Bacillus subtilis* colonies, and the other causes them to express lipase, a fat-digesting enzyme. Then, by the means of food coloring, he tests how long it takes for transplanted bacteria to replace the existing bacterial flora in the pigeon's gut, hence addressing today's research into microbiome transplantation. In addition, Van Balen has designed two pseudo-functional objects: a pigeonry to be attached to the windowsill that allows pigeons to be fed, and an interface for parked cars, encouraging pigeons to land and defecate on the windscreen. Beyond its neo-Dadaist spirit the work needs to be analysed upon its epistemological subtexts: the pigeons themselves are not genetically altered; the bacteria in their guts are, the pigeons being mere 'messengers' of the transgenic. Van Balen alludes to a research paradigm called metagenomics which studies not only the DNA sequences of individual organisms but also their symbiotic or parasitic interactions with other inhabitants of their environment. In addition, the design of speculative artefacts metaphorically echoes the hyperbolic engineering language of synthetic biology, which speaks of the design of genetic 'circuits,' 'modules,' 'standardized parts,' and 'chassis,' instead of referring to organisms as living beings. Inspired by the concept of orthogonality imported from computer sciences, synthetic biology presupposes that, unlike in most living systems, one effect produced by one technical component does not create mechanical side effects on other technical components of the system, 'just like in a car adjusting the rear-view mirror does not affect the steering' (Billerbeck & Panke 2012, 23).

Such epistemic shifts, especially with regard to biological media in the sense of *means* of transformation, generation, or communication, can also be observed within the evolving practices of pioneers in the field of bio media art themselves, Eduardo Kac and Joe Davis.

In retrospect Eduardo Kac's well known transgenic installation *Genesis* (1999) appears as a typical case for art centrally using bacteria as mere breeding containers for genes. Kac focuses on a synthetic gene, generated for artistic purposes as a transcription of a sentence from the biblical book of Genesis into Morse code, then converting the Morse code into DNA base pairs and cloned into the plasmid DNA of bacteria, which are then being submitted to UV light as a mutagenic factor in order to potentially alter the encoded extra-biological information. The bacteria, genetically altered themselves to fluoresce either blue or yellow, serve as biomarkers to indicate the difference between colonies with and without the artistic gene and, by making contact with each other, turn green and indicate mutation in the plasmid DNA while 'co-authoring' the genetic text. The main focus is put on the change of the genetic sequence as a paralinguistic entity according to the *Genesis* narrative. Less than a decade later, in *Specimen of Secrecy about Marvelous Discoveries* (2006), Kac makes microbial activity the core of a series of canvas-like three-dimensional biotopes, technically conceived as self-sustaining Winogradsky columns, expressing different colors and shapes pre-arranged by the artist and then left to arbitrary metabolic exchange over time. Here, abiotic factors such as temperature, humidity, airflow, water, and light levels, as well as earth and nutrient-rich media as *milieus*, pave the way for metabolic activity of a large variety of aerobic and anaerobic microbial life in order to produce constantly evolving living motifs. Kac addresses ecological and evolutionary issues to challenge a human-only perception of the biotopes: 'We shall not confuse our ability to describe a living entity in a given manner [...] with the phenomenological consideration of what it is like to be that entity, for that entity.' In this work, a dialogical situation between the observer and the microbial life is encouraged:

> The biotope [...] is an artwork that asks for the participation of the viewer in the form of personal care. Like a pet, it will keep company and will produce more colours in response to the care it receives. Like a plant, it will respond to light. Like a machine, it is programmed to function according to a specific feedback principle [...]. Like an object, it can be boxed and transported. Like an animal with an exoskeleton, it is multicellular, has a fixed bodily structure and is singular. What is the biotope? It is its plural ontological condition that makes it unique (Kac 2007, 92–96).

In the same vein, Joe Davis, who has synthesized and inserted extra-biological textual and graphic information into bacterial cells as material carriers of genetic information since 1986, has shifted toward artistic practices in which bacteria are ascribed agency as organizers and tool-builders. In his early genetic art work such as *Microvenus* (1986) recombinant techniques of molecular biology are used to encode a graphic icon resembling both an ancient Germanic rune as a fertility symbol and an iconic outline of the external female genitalia into *E. coli* bacterial plasmid DNA (Davis 1996). Whilst its visual appearance is almost negligible—a banal Petri dish, without any microbial effects of color or shape—its complex conceptual resonances are concerned with attempts to communicate with extra-terrestrials, a philosophical reversal of the panspermia hypothesis which stipulates that all terrestrial life originates from microscopic life-forms existent in the Universe. But 25 years later, within the context of today's Synthetic Biology, Davis' *Bacterial Radio* (2011) subversively acknowledges bacterial agency, beyond the genetic code, to act as *means* of transformation, generation, or communication. The piece features in its production phase a living, conductive radio circuit made of and by genetically altered *E. coli* bacteria, containing a modified gene sequence from the orange marine puffball sponge *Tethya aurantia*. Naturally, this gene forms the sponge's glass skeleton of silicatein needles through bio-encapsulation. Instead, it is now capable of making the engineered bacteria metabolize metals to plate electronic circuits (Davis 2012, 110–13). Davis refers to the once popular hobby of making crystal radios—that can work without any electrical source because the necessary power is generated by the electromagnetic waves of the emitter the radio receives. Davis is interested, here, in the transition of physical to biological systems, and in alternative knowledge generated by do-it-yourself practitioners as a subversive counter-pole to the profit-based industry. The artist takes synthetic biology's engineering metaphors literally, but ironically reverses their goal: Instead of applying principles of electronic engineering to biology, he applies biological principles to electronic engineering, and 'trains' bacteria to grow a totally anachronistic old-fashioned radio.

Contemporary biotechnology is fascinated with the technical possibilities of prokaryotic organisms in many ways. On the one hand, research on minimal organisms tests biological units reduced to their minimal functions necessary for survival, for example when bio-entrepreneur Craig Venter transfers a synthesized genome into the bacterium *Mycoplasma capricolum*, which becomes a pure 'chassis' for 'circuits,' 'modules or 'standardized parts' of synthetic biology. On the other, technical features of unicellular organisms can be implanted and enhanced, or extracted through biomimetic strategies. For example, the CRISPR (clustered regularly interspaced short palindromic repeats) gene-editing system consists of an adaptation of a naturally occurring immune mechanism in bacteria and archaea whose DNA contains genetic sequences from viruses having previously infected the prokaryote. Third, the field called directed evolution places strong selective pressure on cultured bacteria in automated systems to speed up microbial evolution and provide fitness advantages by which resulting mutated bacteria produce new proteins. Artists today translate the underlying epistemological mind-sets philosophically and aesthetically. Some appropriate or create *biobricks*, collected in the Registry of Standard Biological Parts set up by MIT and presented in popular events such as the International Genetically Engineered Machine Competition (iGEM), for example Daisy Ginsberg and James King's *E. chromi*

(2009) wherein engineered bacteria secrete colored pigments to potentially serve as purposeful bio-indicators to sense toxins in home-diagnosis of stool (Ginsberg 2014, 123–25), as *media of biology* employed to *measure* and *observe* biological systems. The Indian group ArtScienceBangalore has created *BBa_K221000: First volume of teenage gene poems* as a *biobrick* (2009) that enables *E. coli* bacteria to artificially produce geosmin, an organic compound responsible for the earthy smell during Monsoon rainfalls.[7] This olfactory sensation is a cultural reference to the Monsoon period as fertile but melancholic, traditionally the stage of first loves, such as depicted by Bollywood movies. ArtScienceBangalore also reminds us that second- and third-world populations not only have suffered from corporate industrial negligence, such as shown by the Bhopal gas leak disaster, but also may become the first and cheap guinea pigs to be exposed to nontested synthetic biology products in the environment. Therefore, their tactical media art and activist citizen project *Searching for the Ubiquitous Genetically Engineered Machine* (2011) offers DIY kits to a network of rural volunteers for grassroots detection and mapping of *biobricks* in soil bacteria,[8] comprising devices for DNA extraction, electrophoresis, and polymerase chain reaction. The art duo C-Lab employs both naturally occurring bacteria to construct a biotechnological mirror, and genetically modified ones to make bacteria express smell perceived as natural. *Living Mirror* (2013) makes use of magneto-tactic bacteria's ability to swim along the earth's magnetic field. Depending on the input, a portrait image translated into a magnetic field, the bacteria reorient their body in real time, causing light to scatter and to create an image in a liquid bacteria culture, drawing on the myth of Narcissus who fell in love with his own image in the water reflection, and thus 'highlights how contemporary science has shattered the idea of our own body by recognising that we are mostly made up of non-human bacterial cells.'[9] In *Banana Bacteria* (2012), the duo genetically alters *E. coli* bacteria, known for their unpleasant smell, to express enzymes to produce savoury banana oil. Meanwhile, Panama disease is threatening the Cavendish, the world's most popular banana, grown as a standardized human artefact in monocultures that cannot sexually reproduce. Artist Orkan Telhan conceives of his *Microbial Design Studio* (2015) as an open source countertop wetlab to genetically engineer bacteria by processes of directed evolution, so that they can synthesize different flavors—for example *Bananaworks* (2016): Bacterial media encapsulated in macro-fluidic gels in various compositions, evoking colorful candy, 'semi-living encapsulations capable of diversifying their taste independently, and of creating infinitely new possibilities that cannot be created by nature-born (wild) bananas or microorganisms alone' (Telhan 2016, 81–82).

With regard to the ambivalent notion of 'bacterial agency' a cluster peculiar to artistic practice emerges: the desire to deconstruct the alleged exclusiveness of human beings as tool-building agents and to counterbalance human prowess with unicellular organisms' quality to be organized and organizing, adapting, evolving, moving, sensing, and process-

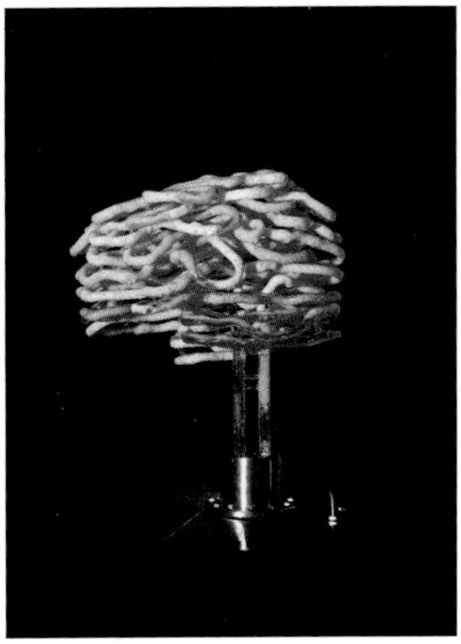

Fig. 1. Thomas Feuerstein, *Pancreas* (2009), photo by the artist. (Plate 27, p. 319)

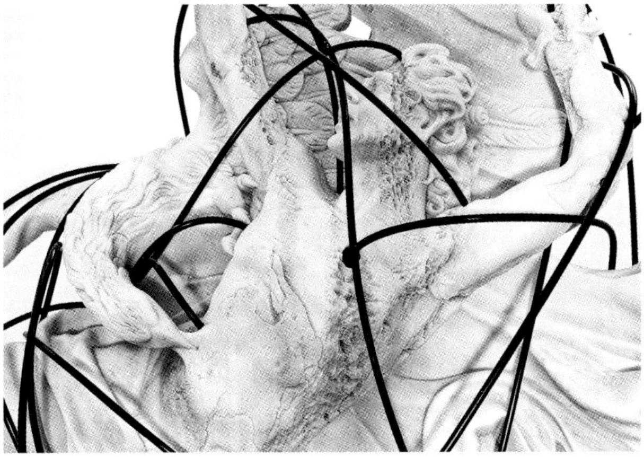

Fig. 2. Thomas Feuerstein, *Prometheus Delivered* (2017), photo by the artist. (Plate 28, p. 319)

ing information. Within an ecological account of natural agency, they partake in a 'cognitive turn in microbiology' considering that 'bacteria are purposive agents, and purposive agency is the mark of cognition' (Fulda 2017, 70–71). Affirmations of bacteria's cognitive ability distance them from just 'biotic systems [that] are analogous to [...] artificial machines: thermodynamic engines, pumps, and information-processing systems. Bacteria are analogous to complex human-made cybernetic systems', because

> the colony of individuals, the social group, gleans information from the environment. They "talk" with one another, distribute tasks, and convert their collective into a huge "brain" that processes information, learns from past experience, and, we suspect, creates new genes to better cope with novel challenges (Ben-Jacob, Shapira, and Tauber 2011, 56 57).

Artists Thomas Feuerstein renders such bio-philosophical considerations as 'metabols,' combining our brains' 'metabolism of signs and symbols' with 'digestive systems, memory and storage in our brain and in the rest of our body alike' (Feuerstein and Hauser 2020). Engineered bacteria are the protagonists of his biotechnological sculpture *Pancreas* (2009), modified to produce glucose by breaking down cellulose from shredded books to feed human neuroglia cells that grow into the shape of a human brain, metaphorically the center of human cognition. The feeding of the artificial brain follows a strict diet, since it exclusively consists of Hegel's *Phenomenology of Spirit:* Metabolically, 'food for thought' becomes 'thought for food.' In *Prometheus Delivered* (2017), endolithic microbes eat rocks to produce nutrients for an organic sculpture grown out of liver cells; a symbolic sculpture meant to last forever is metabolically transformed into a biotechnological artefact. Feuerstein employs chemolithoautotrophic bacteria, living in rocks and feeding on inorganic matter by chemosynthesis, to first decompose a marble sculpture (a replica of *Prometheus Bound* by Nicolas Sébastien Adam from 1762) as the very icon of Western cultural heritage—man striving for techno-scientific knowledge—and then recompose a 'live(r) sculpture' (Adler 2018, 12) to reference Prometheus' immortal liver: The organ considered central to knowledge and life, eaten every day by an eagle but regenerating every night, as 'a cautionary symbol to humans of both emancipation and hubris' (Koos 2018, 45).

In this context, the threat microorganisms represent to cultural heritage seen from the standpoint of bio-deterioration and conservation studies needs to be re-examined under a different light. Even though bacteria, algae, fungi, cyanobacteria, mosses, and plants cause stains, crusts, cracks, patinas, biofilms, or corrosion, at the same time microorganisms also assist in cleaning, bio-removal of unwanted substances, or, what's more, help conserve cultural heritage by bio-calcification to consolidate supports such as limestone (Caneva, Nugari, and Salvadori 2008). Among the group of bio-deteriorating microorganisms, the cyanobacterial genus *Leptolyngbya* has been found to be responsible for the degradation of frescos of the Domus Aurea in Rome, built by the emperor Nero around 64 BC (Albertano and Caiola 1988). However, human presence is equally responsible because the permanent illumination of the frescoes for tourists attracts the red- or green-pigmented photo-tactic *Leptolyngbya* which move toward light sources. Artist Edgar Lissel, in *Domus Aurea* (2005), employs them to remediate what they deteriorate, by first taking photographs of the frescos and then projecting the negative image onto large rectangular plates coated with agar-agar medium while inoculating them with bacteria sampled from the frescoes. The red *Leptolyngbya*—which have conserved their photo-tactic qualities from their green, photosynthesizing precursors but evolved to also heterotrophically feed on sugars under low light conditions—now recreate the image with their pigments. Similarly, Sabine Kacunko has sampled living bacteria and fungi from the Colosseum in Rome, including the Gram-negative *Bacillus cereus* known for its capacity to consolidate stone, while the Colosseum's building material itself consists dominantly of travertine, a terrestrial sedimentary and porous rock, solidified by bacteria via bio-calcification (Ranalli and Sorlini 2008). In *Invincible* (2015), Kacunko projects in real time a large circular image showing the bacterial biofilm cultured in a Petri dish onto the walls of the world-famous icon of cultural heritage, thanks to which it exists and survives. The colorful pigments produced by the bacterial metabolisms turn the spotlight on to the overlooked fact that 'the micro-organisms protect the monument from destruction caused by harmful environmental influences and thus secure the transmission of our cultural memory' (Kacunko 2016, 9), whilst microbiologists themselves address bacteria as 'ancient architects' capable of building 'highly patterned

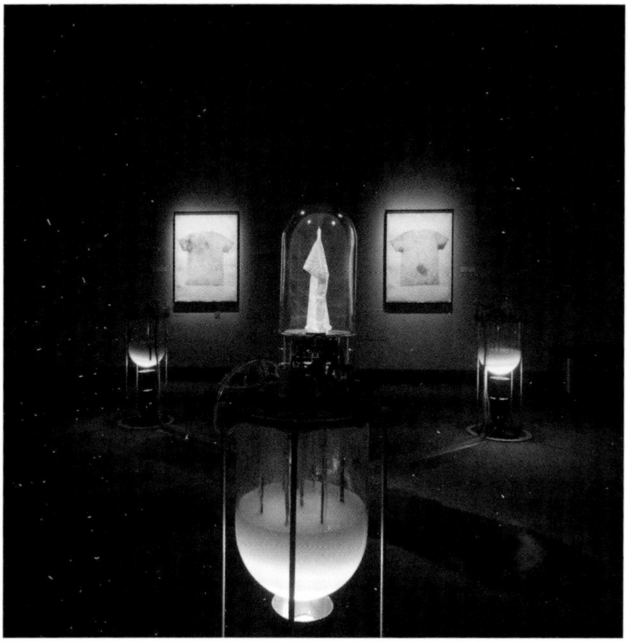

Fig. 3. Paul Vanouse, *Labor* (2019), Burchfield Penny Art Center, photo by Tullis Johnson.
(Plate 29, p. 320)

dwellings' via 'interspecific sensory and coordinated physiological activity ("consciousness" [...] and "design") to challenge the preconceived idea that conscious human architecture constitutes an exceptional feature' (Krumbein and Asikainen 2011, 63–65).

Although from a very different angle, these challenges are complemented by contemporary microbiome studies, including research around the so-called gut-brain axis which questions the division of metabolic and cognitive activity, as well as the existential role of microbial diversity for human health (Blazer 2014), learning, social interaction, or moods, and the notion of human individuality as such—'*Mind the Gut* (Bencard & Whiteley 2018)!'[10] Correspondingly, a growing number of art projects address human contact with their microbial companions, such as Sonja Bäumel's *Expanded Self* (2012) as microbial self-portraiture via nonhuman life forms, growing skin microbes of her body in large-sized Petri dishes. Jean-François Lapointe's performance *1000 Handshakes* (2014) enacts metagenomic self-portraits, *Microbiome Selfies*, demonstrating the gradual change in 'his' microbial community by pressing palms with people. Bringing matters to a head even further, in Paul Vanouse's micro-performative installation *Labor* (2019) human bodies are fully eliminated while a complex interplay of microbial activity creates scents typically associated with human exertion. *Staphylococcus epidermidis*, *Corynebacterium xerosis*, and *Propionibacterium avidum* are cultured in three connected bioreactors, metabolize sugars and fats, and then infuse a wearer-less white t-shirt as the nostalgic olfactive indicator of the progressive disappearance of work and workers as we knew it. In times of algorithmic finance and high frequency trading, *Labor* produces the sweaty smell of labor not as by-product but as end-product. The instrumentalization of sequential micro-gestures on the industrial production line is continued here in a self-regulating installation to embody society's shift from human and machine labor to increasingly pervasive forms of microbial manufacturing and computerized bio-optimization.

Rehabilitating bacteria as media, subject of inquiry, and models in disciplines of the humanities increasingly aware of the impact of techno-scientifically shaped worldviews bears witness not only to a general change in attitude toward prokaryotic single-cell organisms within the trend of a post-anthropocentric self-understanding and in times of major ecological crises, it also indicates the urgent need for hybrid trans-disciplinary vectors of thought in a period when research is characterized by excessive hyper-specialization. After the paradigm shifts brought about by the *linguistic, performative, and pictorial turns*, an *epistemological turn* emerges: Art here is no longer concerned with the aesthetic transposition merely of knowledge, but of how knowledge is being produced. While the techno-sciences today have themselves become powerful producers of aestheticized images, artistic strategies call for an analysis that is not based primarily on imagery but on material media and epistemic connections. Phenomena that once assumed the form of artistic images are being translated, scattered, and fragmented, here, into a variety of instances of biomediality—not as a means to an end, but as fully integrated elements of the aesthetic object. Merging initially disconnected disciplines and methods of inquiry, the conjunction of research areas makes bacteria an extremely prolific epistemological art/science interface.

Notes

1. This chapter is related to research conducted within the framework of our Big Bacteria Research Network at the University of Copenhagen, aiming at establishing an interdisciplinary platform between the natural and the health sciences, the arts and the humanities. Big Bacteria examines the role of bacteria as epistemic tools; it shifts the attention from the paradigm of Big Data with its immeasurable sequential data to Big Bacteria, emphasizing the importance of relational networks essential for interdisciplinary research. https://bigbacteria.ku.dk.
2. The SO_3 exhibition, curated by the author in 2015 at the Espace Multimédia Gantner, Bourogne/France, presented 'three tender significant others' by artists Adam Brown, Tagny Duff, and Paul Vanouse as a triptych of works encouraging 'affectionate experiences with our gentle but usually despised biological better halves—viruses, bacteria and plasmids—as today's science focuses on the benefits of bacteriophages, microbiome studies, or decontaminating bacteria.'
3. The point of departure is the atomistic principle of loosely coupled, independent physical units that, according to the early media philosophy of Fritz Heider, make media functions, such as transmission, mediation and processing, and their perception possible. Cf. Heider, Fritz.

1926.'Ding und Medium.' *Symposion. Philosophische Zeitschrift für Forschung und Aussprache*, Vol. 1, Berlin, 109–57.

4. In Hofmann, Werner. 1996. "Avantgardistische Skepsisverweigerung. Fortschrittseuphorie in den Lebenswelten der Kunst." In *Vom Weltbildwandel zur Weltanschauungsanalyse*, edited by Volker Drehsen and Walter Sparn, Berlin: De Gruyter, 38.
5. Moreira, Ines: Why Decon at petit CABANON? Gallery announcement from November 2007, published within the framework of Marta de Menezes' exhibition at the petit CABANON in Porto from November 17, 2007 to January 10, 2008.
6. From a curatorial standpoint, this means that a large part of an exhibition budget is dedicated to constantly re-growing rotting ephemerals. During the *sk-interfaces* exhibition in Luxembourg (2009–10) Takita's moss sculpture and the Tissue Culture and Art Project's *Victimless Leather* jackets had to be re-cultured repeatedly to be able to be shown for 15 weeks.
7. http://parts.igem.org/Part: BBa_K221000:Design.
8. Interview with C-Lab (Laura Cinti and Howard Boland), *We-make-money-not art*, October 2, 2013; https://artscienceblr.org/index.php/searching-for-the-ubiquitous-genetically-engineered-machine-igem-2011/.
9. http://we-make-money-not-art.com/the_living_mirror/.
10. *Mind the Gut!* is the title of an art and science exhibition since 2017 on the microbiome and its interaction with brain functions at the Medical Museion in Copenhagen, with which the author is affiliated.

References

Adler, Sabine (ed.). 2018. *Promotheus Delivered. Thomas Feuerstein.* Munich: ERES.

Albertano, Patrizia, and Maria Grilli Cailoa. 1988. "Structural and ultrastructural characters of a red biodeteriorating Lyngbya sp. in culture." *Archiv für Hydrobiologie* Suppl. 80, 1–4: 55–57.

Badura, Michael. 1977. "Manifest dem Laubvogel gewidmet." In *Michel Badura—Die Eingeweckte Welt und andere Arbeiten aus den Jahren 1966–1977*, edited by Klaus Ulrich Düsselberg, 33. Krefeld: Sassafras (first published in *Bric à Brac*. Barlisssen, 1967).

Ben-Jacob, Eshel, Yoash Shapira, and Alfred I. Tauber. 2011. "Smart Bacteria." In *Chimeras and Consciousness* edited by Lynn Margulis, Celeste A. Asikainen, and Wolfgang E. Krumbein, 55–62. Cambridge MA: MIT Press.

Bencard, Adam, and Louise Emma Whiteley. 2018. "Mind the Gut—displaying microbiome research through artistic collaboration." *Microbial Ecology in Health and Disease* 29(2): 1555433.

Billerbeck, Sonja, and Sven Panke. 2012. "Synthetische Biologie—Biotechnologie als eine Ingenieurswissenschaft." In *Leben schaffen? Philosophische und ethische Reflexionen zur Synthetischen Biologie* edited by Joachim Boldt, Oliver Müller, and Giovanni Maio, 19–40. Paderborn: Mentis.

Blazer, Martin J. 2014. *Missing Microbes: How the Overuse of Antibiotics Is Fueling Our Modern Plagues*. New York: Henry Holt & Company.

Caneva, Giulia, Maria Pia Nugari, and Ornella Salvadori (eds.). 2008. *Plant Biology for Cultural Heritage. Biodeterioration and Conservation*. Los Angeles CA: Getty Conservation Institute.

Davis, Joe. 1996. "Microvenus." *Art Journal, Contemporary Art and the Genetic Code* 55(1): 70–74.

Davis, Joe. 2012. "Bacterial Radio." In *Prix Ars Electronica: CyberArts 2012* edited by Hannes Leopoldseder, Christine Schöpf, and Gerfried Stocker, 110–13. Ostfildern: Hatje Cantz.

Dunn, Rob. 2010. "Painting with Penicillin: Alexander Fleming's Germ Art." Smithsonian Magazine, July 2010; http://www.smithsonianmag.com/science-nature/painting-with-penicillin-alexander-flemings-germ-art-1761496/#zx3op7s8FrK2e14Z.99.

Fleming, Alexander. 1937/2007. "The Growth of Microorganisms on Paper." In *Signs of Life. Bio Art and Beyond* edited by Eduardo Kac, 345–46. Cambridge MA: MIT Press. First published in *Report of Proceedings. International Society for Microbiology. Second International Congress for Microbiology* edited by Ralph Terence St. John-Brooks. London, 1937.

Fulda, Fermín C. 2017. "Natural Agency: The Case of Bacterial Cognition." *Journal of the American Philosophical Association* 3(1): 69–90.

Feuerstein, Thomas, and Jens Hauser. 2020. "From Metaphors to Metabols, Conversation Jens Hauser & Thomas Feuerstein" in *Stofsk(r)ifter: Metabolic Machines by Thomas Feuerstein* edited by Adam Bencard, Martin Grünfeld and Jens Hauser. Copenhagen: Medical Museion, 50–57.

Hauser, Jens. 2007. "The Grammar of Enzymes." In *Wim Delvoye. Cloaca 2000–2007*, 26–35. Luxembourg: Casino Luxembourg Forum d'art contemporain.

Hauser, Jens. 2013. "Remediating Still Life, Pencils of Nature and Fingerprints." In *Relive. Media Art Histories* edited by Sean Cubitt and Paul Thomas, 275–307. Cambridge MA: MIT Press.

Hauser, Jens. 2014. *Biotechnologie als Medialität: Strategien organischer Medienkunst*. Bochum: Ruhr-Universität Bochum.

Hauser, Jens. 2016. "Biomediality and Art." In *Recomposing Art and Science: Artists-in-Labs* edited by Irene Hediger and Jill Scott, 201–19. Berlin: De Gruyter.

Heider, Fritz. 1926. "Ding und Medium." *Symposion. Philosophische Zeitschrift für Forschung und Aussprache* 1, 109–57.

Hofmann, Werner. 1996. "Avantgardistische Skepsisverweigerung. Fortschrittseuphorie in den Lebenswelten der Kunst." In *Vom Weltbildwandel zur Weltanschauungsanalyse* edited by Volker Drehsen and Walter Sparn, 31–38. Berlin: De Gruyter.

Hoffmann, Peter Gerwin. 1987/2007. "Kunstwesen. Galerie-Notiz aus Graz vom 9.9.1987." Published as "Kunstwesen" in *Animal Art* edited by Richard Krische. Graz: Steirischer Herbst. Newly published in *Peter Gerwin Hoffmann* edited by Günther Holler-Schuster, 142–43. Köln: Walther König.

Kac, Eduardo. 2007. "Specimen of Secrecy about Marvelous Discoveries." In *Eduardo Kac* (exhibition catalogue), 92–96. Valencia: Instituto Valenciano de Arte Moderno.

Kacunko, Slavko (ed.). 2016. *Sabine Kacunko. Bacteria, Art and other Bagatelles*. Wien: Verlag für moderne Kunst.

Kaprow, Allan. 1971/1993. "The Education of the Un-artist, Part I." *Art News* 69(10), 1971: 28–31. Newly published in *Essays on the Blurring of Art and Life / Allan Kaprow* edited by Jeff Kelley, 97–109. Berkeley CA: University of California Press.

Koos, Anuschka. 2018. "The Myth of Prometheus." In *Promotheus Delivered* edited by Sabine Adler, 45–47. Munich: ERES.

Krumbein, Wolfgang E., and Celeste A. Asikainen. 2011. "Ancient Architects." In *Chimeras and Consciousness* edited by Lynn Margulis, Celeste A. Asikainen, and Wolfgang E. Krumbein, 3–70. Cambridge MA: MIT Press.

Landecker, Hannah. 2006. "Microcinematography and the History of Science and Film." *Isis* 97: 121–32.

Lippard, Lucy R. 1973. *Six Years. The dematerialization of the art object from 1966 to 1972*, New York: Praeger.

Porath, Erik. 2008. "Begriffsgeschichte des Mediums oder Mediengeschichte von Begriffen? Methodologische Überlegungen." In *Begriffsgeschichte der Naturwissenschaften. Zur historischen und kulturellen Dimension naturwissenschaftlicher Konzepte* edited by Ernst Müller and Falko Schmieder, 253–72. Berlin: De Gruyter.

Ranalli, Giancarlo, and Claudia Sorlini. 2008. "Bioremediation." In *Plant Biology for Cultural Heritage. Biodeterioration and Conservation* edited by Giulia Caneva, Maria Pia Nugari, and Ornella Salvadori, 340–46. Los Angeles CA: Getty Conservation Institute.

Sarasin, Philipp. 2007. "Die Visualisierung des Feindes. Über metaphorische Technologien der frühen Bakteriologie." In *Bakteriologie und Moderne. Studien zur Biopolitik des Unsichtbaren 1870–1920* edited by Philipp Sarasin, Silvia Berger, Marianne Hänseler, and Myriam Spörri, 427–61. Frankfurt am Main: Suhrkamp.

Schult, Hans-Jürgen. 1969. *BIOkinetische SITUATIONEN*. Catalog published in conjunction with an exhibition at Schloß Morsbroich. Leverkusen: Städtisches Museum.

Telhan, Orkan. 2016. "Biorealize: Microbial Design Studio and Bananaworks." In *WETWARE: Art – Agency – Animation* edited by David Familian and Jens Hauser, 80–89. Irvine CA: Regents of the University of California.

Thacker, Eugene. 2004. *Biomedia*. Minneapolis MN: University of Minnesota Press.

Exceptional Matters

Adam W. Brown

Cartesian dualisms have shaped our thinking throughout the centuries, resulting in the creation of artificial distinctions between ideas such as organic and inorganic, alive and dead, and human and animal. In this essay, three artworks are examined that intend to decenter human exceptionalism by questioning how human entanglements with matter and 'living' species shape our histories, stories, and myths and therefore redefine our ontological status.

The largest and potentially the most problematic binary is the construction of the human versus nature divide, which delineates a false opposition between the artificial human-made and the 'natural,' nonhuman entities of the world. These dichotomies, which flourished during the period of Enlightenment and intensified throughout the Industrial Revolution, oversimplified complex issues and in turn have distorted human thinking by extracting humans from nature. Each artwork in turn attempts to rattle the foundations of anthropocentricism by questioning inherent and associated dichotomies relevant to the natural vs artificial polemic. The ongoing series titled *Origins of Life* challenges the notion that the constituents of aliveness are different than dead matter by investigating agency inherent in all elemental matter. The *Great Work of the Metal Lover* unravels the idea that humans possess cognitive superiority above all other species by demonstrating that microbes can solve the ancient alchemical problem of transmuting base matter into noble gold, which human beings have yet to manage. Finally, *[ir]reverent: Miracles on Demand* proves that microbes can perform 'miracles,' too, challenging our understanding of historical events, organized religion, and the supernatural.

Begin with our understanding of the origins of life. How life began and exactly what life is have been religious, philosophical, and scientific puzzles at least since the beginning of recorded history. Religious explanations of life's origins are generally teleological. For example, the Judeo/Christian myth of Genesis declared that life started through divine intervention, that nature was developed for man, and that man himself was made in the image of God. One of the more enduring philosophical arguments was that life formed through vitalistic and supernatural means, arising from nonliving matter through a process known as spontaneous generation. It was hypothesized that maggots and funguses arising from spoiled food provided evidence for this everyday common occurrence. Not surprisingly, science has had a profound impact on these myths in its attempt to replace them with a rational account of origins. Darwin undermined the Biblical account of Genesis by introducing the concept of evolution by natural selection at the same time as Louis Pasteur proved the impossibility of spontaneous generation. The new dogma that emerged stated that all life springs from previous life. But that, too, created a teleological conundrum. If spontaneous generation is impossible, how did the first living organisms arise? How did 'dead' matter become 'living' without a god and in the absence of any 'vital force'?

Around 1950, the physicist Harold Urey proposed that it might be possible to evolve biological molecules from chemical ones, bridging the gap between living and nonliving. He proposed to recreate the atmosphere of the Earth and add an energy source such as lightning to the mix. A graduate student by the name of Stanley Miller heard one of Urey's talks on this subject and proposed to Urey that they carry out such a test. Urey was reluctant, but Miller convinced him that it would make a good PhD thesis. Using the best knowledge about the Earth's primordial atmosphere available at the time, Miller set up an experiment in which methane, hydrogen, ammonia, and water vapor were mixed in the presence of an intermittent electrical discharge. Within days, the experiment yielded the amino acid glycine, and within a week alanine and urea as well. No previous experiment had yielded even one amino acid, let alone several, so Miller's experiment represented a major breakthrough. He published his results in 1953 in the journal *Science* and it quickly became a scientific and public icon of origins of life experimentation (Miller 1953).

Origins of life: Experiment #1 is a continuation of Miller's 1953 experiment. This experiment has been replicated only a handful of times, by Hough and Rogers in 1956 and Dr. Jeffrey Bada in 2013, and it yielded the same compounds as Miller's original (Rogers 1956; Bada 2013). No one has ever run the experiment through multiple cycles of synthesis to see if, after several weeks or months, small peptides, sugars, or other important compounds result. Moreover, the number of variations run on Miller's experiment is very limited, despite the fact that the primordial atmosphere was almost certainly not the one that he replicated in his experiment. And now a great deal is known about the atmospheres of other planets and their moons. Thus, there are a huge number of permutations of Miller's experiment that can fruitfully be run by making this into an 'open-source' experiment in which many people can contribute and participate.

Adam W. Brown and Dr. Robert Root-Bernstein transformed Miller's experiment into an open-source performative re-enactment and continuation of working sculptural and performative art piece called *Origins of Life: Experiment #1.x* that added the means to resupply the system with fresh gases, water, and minerals so as to over-populate the chemical species in the system. This over-population drove the chemical syntheses in new directions, so that we were able to obtain not only amino acids, but sugars, fatty acids, and lipids, and even nucleic-acid bases and the organic chemical that is a building block to RNA and DNA as well as providing energy to drive many processes of living cells, Adenosine triphosphate (ATP). These represent the entire range of chemical compounds required to build the constituents of a cell!

The artwork probes the historical, philosophical, and scientific debate about origins of life and the dualistic argument between vitalism and the mechanistic or deterministic materialism view of the workings of life. Historically, the vitalists of the twentieth century believed that 'life' consisted of something 'special' or unique that existed outside the material world and could not be found by taking a living thing apart (wholism), while the mechanists believed that 'life' could be explained by 'physico-chemical' processes that were deterministic, mechanistic, and could be understood from the study of its parts.

These Cartesian binaries have shaped our thinking throughout the centuries, resulting in the creation of artificial distinctions between ideas such as organic and inorganic, alive and dead, and human and nonhuman. Is there a middle ground, a paradigm that does not make such distinctions? Is there a model that takes into account emergent processes, complexity, energy, materiality, and evolution? Do the laws of physics and chemistry contain in them the potential to give rise to the autopoetic, self-emergent, replicating entities that we call living?

Such a system would have to evolve, and to evolve by means of natural selection the system would have to integrate five processes. The first is a means of creating variations among the elements of the system. The second is the over-production of these elements so that not all can survive. The third is a way to nonrandomly select among the variations so that those elements that do survive have a different distribution than the elements that were originally present. The fourth is a method to reproduce the surviving variants so that that population of elements changes over time. Finally, this process must be reiterated over and over again so that the elements are modified with each passing generation. The result of such an integrated set of processes is a continuous cycle of over-production-selection-reproduction that results in adaptation over time—in short, evolution.

ReBioGeneSys is our answer to the dead-end nature of the Miller experiment, which can only produce biologically-related chemicals, but has no ability to evolve. *ReBioGeneSys* itself evolved from Origins of Life, 1.x by an iterative process of trial and error. *ReBioGeneSys* is augmented by several features that provide multiple

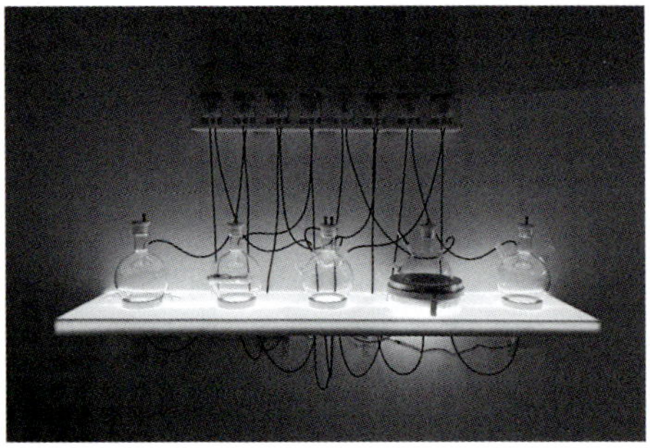

ReBioGeneSys, 2015. *ReBioGeneSys* is a fully functioning scientific experiment capable of being reconfigured into any real or imaginary world, be it Venus, Titan, our prebiotic Earth, or Middle Earth. By combining all the scientific research on the origins of life in one set of integrated processes, *ReBioGeneSys* creates 'mashed-up' extreme minimal ecosystems theoretically capable of forming the self-organizing chemistries necessary to produce semi-living molecules and perhaps even protocells. (Plate 30, p. 321)

means of chemical production and processes for carrying out nonrandom selection. All known natural conditions such as desiccation-hydration, freeze-thaw, and day-night cycles, exposure to UV light, electrical energy, and heat are incorporated into one system. Any possible environment can be simulated or invented. Prebiotic compounds are not only synthesized, but selectively destroyed, some surviving to participate in further evolution, others going extinct.

Every permutation of every environment that might support life is incipient in *ReBioGeneSys*. What evolves is simply a matter of how many permutations need to be explored and how long it takes to explore them. If life can evolve, then it will. The evolution of the sculpture is inherent in its materiality and the processes it carries out, independently of the human beings that sculpted it, unaffected by the transience of the audiences that view it. An incarnation of new materialist philosophy that nonetheless is designed to produce a vitalistic outcome, *ReBioGeneSys* is an artifice designed not to mimic, but to produce nature itself through self-sustaining, ever-recursive autopoeisis. What makes *ReBioGeneSys* different from all previous attempts to reproduce life, whether through art or science or both, is that it contains within itself the entire cycle of self-evolving processes. The apparatus simply makes manifest the intrinsic agency of matter itself.

If the materiality of chemical reactions bridges the nonliving to the living world then the next artwork reverses the equation to bridge the living world back to the chemical. *The Great Work of the Metal Lover* is a neo-alchemical artwork that solves the ancient problem of the philosopher's stone or the transmutation of base materials into gold. A bioremediation process with extremophilic bacteria, made in collaboration with Dr. Kazem Kashefi, titled *The Great Work of the Metal Lover* is a 2012 artwork that re-examines the problem of transmutation through the use of modern microbiological practice. *The Great Work of the Metal Lover* solves the unobtainable ancient human riddle of transmutation of gold.

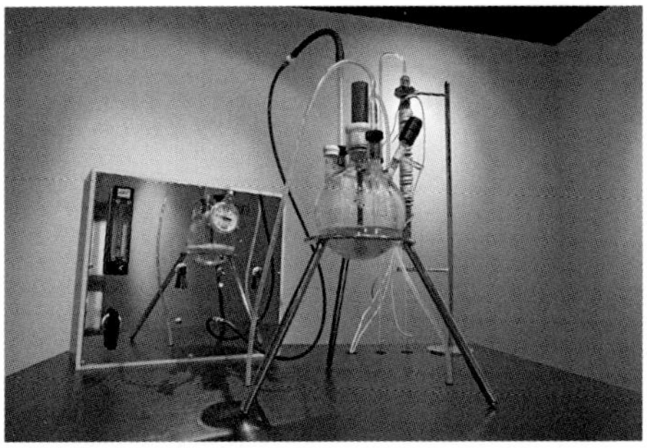

The Great Work of the Metal Lover, 2012. Gold production is accomplished by the pairing of a highly specialized metallotolerant extremophilic bacterium (*Cupriavidus metallidurans*) and an engineered atmosphere contained within a customized alchemical bioreactor. The extreme, minimal ecosystem within the bioreactor forces the bacteria to metabolize high concentrations of toxic $AuCl^3$ (gold chloride), turning soluble gold into usable 24K gold. (Plate 31, p. 321)

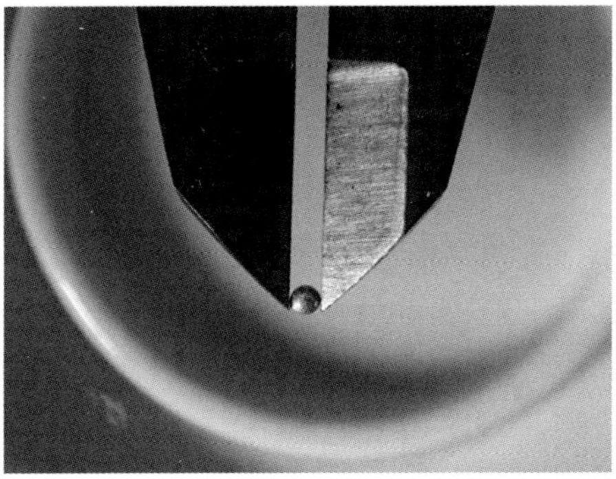

The Great Work of the Metal Lover, 2012. 150 mg of 24K gold(Au) produced after 2 weeks of running the bioreactor. (Plate 32, p. 322)

The alchemical attempt to transmute matter (best exemplified by turning lead into gold) is an example of Man's hubristic belief that he could emulate God and create from dross the most valuable things, just as God turned dust into men. Human hubris, however, has been stymied by the laws of nature, which seem to forbid such transmutations. Our human exceptionalism has been further undermined by the discovery of microbes that, if not exactly capable of transmuting one element into another, can do something very close to that which still eludes human capability, and that is to separate out gold from dross. *The Great Work of the Metal Lover* has harnessed these microbes to achieve the dream of creating gold, in this case from a water solution, fulfilling the alchemical dream, but only by displacing human beings from center in the cosmic equation.

It stands to reason that the earliest life forms to evolve on earth were extremophiles, since life certainly emerged in conditions that we would consider poisonous to human life. These extremophiles have adapted to thrive in environments of intense heat, acidity, salt, and high concentrations of heavy metals, and have been discovered everywhere from hydrogen sulfide-powered hot springs to deep sea thermal vents to industrially designed toxic heavy metal plating containers. In fact, discoveries made in the last decade indicate that extreme life is abundant deep below earth's surface, constituting substantially more biomass than all of the humans on the planet. Just as the bacterial microflora play a central role in the human body's ecological make-up, so too does the *microbial dark matter*, regulating the material exchange of minerals via earth's metabolic processes (Magnabosco 2018). In this paradigm the distinction between 'dead' matter and life become murky, overthrowing previously held beliefs about what constitutes aliveness. Indeed, principles of evolution drive microbes to develop novel adaptations to utilize metals and minerals as part of their metabolisms.

Gold production is accomplished by the pairing of a highly specialized metallotolerant extremophilic bacterium, *Cupriavidus metallidurans,* and an engineered atmosphere contained within a customized alchemical bioreactor. The extreme minimal ecosystem within the bioreactor forces the bacteria to metabolize high concentrations of toxic $AuCl_3$ (gold chloride), turning soluble gold into usable 24K gold. Through a reduced artificial anthropogenic environment, the liquid medium is purified by a collaboration with a colony of living bacteria. The reaction is observable both by the purple color that the gold precipitate creates in the reaction vesicle and also by electron microscopy, which reveals the nuggets of gold that each bacterial colony produces. Thus, the all illusive and unachievable goal of the alchemist's magnum opus is solved not by human genius alone, but by help by a collaboration with another species.

If the abilities of Earth's smallest life forms such as *C. metallidurans* to extract gold from Earth's water supplies appear miraculous, microbes turn out to have other miraculous abilities that have literally fooled human beings into believing in divine intervention. For centuries, the great monotheistic religions such as Catholicism proliferated stories about autonomous human mastery over all other organisms. Touting that man was made in God's image has blinded us to the fact that we are indeed interconnected to nature and that our histories have been influenced by nonhuman species. *[ir]reverent: Miracles on Demand* is an artwork that examines the impact of invisible microbial agents on the course of human history and belief systems.

> … as nothing happens in nature which does not follow from her laws, and as her laws embrace everything conceived by the Divine intellect, and lastly, as nature preserves a fixed and immutable order; it most clearly follows that miracles are only intelligible as in relation to human opinions, and merely mean events of which the natural cause cannot be explained by a reference to any ordinary occurrence, either by us, or at any rate, by the writer and narrator of the miracle. (Baruch Spinoza, *Theologico-Political Treatise*)

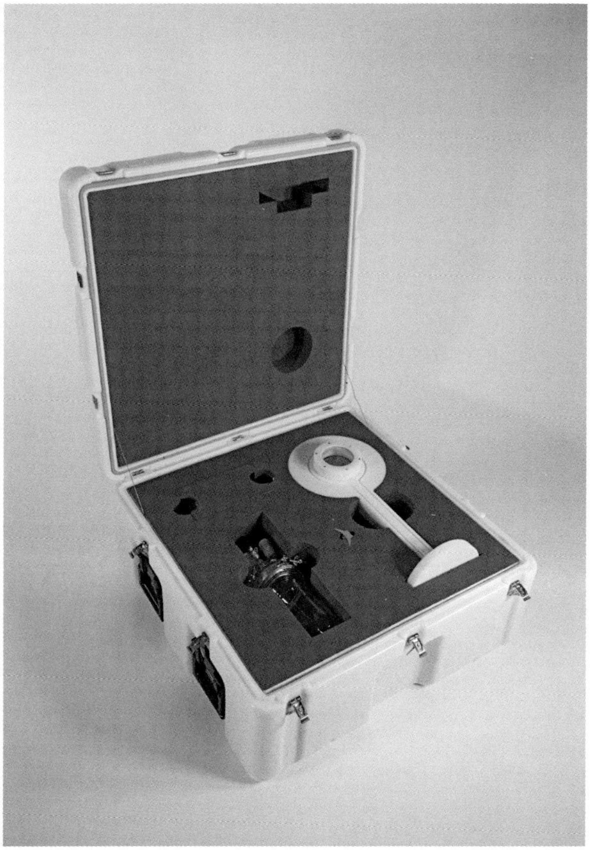

[ir]reverent: Miracles on Demand Kit, 2019. Portable Kit containing equipment necessary to create a miracle on demand. Items include custom PTFE (Teflon), glass and stainless-steel Monstrance/incubator, bioreactor, stainless-steel Luna and control system. White ABS shell case with red velvet interior.

The seventeenth-century philosopher Baruch Spinoza succinctly discusses miracles. According to Spinoza, miracles are not a transgression of natural and scientific laws, but rather events that reveal epistemological 'gaps' in our knowledge. In other words, miracles are events that surpass our limited human comprehension. Moreover, nothing would appear miraculous if we had a perfect understanding of the laws of nature.

Countless historical accounts of blood flowing from bread were recorded during the classical period of antiquity, the middle ages, as well as modern times. Blood miracles were used as a complex anthropogenic currency to justify everything from corrupt human conquests and evidence of divine agency to anti-Semitic persecution.

A good example of how such supposedly supernatural miracles could be used for religious and political purposes occurred over 2,000 years ago in the ancient Phoenician (now Lebanese) city of Tyre. Tyre was a thriving commercial center featuring a prominent Mediterranean harbor. The island city, located one-half mile from the Phoenician shore, was heavily fortified, surrounded by massive defensive walls, making it a key naval base. In 322 BC when Alexander the Great on his famous march was denied entry to Tyre, he decided to besiege the city. The Macedonians under Alexander worked for seven months to build a causeway to the island from the mainland so that their catapults could breach the walls. The weary Macedonians, laboring to finish the causeway under a constant rain of arrows and stone balls, were going to give up if it were not for an omen. When the Macedonian soldiers broke bread, drops of blood trickled out and dripped onto the floor. The presence of blood was read as an omen prophesying Tyre's destruction.

Quintus Curtius Rufus, the Roman historian, in his *History of Alexander the Great* written in the first century AD, elaborates as follows:

> Among the Macedonians, when some of the soldiers happened to break bread, they noticed drops of blood trickling out, and when the king was alarmed, Aristander, the most skilled of the seers, declared that if the blood had flowed from without, it would be an unfavorable omen for the Macedonians, but on the contrary, since it had flowed from an inner part, it foretold destruction for the city which they had determined to besiege.

The Macedonians completed the causeway, captured Phoenician ships, stormed Tyre, killed 8,000 Tyrians, sold another 30,000 into slavery, and left Tyre in ruins (Rufus 1984). This is just one example of many reported throughout antiquity citing the miraculous appearance of blood on food.

In late medieval Europe, stories of miraculous transformations of living matter became a focus of devotion and worship. Reports of statues, cloth, and eucharist hosts that oozed blood operated as powerful tropes giving proof of the transcendence of Christ and thus hope for ordinary humans (Bynum 2012). In an odd and paradoxical confluence of proof of Christ's transcendence and anti-semitic propaganda, blood miracles were often used to justify religious persecution. Jews were accused of stealing and stabbing consecrated eucharist bread so that they could use the flowing blood in cult practices. Thus, the action of stealing and desecrating hosts simultaneously sanctified semitic persecution while proving the existence and power of Christ. This host-desecration libel was exemplified by the massacre at Pulkau, Austria, on April 23, 1338. According to several Austrian and Bohemian chroniclers, on that day a crowd had gathered in the town center in response to rumors that a desecrated host had been found on the doorstep of a local Jew; passions, we are to infer, flared into violence; and most, possibly all, of the town's Jews, numbering about 150 persons, were seized, brutalized, and massacred and their possessions plundered. Reports

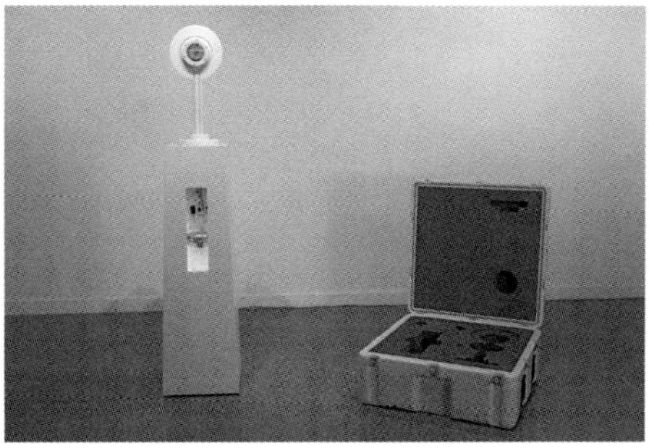

[ir]reverent: Miracles on Demand, 2019. Full Installation view. 'Bleeding' Eucharist Host, *Serratia marcescens*; Custom PTFE (Teflon), glass, and stainless-steel Monstrance/incubator, bioreactor, stainless-steel Luna and control system; White acrylic polymer pedestal with glass window; White ABS shell case with red velvet interior. (Plate 33, p. 322)

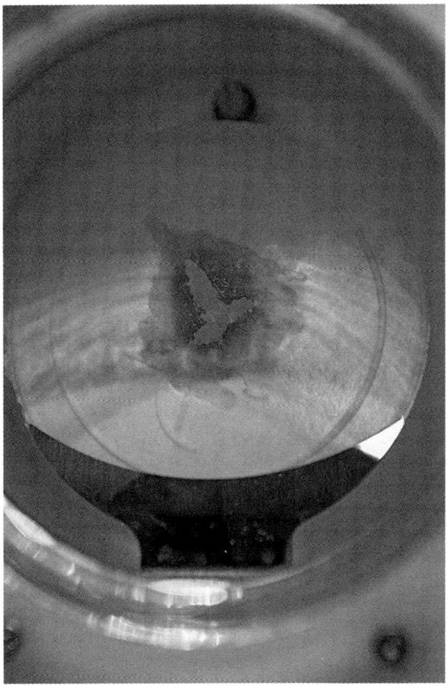

[ir]reverent: Miracles on Demand, 2019. Detail. 'Bleeding' Eucharist Host, *Serratia marcescens* and stainless-steel crescent-shaped Luna. (Plate 34, p. 323)

indicate that victims were burned alive and reduced to dust. Some of the victims, unable to flee from their persecutors, committed ritual suicide (a form of martyrdom Hebrew sources call kiddush ha-shem, 'sanctification of the divine name') in order to avoid forced baptism. The event was commemorated in two paintings depicting a Jewish man committing blood libel by desecrating the host and then hiding the evidence that now appears on the central altarpiece within the Church of Holy Blood in Pulkau (Merback 2005).

Many microbiologists agree that it is likely that the appearance of 'blood,' at least in part, was due to a microorganism, invisible to the eye, named *Serratia marcescens*. This bacterium produces a red, blood-like pigment called mucilaginous prodigiosin. Ubiquitous in our environment, this microorganism exists in the soil and is responsible for the pink-colored material that one finds in the shower.

[ir]reverent recreates these 'miracles' on demand by employing *S. marcescens*, a microorganism that grows on bread and creates a viscous fluid that has an uncanny resemblance to blood. Physically, the installation consists of a bio-incubator inspired by medieval Roman Catholic Monstrance, a sacred vessel (still used today) intended to exhibit the blessed bread after being blessed by a priest. A Eucharist host (bread) is placed in the oculus, which is a small window under the monstrance designed to

'display' the host; then micro-droplets of media containing a strain of *S. marcescens* are delivered to the host via a bio-incubator. Over the course of two days the inoculated host incubates the microorganism, causing the bread wafer to 'bleed,' thus producing a blood 'miracle on demand' similar to those seen in churches centuries ago.

Taken as a whole, these artworks can be seen as a logical continuation of the Darwinian revolution that took human beings out of the center of nature and made us into just another naked ape. But the work goes further to question whether this naked ape has any special status in nature at all. It appears that the distinctions between live and dead matter results in an artificial and false dichotomy. It stands to reason that if matter has rules about how it forms via hydrogen and covalent bonds etc. then matter itself embodies agency and its properties contingent to environment and conditions. All living things are manifestations of this material agency, and as special as we would like to think ourselves, the more we know about nature the more we see that the manifestations of material agency have properties and possibilities that we human beings may aspire to but have yet to achieve. If we are at all exceptional, it is only in being able to be cognizant of our humble place in nature and being able to recognize the benefits of working as part of nature.

References

Bada, Lazcano A. 2013. "New insights into prebiotic chemistry from Stanley Miller's spark discharge experiments." *Chem. Soc. Rev.* 42(5): 2186–96.

Bennett, J.W. 2000. "Seeing Red: The Story of Prodigiosin." *Advances in Applied Microbiology* 47: 1–32.

Bynum, Caroline Walker. 2012. *Christian Materiality: An Essay on Religion in Late Medieval Europe.* New York: Zone Books.

Magnabosco, C. 2018. "The biomass and biodiversity of the continental subsurface." *Nature Geoscience* 11: 707–17.

Merback, Mitchell B. 2005. "Fount of Mercy, City of Blood: Cultic Anti-Judaism and the Pulkau Passion Altarpiece." *The Art Bulletin* 87(4): 589–642.

Miller, S.L. 1953. "A production of amino acids under possible primitive earth conditions." *Science* 117: 528–29.

Rogers, Hough L. 1956. "Synthesis of amino-acids from water, hydrogen, methane and ammonia." *J. Physiol.* 132(2): 28–30.

Rufus, Quintus Curtis. 1984. *History of Alexander.* London: Penguin Group.

From Soft Sculpture to Soft Robotics
Retracing Entropic Aesthetics of the Life-like

Jonas Jørgensen

Introduction

Around the late 1960s a group of artworks were produced that have been historicized under the heading of *soft sculpture*. The epithet refers to Claes Oldenburg's Pop Art sculptures, but also to works associated with Postminimalism, Process Art, and Anti-Form. Straddling several art historical traditions, soft sculpture as a category collects together an ensemble of works that is internally divergent in many respects. And soft sculpture operates with several different notions of softness: Some works index a physical process involving molten or plastically deformed parts, others use pliable, yielding, or lightweight materials, some merely have visual cues that indicate a possibility of deformation, while others still are made of elastic rubbers.

When reading descriptions of works of soft sculpture, it is remarkable to notice that their similarities with living organisms and bodies are so frequently emphasized as a central tenet of their aesthetic, though most are nonfigurative and motionless.[1] Oldenburg's sculptures, for instance, are said to 'take on their own life' (Babington and Owens 2009). The felt compositions of Robert Morris 'allude to the human body through their response to gravity and epidermal quality' (Blessing n.d.) and have 'human qualities' (MoMALearning n.d.). Richard Serra's rubber strips in *Belts* (1966–67) are imbued with an 'anthropomorphic quality' (Spector n.d.), and Lynda Benglis' polyurethane foam accumulations attributed biological drives of their own when claimed to embody an 'erotics of Anti-Form' and 'movement of a (…) Dionysian sort' (Edelman 2004).

But why is soft sculpture so frequently described as life-like? Might this be a direct consequence of or somehow related to its soft materiality or is something else at play? Soft and pliable materials indeed figured prominently within an earlier tradition of crafted objects that explicitly sought to not just imitate, but simulate organic life, namely the fraction of

European automata that Jessica Riskin refers to as 'eighteenth-century wetware' (Riskin 2003). As Riskin writes,

> These machines all reflected the assumption that an artificial model of a living creature should be soft, flexible, sometimes also wet and messy, and in these ways should resemble its organic subject (Riskin 2003, 112).

The intuition that soft matter plays a central role for the functioning of living bodies has, moreover, recently started to assert itself within the technical research field of *soft robotics*. Here, efforts are made to construct robots from materials such as silicone, often by drawing explicit inspiration from the bodily mechanics, control strategies, and abilities of soft or partially soft animals.

In this essay, I interrogate an apparent contiguity between softness and what, for lack of a better term, I will refer to as *the life-like*. By this, I designate experiences and descriptions of nonliving entities and material processes that ascribe to these physical dynamics and qualities characteristic of living organisms.[2] The line of questioning pursued originates from a syncretic ambition, as it seeks to integrate the historical and aesthetic perspectives of media art history with the description of life within physics and with recent technical research on soft robotics. The extensive question that the essay revolves around is, thus, what connections can be drawn between softness and the process of life, beyond the obvious observation that most life forms (that we know of) are composed of soft matter. This question is approached by way of analyses of a number of selected works of soft sculpture and robotic art. In terms of the wider interests of this anthology, the essay can thus be said to deal with softness, and its changing status as an interface for an aesthetics of life.

To begin with, I focus on postminimalist soft sculpture and its conceptual underpinnings and reception to explore the different ways in which such works conjure up impressions of the life-like.[3] In doing so, I aim to explicate the kinds of liveliness postminimalist soft sculpture can be seen to enact. Secondly, I proceed to two recent artworks from a burgeoning strand of contemporary media art that appropriates soft robotics technology for an artistic end. And finally, as a way of concluding, I discuss how the relations between softness and life, unearthed through the analyses, relate to measures currently taken in order to imitate natural organisms within technical soft robotics research.

The underlying thesis of the essay is that the concept of *entropy* and the physical description of life as a process of entropy evasion can be used to analyze the different notions of life evoked by soft sculpture. And I argue that accounts of postminimalist soft sculpture as life-like actualize what I term *an entropic aesthetics of the life-like*, wherein life is envisioned as processual and the living organism's physical intertwinement with its environment is stressed. This aesthetics, I further argue, can be extended to also encompass two contemporary soft robotic artworks through a historically bound reworking of the entropy concept that emphasizes an understanding of entropy as sameness.

Forces Become Process

The question of why soft sculpture is so frequently described as life-like obviously has more than one answer. In some works of soft sculpture we find so-called natural materials such as fur, wool, jute, or hemp, that have connotations of life because they derive from living organisms. In others, the belts by Serra for instance, there is something about the scale and proportions that comes close to the human body. But in the subgroup of works where softness is introduced via pliable materials or physical deformation, the sense of life-likeness also appears to hinge on a reflexivity between the sensuous apprehension of soft matter that is part of the viewer's own body and soft matter within its proximity. In Lucy Lippard's 'Eccentric Abstraction' essay, which mentions several works of soft sculpture, she insightfully remarks that for sculpture

> The use of a flexible instead of a fixed medium opens up an area somewhere between kinesthetic and kinetic art in which moving or movable elements are extremely understated (Lippard 1966, 106).

In flexible sculpture, movement is thus arguably implied in a subtle manner that seems to implicate the kinesthetic. Kinesthetic experience relates to the sensations we have of our own living bodies. They involve an embodied knowledge of the effects that physical forces and gravity have upon the type of matter we inhabit from the inside, namely soft matter. But from Lippard's observation the inference can be drawn that soft sculpture also appears to trigger kinesthetic responses, hence, this might in part be what compels the viewer to empathetically attribute to soft sculptures a liveliness of its own.

The coexistence of movement and nonmovement, observed by Lippard, also brings up how soft matter in general seems to exist in between set categories. Soft matter mediates between not just the fixed and the moveable, but also between liquids and solids, the stable and the unstable, the structured and the unstructured. In many respects, soft bodies are characterized by being receptive to forces coming from their outside, and hence they are also tied to process.

In his programmatic text 'Anti-Form,' written in 1968, against the minimalist aesthetic of 'object-type art' which he had previously pursued, Robert Morris famously announced an aesthetics of process that was to become central to postminimalism. For Morris, the painting practices of Jackson Pollock and Morris Louis provided a point of departure for

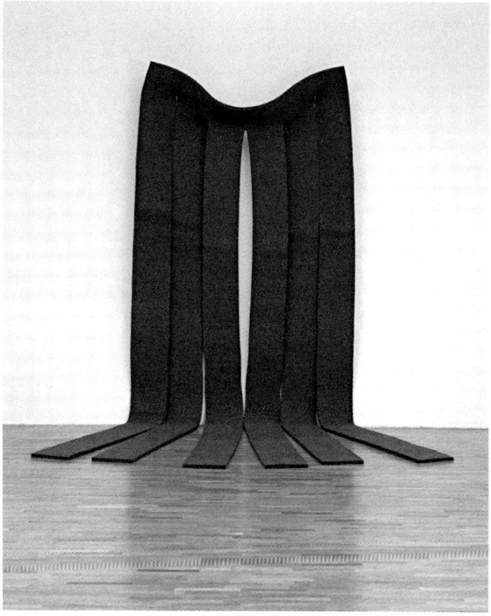

Fig. 1. Robert Morris, *Untitled* (1969). New York, Museum of Modern Art (MoMA). Draped 1' (2.5 cm) thick, gray-green felt. 15' 3/4' × 6' 1/2' × 1' (459.2 × 184.1 × 2.5 cm). The Gilman Foundation Fund. Acc. n.: 378.1975. © 2019. Digital image, The Museum of Modern Art, New York/Scala, Florence. © Robert Morris / VISDA.

rethinking artistic creation. 'The stick that drips paint,' Morris wrote, 'acknowledges the nature of the fluidity of the paint' (Morris 1968, 43). Hence, compared to the brush 'it is in far greater sympathy with matter because it acknowledges the inherent tendencies and properties of that matter' (Ibid.). Against an idealist focus on form, which Morris took to be a 'conservative enterprise' (Ibid. 45), he proclaimed process a 'more direct revelation of matter itself' (Ibid. 44). Hence in his works from the period, Morris explored a balance of flexibility and stiffness in materials that included soft industrial felt.

The credo of Morris and other postminimalists, such as Richard Serra, that sculptural form should be derived from the inherent properties of materials, is related to the notion of 'truth to materials.' Yet in postminimalist practice, this notion was radicalized and became an ambition of bringing matter and form into a condition of mutuality: The forms and the physical relations between parts should be generated from and made possible by the specific physical characteristics of the materials used. This is evident in Serra's bent sheet of vulcanized rubber upheld by forces of friction in the work *To Lift* (1967). But even more elegantly displayed in *Right Angle Prop* (1969) where a bent piece of lead alloy is used to balance another piece of the same material against a wall, something that is made possible by the specific density, friction, and stiffness of the material.

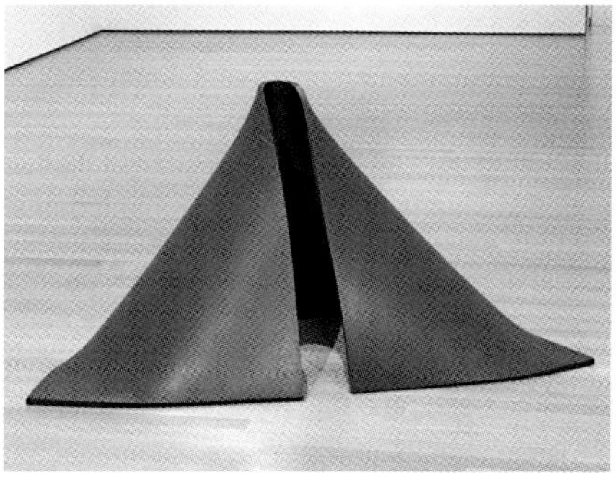

Fig. 2. Richard Serra, *To Lift* (1967). New York, Museum of Modern Art (MoMA). Vulcanized rubber, 36" × 6' 8" × 60" (91.4 × 200 × 152.4 cm).
Gift of the artist. Acc. n.: 458.2007. © 2019. Digital image, The Museum of Modern Art, New York/ Scala, Florence. © Richard Serra / VISDA.

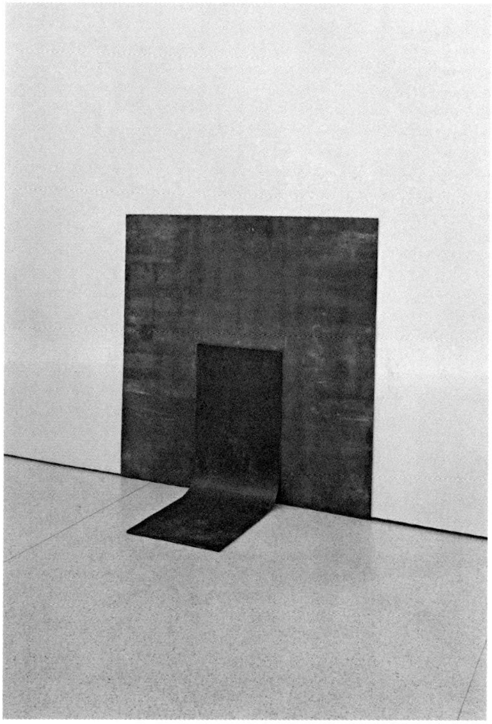

Fig. 3. Richard Serra, *Right Angle Prop* (1969). New York, Solomon R. Guggenheim Museum. Lead antimony. 72 × 72 × 34 inches (182.9 × 182.9 × 86.4 cm).
Purchased with funds contributed by The Theodoron Foundation, 1969 © 2019. The Solomon R. Guggenheim Foundation/Art Resource, NY/ Scala, Florence © Richard Serra / VISDA.

Entropy and Life

In Morris' and Serra's works, the force of terrestrial gravity enters the artwork to become a co-constitutive element of sculptural composition. In this way, an indispensable link between the artwork and its surroundings is established that echoes the subjection to gravity that all earthly organisms must endure. In other postminimalist works, a physical connection between organism and environment is also thematized, but in a different way that can be more directly brought to bear on a physical processual view of life.

Fig. 4. Eva Hesse, *Tori* (1969). Philadelphia Museum of Art. Polyester, resin, and fiberglass on wire mesh. Largest of nine units 119.4 × 43.2 × 38.1 cm. Purchased with funds contributed by Mr. and Mrs. Leonard Korman, Mr. and Mrs. Keith Sachs, Marion Boulton Stroud, Mr. and Mrs. Bayard T. Storey, and with other various funds, 1990, 1990, 121-1-9, © The Estate of Eva Hesse. Courtesy Hauser & Wirth. (Plate 35, p. 324)

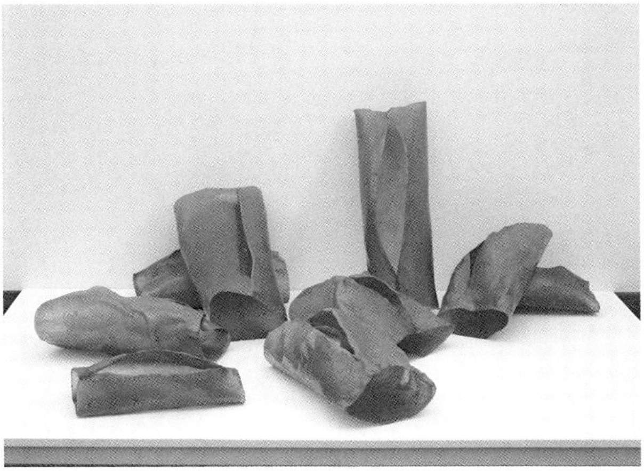

From a physical perspective, acting as a catalyst of process can be interpreted as letting entropy lead. As a concept, entropy originates in thermodynamics, but it also became a frequent reference of late 1960s art discourses and is explicitly mentioned by Morris in 'Anti-Form.' Defined technically in statistical physics as related to the number of microstates of a given system, and thus an expression of its number of degrees of freedom, entropy is often interpreted as a measure of disorder, even if such an interpretation does not always hold (Piazza 2011, 80).[4] Entropy is relevant in relation to the life-like because life seems to evade the second law of thermodynamics, which implies that all physical processes yield a net increase in entropy. Given the physical description of life as being, at its core, anti-entropic, it might seem paradoxical that postminimalist and Anti-Form works with soft materials that are based on letting physical processes run their due course, and usually appear somewhat disorganized or haphazard, are frequently described as possessing qualities shared with living organisms. However, Erwin Schrödinger (2012[1944]), and before him Ludwig Boltzmann (1974[1875]), have pointed out that living organisms only evade the decay to thermodynamical equilibrium by continually exporting entropy to their surroundings. Within the physical description, the locus of life is therefore not just the living organism itself. Instead, life is seen to inhere in the relation that the organism continually sustains with its environment. The living organism is considered an *open system* (in the physical sense) that exists in a state of continual exchange of entropy, energy, and matter with its environment. Interestingly, this ecological perspective on life also runs as a latent current through the associations conjured up by postminimalist soft sculptures by Eva Hesse and Lynda Benglis.

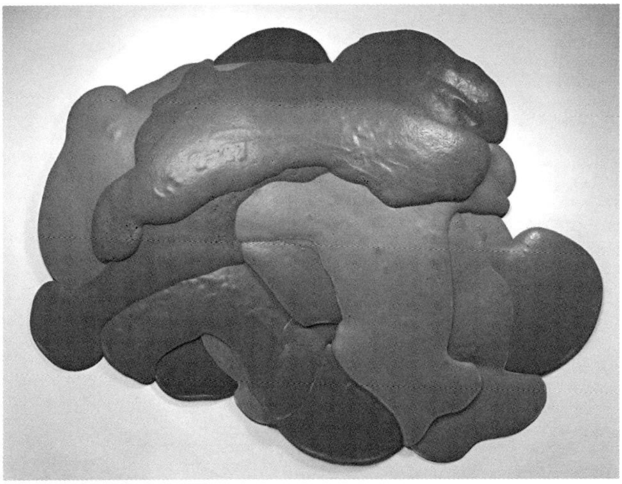

Fig. 5. Lynda Benglis, *NIGHT SHERBET A* (1968). Dayglo pigment, phosphorescence and poured polyurethane foam. 5 × 48 × 60 1/2 inches (12.7 × 121.9 × 153.7 cm). © Lynda Benglis / DACS, London / VAGA, New York / VISDA, Copenhagen. Courtesy Pace Gallery and Thomas Dane Gallery. (Plate 36, p. 324)

In the text on Hesse's work *Tori* on Philadelphia Museum's webpage, for instance, it is stated that Hesse 'imbued a subtle eroticism and a sense of human presence in her sculptures' (Philadelphia Museum 2018). The nine forms that make up the work are, moreover, described as 'organic and rigid, fleshy and repellent, corporeal and otherworldly' (Ibid). And the organization of the piece is referred to as a 'casual arrangement' that looks 'more like something discovered by chance than deliberately set in place' (Ibid).

In the description, the nine sculptural elements are thus likened to a group of organisms or parts thereof (being 'fleshy,' 'corporeal'). But they also take on the role of a kind of elements in the environment that appear as if organized, perhaps for some purpose or from some previous activity, and hence their organization also comes to index the past presence of one or more living creatures. It is thus both the frailness of the parts and the balanced disorder with which they are arranged that imbue the work with a life-like quality. The elements in *Tori* also resemble the shed skin of a snake. Within the reception, Lynda Benglis' polyurethane *pours* are frequently associated with a comparable discarding of bodily matter and bodily fluids that are imagined to leave the body to be cast off to the environment (Chadwick 1996).

In such readings, the works come to reflect a process of *abjection*, described by Julia Kristeva as the expulsion of bodily matter, a shedding of 'the me that is not me' (Kristeva 1982). When read through this prism, Hesse and Benglis' works allude to the fact that entropy does not just vanish in organisms but is exported to the environment through a transfer of matter and energy. They acknowledge that life is not simply anti-entropic, but rather a modulation of entropic flows that envelop both the organism and the environment. Matter must continually enter the body and leave it again in a more entropic state in order for the body to maintain its structured organization. And we can take Kristeva literally when she writes that 'These body fluids, this defilement, this shit are what life withstands (...) Such wastes drop so that I might live' (Kristeva 1982, 3).

In line with the above reading, author Darian Leader has noted a tension between a kind of unruly matter that constitutes the living body and this body's highly organized character as a general theme within Hesse's work. As Leader writes,

> When we look at Hesse's three-dimensional work, we are looking at ourselves (...) Hesse shows us that the body itself is a tension between imposed, regulating structures and a substance that is never entirely subsumed by them (Leader 2002).

To summarize and conclude, the critical reception illustrates how soft sculptures by Morris, Serra, Hesse, and Benglis can come to appear life-like, in the sense that they may be read as analogies that parse different aspects of the relation through which a living organism is inscribed into its physical environment. Although they are explicitly preoccupied with process, Morris' and Serra's works predominantly channel this relation via a focus on physical forces, notably gravity, and their effects on soft materials, and analogously soft bodies, as well as the potential of such forces to generate specific inherent organizations of different kinds of soft matter. Hesse's and Benglis' work is instead associated with a transformation of biological matter, and processes wherein matter is passed between organism and environment. Albeit rooted in an analogy between artwork and organism, the perception of soft sculpture as life-like evoked by these works differs from the Romanticist notion of the artwork as a *quasi-organism*. For within that analogy, it was the holistic unity of all the parts of the artwork, and the resultant 'self-organization,' that was essential and proclaimed a source of the artwork's beauty (Gorodeisky 2016). Postminimalist soft sculpture, in contrast, propels *an entropic aesthetics of the life-like*. Within it, life is portrayed as processual and as intimately tied to its physical surroundings. Organisms and bodies are rendered permeable to outside forces and enrolled in material flows and transformations, within a dynamism of forces of chaos and order. Following Rosalind Krauss' writings on *the formless*, this aesthetic can be described as entailing a blurring of the boundary between organism and milieu, or in art historical terms—a dissolution of the figure/ground relationship (Krauss 2000, 75).

Within a number of recent soft robotic artworks, the theme of physical intertwinement between organism and environment also recurs. Yet in these works the viewer-become-user is confronted not just with a soft object or a sculpture, but surroundings that have themselves become life-like. In the following sections, two such works will be analyzed.

BRALL

The Breathing Wall (BRALL) (2015) is an interdisciplinary project by artist Ece Polen Budak, engineers Onur Zirhli and Ozge Akbulut, and soft roboticist Adam A. Stokes. The installation consists of nine silicone foam tiles mounted side by side on an upright panel. The tiles have different organic surface structures and a porous sponge-like surface—a few are endowed with a layered appearance reminiscent of Benglis' foam pours, while others are characterized by saggy folds. Each tile possesses its own air compartment and has a separate sensor implemented underneath. The structure as a whole 'responds to touch by modulating its breathing' (Budak et al. 2016), and when one of the silicone tiles is touched, the tile will start to inflate and a playback of recorded sounds of human breathing is triggered.

In their paper on *BRALL*, Budak and her collaborators state that the work 'investigates the tactile possibilities of human interaction with synthetic biomorphic surfaces' (Budak et al. 2016, 162). And they claim that by way of its responsiveness to touch and its 'breathing,' the work 'engages the viewer in a similar fashion to that of a living organism' (Ibid. 163). They believe this 'foster[s] greater responses on the part of the user,' so that 'the interaction [with the installation] will be closer to that of organic life' (Ibid.). And they further quote research to support the notion that the presence of biomorphic shapes in an environment 'can serve to enrich human emotional experience' (Ibid. 162).

Like the soft sculptures discussed earlier, *BRALL* evokes an entropic aesthetics of life, both by implying exposure to the forces of the surroundings and by way of the installation's use of breathing. From an art historical perspective, the gesture contained in *BRALL* of rendering a wall soft can be seen to reiterate Oldenburg's practice wherein rigid objects,

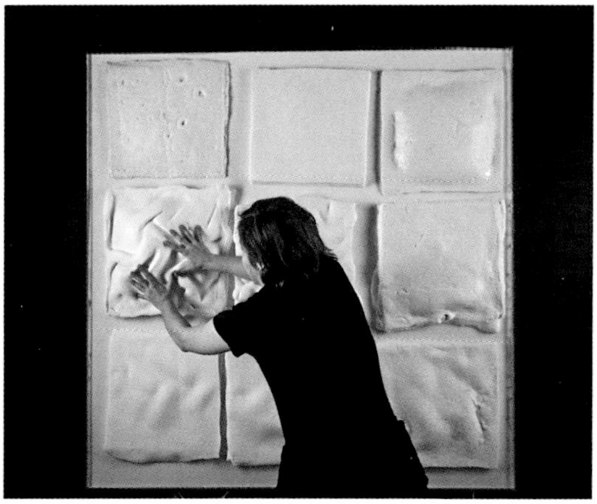

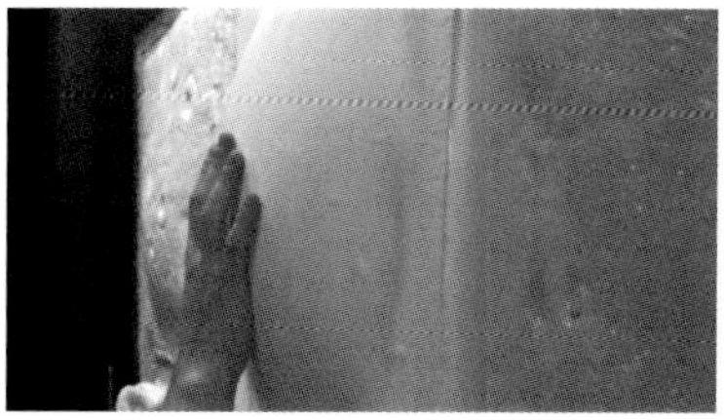

Fig. 6. Ece Budak and Ozge Akbulut, *BRALL* (2015). Silicone on polycarbonate panel, 145 cm × 145 cm. © Ece Budak and Ozge Akbulut. Photo and video still © Baris Dervent, Murat Ugurlu. Courtesy of the artists. (Plate 37, p. 325)

such as a fan or a typewriter, were replicated in pliable materials. The resulting soft versions of these objects would often collapse under their own weight, and hence came to appear blatantly nonfunctional, and softification thus became a means of effectuating a defamiliarization or estrangement from otherwise well-known mass-produced commodities and technologies with a fixed function. In relation to a wall, the operation of rendering soft perhaps even more directly negates the intended functionality. That is, it effectively cancels out the wall's traditional architectural functions of supporting the roof of a building and of shielding against the outdoors. If a wall is soft, instead of a clear separation between the domestic interior and the outdoors, one gets a correspondence and connection. The wall becomes permeable and the human dwellers become exposed to the forces of nature.

The simulated breathing used in the work is a recurrent feature of artworks with soft robots.[5] Breathing is the definitive sign (or index) of life, 'life depends on breathing,' as Despina Kakoudaki puts it (Kakoudaki 2014, 93). However, breathing is equally one of the basic mechanisms through which it becomes evident that we are never just ourselves, alone, in isolation from our surroundings. It is an occurrence that transgresses on the borders between the inside and the outside of the body and connects the organism with its environment in an intimate manner. The body is literally reproduced differently with each breath it takes, as oxygen molecules from the atmospheric air enter the lungs and from there the bloodstream, and as entropy is exported via respiration. The way the breathing mechanism is accomplished in *BRALL*, where exhalation is achieved by passive elastic contraction of the tiles, is also in itself an entropic mechanism. For elasticity, the property that in general seems to endow soft elastomers with a life of their own, has an entropic origin. The inherent tendency of a rubber band to return from a stretched state is predicted by the second law of thermodynamics within statistical physics, as the contraction maximizes the number of possible configurations of the polymer chains of which it is composed, thereby increasing entropy (Piazza 2011, 83).

Synthetic Seduction

Synthetic Seduction is a collaborative exhibition by Stine Deja and Marie Munk that consists of individual works by the two artists assembled as a joint installation.[6] The installation is described as 'a futuristic laboratory setting, simulating the proverbial hospital room' (SixtyEight Art Institute n.d.). Within this setting, two video works by Deja and four interactive sculptures by Munk are on display. The largest sculpture, titled *Skin-to-skin* (2018), consists of a soft slouching circular base on which smaller oblong parts are also lying. Some parts are attached to the base with umbilical cord-like tethers and the different parts all share an uncannily illusionistic appearance that makes their surfaces resemble those of

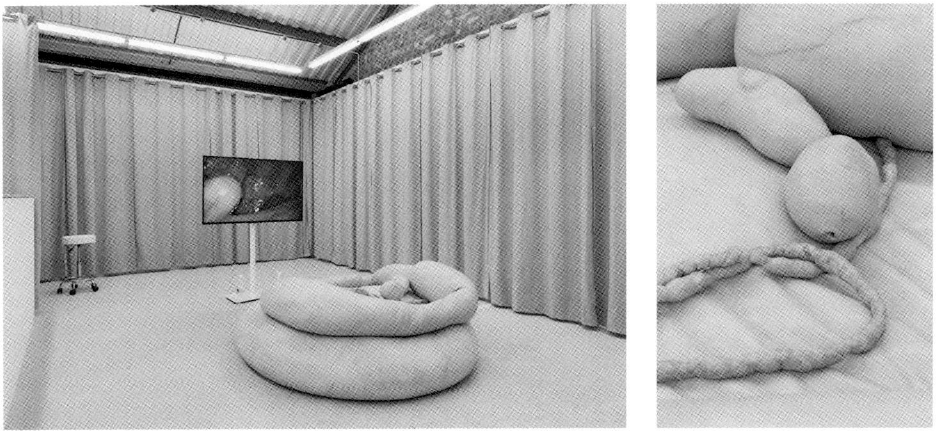

Fig. 7. Stine Deja and Marie Munk, *Synthetic Seduction* (2018). Installation view, Annka Kulty's Gallery, London.
© The artists. Courtesy of the artists and Annka Kulty's Gallery. (Plate 38, p. 326).

human bodies—bluish vein-like protrusions are visible under the skin-like surfaces as well as belly button-esque crevices.

The sculptural pieces are constructed from silicone stuffed with a soft foam material that makes them yield to the touch. And *Skin-to-skin* is positioned so as to function as a recliner to sit on or lie on while wearing headphones and watching Deja's video work *The Intimacy Package* (2018) on a large screen.[7] Two of the smaller sculptures produce heartbeat-like sounds when compressed, and some of *Skin-to-skin*'s tethered elements also have moving parts below their 'skins' that are reminiscent of the kind of rigid wheels mounted on rotating spokes found in massage chairs (Marie Munk, email to author, November 14, 2018).

Despite possessing visual and haptic qualities that call to mind the soft, moist, and wet character of biological life, *Synthetic Seduction* perhaps less obviously aligns with the thematic of physical exchange between organism and environment that connects *BRALL* with the earlier works of soft sculpture. The installation thematizes artificial corporeality and softness as a means of achieving physical (artificial) intimacy (SixtyEight Art Institute n.d.). Through allusions to the mother-infant relationship (umbilical cords, skin to skin contact, the breast-like appearances) notions of physical dependence or symbiosis with an enveloping artificial (m)other are evoked. The relationship between infant and mother is a symbiotic relationship between two organisms, but it is not symmetrical—in terms of survival, the child obviously needs the mother more than she needs it. In a sense, the mother can thus be said to constitute the child's immediate physical surroundings that sustain it. However, within the installation as a whole, the mother-like interactive sculpture *Skin-to-*

skin attains a strange kind of presence. On the one hand, it is a center and the anchor piece, and its appearance, as well as its movement and sound, arguably draws attention to it and encourages active physical exploration on the part of the viewer. Yet when it is used to sit or lie on in order to watch the video work, it is relegated to having only a kind of ambient presence and to exist at the periphery of perceptual awareness, as a piece of furniture. *Skin-to-skin* thus oscillates between encountering the viewer via what Don Ihde terms an *alterity relation* and a *background relation* (Ihde 1990). Within the former the sculpture attains a subject-like status as a *quasi-other*, whereas in the latter it functions as a context for human existence, but is not at the center of experience itself. The soft settings envelop the viewer as she lies down and then become less noticeable as she is absorbed by the video, thus evoking an experience of having merged with an artificial (m)other.

Entropic Aesthetics and Twenty-First-Century Soft Robotic Media Art

In the previous two sections I have sought to articulate in which respects two more recent soft robotic artworks may be said to extend the entropic aesthetics of life latent in post-minimalist soft sculpture. Yet, in *BRALL* and *Synthetic Seduction*, we can also discern a reformulation of the entropic aesthetic. To unpack it, we might start from the theorization of entropy that occurs within the writings of Robert Smithson, who extends the entropy concept from the physical domain to also encompass aesthetic and cultural phenomena. In Smithson's writings, which were closely tied to his artistic practice, the interpretation of entropy as 'sameness' dominates (Smithson 1996). This is a meaning that originates in thermodynamics, where the second law predicts that when two closed systems are put into contact they will gradually attain the same temperature. But also that this phenomenon will eventually manifest on a much vaster scale, resulting in the so-called *heath death*, a condition of maximum entropy wherein the Universe will have attained an even temperature and all macroscopic movement ceases to exist. Smithson reinterprets and applies this gloomy teleology on a cultural scale with a critical bent, as entropy becomes tied to a diagnosis of cultural and aesthetic uniformity but also to an affirmation of facticity within contemporary minimalist art that he sees as taking part in developing a new kind of monumentality.

But where can this entropic sameness be located today? Given that soft robotics can be conceived of as an intermingling of computing, sensing, and actuation technologies with soft matter, it seems logical to start by looking at the physical computing technologies that the soft sculptural medium is augmented with as it transforms into soft robotic art. From a contemporary perspective, Smithson's notion of entropy, as descriptive of a qualitative sameness that pertains to facets of cultural practices, art, and the built environment, is also obviously missing a crucial component, namely an account of the effects that the proliferation and implementation of computational technologies on a planetary scale over the past couple of decades have had. The contemporary lifeworld is permeated by and saturated with computational media, software programs, algorithms, as well as networks of sensors and actuators assembled as a still expanding *internet of things*. Terms such as *pervasive computing* and *ubiquitous computing* (*ubicomp*) are often used to describe this phenomenon (Ekman et al. 2015). A number of theorists have taken note of how this development has rendered computational processes ambient and atmospheric, and consequently established a dispersed environmental computational agency. Mark B.N. Hansen, for instance, argues that

> Through the distribution of computation into the environment (…) space becomes animated with some agency of its own (Hansen 2012, 33).

And that this sets new requirements for adequately thinking through the relationship between humans and technology. As he puts it,

> We must reconceptualize the coupling of human and technics beyond the figure of the 'technical object' (Ibid. 51).

The animation of space of which Hansen speaks can arguably be seen to have endowed the environment as a whole with life-like qualities, in the sense that the surroundings have attained what appear to us as both intelligences and autonomous agencies of their own. That is, within experience, the physical environment may be attributed abilities and a mode of being that were traditionally reserved for living beings within the modern scientific worldview. And, as Hansen argues, the fact that the forms of intelligence and agency the surroundings manifest can no longer be traced back to a single origin (a subject or an object), but are dispersed and networked in character, makes them troubling and hard to adequately conceptualize and come to terms with.

BRALL and *Synthetic Seduction* are emblematic of this contemporary condition of ubiquitous computing, I argue, wherein architectural elements (a wall) and domestic settings (a couch) via technological augmentation come to attain phenomenal qualities that were previously the preserve of living beings, and as a consequence start to appear semi-living. In these works we are presented with sessile technical entities that are embedded into an environment, yet resemble living bodies visually, haptically, and auditorily. The living figure has merged with the ground, one might say, to extend Krauss' phrasing. The works thus take part in a dramatization of the sameness and indistinction between humans, technology, and environment that ubicomp has brought about, i.e. the interlacing of human existence with a set of intelligent and agential technologies that have begun to spill out into all parts of the human habitat. In doing so, the artworks attain an uncanny quality in the sense that they seem to reactivate a suppressed knowledge of the increasingly central role played by technology and technological agencies within everyday life. In a hyperbolic manner, they appear to give a missing body to ubiquitous computational processes, intelligences, and agencies that, despite the recent focus on materiality and infrastructure within media theory, often tend to be conceived as abstract and dematerialized. Compared with the sophistication of the machine-learning algorithms that have intertwined with contemporary everyday life, they appear strangely analogue and anachronistic, donning an almost vulgar corporeality. Their artificial soft bodies appear crude, perhaps even unintelligent, as they are reduced to only displaying basic bodily functions similar to those controlled by the autonomic nervous system (breathing and intra-bodily motion). It is tempting to think of the works as reversals of Mark Weiser's original vision of ubiquitous computing as the becoming invisible of computing technologies (Dourish 2004; Weiser 1991) and the subsequent notion of unobtrusive *calm technology* he took part in developing (Brown and Weiser 1997). Yet *DRALL* and *Synthetic Seduction* derive their aesthetic effects from a calming and casualizing of human interactions with strange embodied technological entities that at first sight might appear repulsive, via switching back and forth between enrolling them and the viewer in background and alterity relations.

Onwards from Soft Sculpture to Soft Robotic Art

This essay has addressed the question of how softness can be seen to afford impressions of the life-like within postminimalist soft sculpture and contemporary soft robotic art, and what conceptions of soft life these two strands of works activate. I have proposed that the concept of entropy can serve to elucidate aspects of this phenomenon as it unfolds within the critical reception of postminimalist soft sculpture. Extending this analysis, I argued that with a reworking and expansion of the entropy concept that emphasizes entropy as sameness, it also has analytical purchase in analyzing more recent artworks wherein soft robotic life-like entities merge with their surroundings.

In conclusion, and in order to return to the broader interest stipulated at the outset about a possible connection between softness and the process of life, the question I raise is: What might soft sculpture and soft robotic art add to soft robotics research that explicitly seeks to replicate soft bodies of living organisms in a technical medium? By posing this question, I do not mean to insinuate that art must measure up to purposefulness as a yardstick, nor that the value of art lies in its potential as a driver of innovations that eventually become useful. Instead, the question is posed from a commitment to transdisciplinarity in acknowledgement of the historicity and contingency of the modern distinction between art and science/technology (Guattari 1995) and the onto-epistemic potential held by artistic practices.

The obvious and straightforward art historical answer to this question is the traditional avant-garde reply wherein the potential of art is postulated to lie in its criticality and ability to subversively scrutinize cultural tendencies and positions. And this answer also resonates with the contextualizing reading I proposed of *BRALL* and *Synthetic Seduction*. Soft robotic artworks, such as *BRALL* and *Synthetic Seduction*, can indeed be said to add vital reflections to more instrumentalist technical research agendas by probing the ramifications of contemporary technologies on human and nonhuman life. But also by going further and introducing a *naturecultural* (Haraway 2003) perspective by drawing attention to the fact that the biophysical process of life can no longer be imagined to unfold within a 'natural' environment, but is set to evolve within augmented technical settings and on a planet whose state and constitution is tied to human culture and activities.

However, other answers to the question are also possible. In retrospect, the postminimalist aspiration of attaining reciprocity between form and material, and the associated intensive probing of how physical properties can engender specific structural organizations and behaviors that are inherent to soft matter, can be seen to anticipate a central conceptual pillar of technical soft robotics research. I am referring to the idea that the implementation of pliable materials can serve to greatly simplify a robot design or make a robot perform some tasks more easily or better (Laschi and Cianchetti 2014). That is, specific material properties can contribute to attaining specific functionalities. This principle, roboticists claim, is already implemented in the soft bodies of natural organisms whose designs have been optimized by the process of natural evolution. The notion of life as being processual and entwined with the physical environment, contained in an entropic aesthetics of life, is also central to research on biodegradable robots occurring at the fringes of soft robotics. Here, a more ecological view of robots than that which is generally the order of the day in robotics prevails, and robots are envisioned to have life-cycles and sustain themselves through exchanges with specific physical environments (Rossiter et al. 2016). Hence, actuators and other parts are constructed from biodegradable materials such as gelatin but also latex, a material that was also favored by Eva Hesse, who was well aware of its eventual degradation over time. However, such overlaps and shared interests between art and soft robotics research seem to have almost vanished within contemporary media art that uses soft robotics technology. This strand of artworks instead tends to work in the vein of the historical tradition of robotic art.[8] That is, the main impetus is usually an interest in an aesthetic of robotic behavior, conceived of as embodied actions and interactions that unfold on a humanly perceptible time scale. This allegiance serves to extend the interests of the tradition of robotic art found within the broader field of media art. But it would be interesting, I posit, to see artists treating soft robotics less as a fixed technology that is an extension of traditional robotics, and more as a medium or affordance in its own right, with singular capacities and modes of composition that must be invented, and continually reinvented, physically and aesthetically from scratch. Perhaps postminimalist soft sculpture can provide some inspiration for undertaking this work.

Acknowledgements

The author is grateful for funding received to cover image rights cost from the Embodied Systems for Robotics and Learning section at the University of Southern Denmark.

Notes

1. A few kinetic soft sculptures from the period exist. These include Keith Sonnier's inflatables (Lippard 1966; The National Exemplar 2018) and Oldenburg's *Ice Bag -- Scale B, Yellow* (1971).
2. I have chosen to use the term 'life-like' as it is less anthropocentric than the discourses on 'anthropomorphization.' It is also preferable to 'animated' as this term usually refers to a qualitative aspect (something that is endowed with life or the qualities of life). The term 'life-like' is better suited for the transdisciplinary interest of this essay, as it is not tied to a qualitative epistemology, but is also compatible with a physical description of life.
3. It should be noted here that the reading of postminimalist soft sculpture that this focus and approach produces is, of course, highly selective and largely neglects the historical and political contexts within which these works were produced.
4. Entropy is a concept that arguably has many different meanings. It has gone through considerable reworking and has been redefined on its travel from thermodynamics to statistical physics, information theory, and beyond. Entropy has previously been applied to art and aesthetics on a number of occasions (see Jones 1973). Gestalt psychologist Rudolf Arnheim's *Entropy and Art. An Essay on Disorder and Order* (2010[1971]) and artist Robert Smithson's writings are probably the most famous instances of this (Smithson 1996). The use of entropy within this essay differs from how the term is used by Arnheim and a number of other theorists' where it serves as a theoretical basis for developing a naturalizing aesthetics rooted in natural science (Rapkine 1970[1945]; Mandelbrojt 1970; Monod 1970). Within these accounts, entropy is associated with disorder and figures as an opposite of order and information that is privileged by way of a classicizing aesthetics. Compared to such theorizing, the scope of this essay is narrower. I do not intend to apply entropy to aesthetics or aesthetic judgement in general, but instead I use the concept as a prism through which to articulate the life-like aspects of soft sculpture and soft robotic artworks. Within this usage, order recurs as a characteristic of life and living bodies, but it is

the dynamic physical imbrication of organism and environment, announced by the negentropic physical description of life, that I take as my leitmotif.

5. See Jørgensen (2017). Breathing automatons were also popular in the late 18th century. Jacques Vaucanson's android *Flute-player* (1738) is an early example that was followed by other famous breathing automatons such as the Jaquet-Droz family's *Lady-musician* (1774) and *Draughtsman* (1772–74) (Riskin 2003).
6. The exhibition has been presented in different formats, first at Annka Kulty's Gallery, London (February 21- March 24, 2018), then at SixtyEight Art Institute, Copenhagen (June 8 – August 4, 2018), and subsequently also at other venues. In my analysis, I refer to the exhibition at SixtyEight Art Institute, which I visited.
7. I was told by gallery staff of this intended function while visiting the exhibition .
8. See Kac (1997), Wilson (2003), Whitelaw (2004), Penny (2013), and Shanken (2014).

References

Arnheim, Rudolf. 2010[1971]. *Entropy and Art. An Essay on Disorder and Order*. First Edition, 40th Anniversary edition. Berkeley CA: University of California Press.

Babington, Brooke, and Emilie Owens. 2009. "Claes OLDENBURG." Soft Sculpture. https://nga.gov.au/Exhibition/softsculpture/Default.cfm?IRN=37808&BioArtistIRN=8947&MnuID=3&ArtistIRN=8947&ViewID=2.

Blessing, Jennifer. n.d. "Robert Morris Untitled (Pink Felt)." Guggenheim. https://www.guggenheim.org/artwork/3016.

Boltzmann, Ludwig. 1974 [1875]. *The second law of thermodynamics (Theoretical physics and philosophical problems)*. New York: Springer-Verlag.

Brown, John Seely, and Mark Weiser. 1997. "The Coming Age of Calm Technology." In *Beyond Calculation: The Next Fifty Years of Computing*, edited by Peter J. Denning, Robert M. Metcalfe, and J. Burke. New York: Copernicus.

Budak, Ece Polen, Onur Zirhli, Adam A. Stokes, and Ozge Akbulut. 2016. "The Breathing Wall (BRALL)—Triggering Life (in)Animate Surfaces". *Leonardo* 49(2): 162–63.

Chadwick, Whitney. 1996. "Balancing Acts: Reflections on Postminimalism and Gender in the 1970s." In *More Than Minimal Feminism and Abstraction in the '70s*, edited by Susan L. Stoops. Waltham, MA: Rose Art Museum, Brandeis University.

Dourish, Paul. 2004. *Where the Action Is*. Cambridge MA: MIT Press.

Edelman, Robert G. 2004. "In Pursuit of the Organic." http://www.artnet.com/Magazine/reviews/edelman/edelman4-12-04.asp.

Ekman, Ulrik, Jay David Bolter, Lily Díaz, Morten Søndergaard, and Maria Engberg (eds.). 2015. *Ubiquitous Computing, Complexity and Culture*. New York: Routledge.

Gorodeisky, Keren. 2016. "19th-Century Romantic Aesthetics". In *The Stanford Encyclopedia of Philosophy* (Fall 2016 Edition) edited by Edward N. Zalta. https://plato.stanford.edu/archives/fall2016/entries/aesthetics-19th-romantic/.

Guattari, Félix. 1995. *Chaosmosis: An Ethico-Aesthetic Paradigm*. Bloomington IN: Indiana University Press.

Hansen, Mark B.N. 2012. "Engineering Pre-Individual Potentiality: Technics, Transindividuation, and 21st-Century Media". *SubStance* 41(3): 32–59.

Haraway, Donna J. 2003. *Companion Species Manifesto: Dogs, People, and Significant Otherness*. Chicago IL: Prickly Paradigm Press.

Ihde, Don. 1990. *Technology and the Lifeworld: From Garden to Earth*. Bloomington IN: Indiana University Press.

Jones, Peter Lloyd. 1973. "Some Thoughts on Rudolf Arnheim's Book 'Entropy and Art.'" *Leonardo* 6(1): 29–35. https://doi.org/10.2307/1572422.

Jørgensen, Jonas. 2017. "Prolegomena for a Transdisciplinary Investigation into the Materialities of Soft Systems." In *ISEA 2017 Manizales: Bio-Creation and Peace: Proceedings of the 23rd International Symposium on Electronic Art*. Caldas: University of Caldas.

Kac, Eduardo. 1997. "Foundation and Development of Robotic Art." *Art Journal* 56(3): 60–67.

Kakoudaki, Despina. 2014. *Anatomy of a Robot: Literature, Cinema, and the Cultural Work of Artificial People*. New Brunswick NJ and London: Rutgers University Press.

Krauss, Rosalind E. 2000. "Entropy." In *Formless: A User's Guide* edited by Yve-Alain Bois and Rosalind E. Krauss, 169–234. New York: Zone Books.

Kristeva, Julia. 1982. *Powers of Horror. An Essay on Abjection*. Translated by Leon S. Roudiez. New York: Columbia University Press.

Laschi, Cecilia, and Matteo Cianchetti. 2014. "Soft Robotics: New Perspectives for Robot Bodyware and Control". *Bionics and Biomimetics 2*: 3. https://doi.org/10.3389/fbioe.2014.00003.

Leader, Darian. 2002. "Eva Hesse, 'A Quantity for Myself'". *Tate Magazine* 2: 78–81. https://www.darianleader.com/art/eva-hesse-a-quantity-for-myself/.

Lippard, Lucy R. 1966. "Eccentric Abstraction." In *Changing Essays In Art Criticism*, edited by Lucy R. Lippard, 98–111. New York: E. P. Dutton.

Mandelbrojt, Jacques. 1970. "Comments on Rapkine's Notes and Monod's Commentary." *Leonardo* 3(3): 354–354. https://doi.org/10.2307/1572346.

MoMALearning. n.d. "Robert Morris. Untitled." https://www.moma.org/learn/moma_learning/robert-morris-untitled-1969/.

Monod, Jacques. 1970. "Commentary on the Aesthetics of Louis Rapkine." *Leonardo* 3(3): 353–53. https://doi.org/10.2307/1572345.

Morris, Robert (1966). "Anti-Form." In *Continuous Project Altered Daily: The Writings of Robert Morris*, edited by Robert Morris, 41–47. Cambridge MA and New York: MIT Press.

Penny, Simon. 2013. "Art and Robotics: Sixty Years of Situated Machines." *AI & SOCIETY* 28 (2): 147–56. https://doi.org/10.1007/s00146-012-0404-4.

Philadelphia Museum (2018). https://www.philamuseum.org/collections/permanent/85791.html.

Piazza, Roberto. 2011. *Soft Matter: The Stuff That Dreams Are Made Of.* Dordrecht and New York: Springer.

Rapkine, Louis. 1970 [1945]. "Notes for a Scientific Theory of Aesthetics." *Leonardo* 3(3): 351–52. https://doi.org/10.2307/1572344.

Riskin, Jessica. 2003. "Eighteenth-Century Wetware." *Representations* 83(1): 97–125. https://doi.org/10.1525/rep.2003.83.1.97.

Rossiter, J., J. Winfield, and I. Ieropoulos. (2016). "Here today, gone tomorrow: biodegradable soft robots." In *Electroactive Polymer Actuators and Devices (EAPAD) 2016. [97981S] (Proceedings of SPIE; Vol. 9798)*. Bellingham: SPIE. https://doi.org/10.1117/12.2220611.

Schrödinger, Erwin. 2012 [1944]. *What Is Life? With Mind and Matter and Autobiographical Sketches.* Reprint edition. Cambridge and New York: Cambridge University Press.

Shanken, Edward A. 2014. *Art and Electronic Media*. Reprint edition. London and New York: Phaidon Press.

SixtyEight Art Institute. n.d. "Synthetic Seduction." http://www.sixtyeight.dk/synthetic.html.

Smithson, Robert. 1996. *Robert Smithson: The Collected Writings*. Edited by Jack Flam. Berkeley CA: University of California Press.

Spector, Nancy. n.d. "Richard Serra. Belts." Guggenheim. https://www.guggenheim.org/artwork/3891.

The National Exemplar (2018). "KEITH SONNIER. 'Inflated Works (1965–1966)' April 4-May 20." http://thenationalexemplar.squarespace.com/sonnier-inflated/.

Weiser, Mark. 1991. "The computer for the 21st century." In *Scientific American*, Sept. 1991: 94–104.

Whitelaw, Mitchell. 2006. *Metacreation: Art and Artificial Life*. Cambridge MA: MIT Press.

Wilson, Stephen. 2003. *Information Arts: Intersections of Art, Science, and Technology*. Cambridge MA: MIT Press.

Natures Mortes, Live Data
The Art Object in the IoT Era

Tsila Hassine, Olga Kisseleva

This chapter follows an exhibition project that is part research, part speculation, on the future of the art object in the era of the Internet of Things. In what follows we describe not only the exhibition itself, but the processes and sources of inspiration and research that fueled it and contributed to its inception.

Introduction

The story of mankind has been inscribed in its artifacts, long before the invention of writing and, subsequently, history. Our man-made objects have been recording and recounting the stories of culture over the centuries. How can our artefacts connect with us and recite the story of our present? McLuhan divides the evolution of communication and technology into three major periods, each defined by the impact of technology on history and culture. The first era, defined by oral communication, spans from the time humans acquired speech up until the beginning of writing, some 5,000 years later. This was followed by the age of literacy: from the invention of writing, and much later the printing press, up until the discovery of electricity and its subsequent use in long-distance communication with the first telegraph. The telegraph marks the beginning of the electric flow of information (McLuhan 1964). A hundred years after the telegraph, we have experienced several major advancements in electronic communication, the most obvious one being the Internet, and very recently the mobile phone. Nowadays, we're on the verge of another revolution (Rifkin 2015; Schwab, Davis 2016) where we go beyond human communication. We no longer use objects simply as means of communication; we actually start to communicate with the objects themselves.

Man-made objects are both historic documents and aesthetic entities. They are the physical manifestation of the thoughts and ideas of a culture. Every artefact (object which is created, as opposed to object which is found) is an assemblage of materials, tools, knowledge, and machines (Caple 2012). Each man-made object is the convergence of cultural and economic factors as well as material properties. An object can be seen to have three forms of identity: a functional identity (the effect it has on the world), a symbolic identity (the meaning of the object and the role it has in cultural code), and a historical one (how objects encapsulate past experiences and carry ideas from the past to the present) (Caple 2006). For the purposes of this chapter we will focus on the object's symbolic identity.

The symbolic meaning of objects can be found in the way objects have been represented throughout history and across different cultures. Looking at art history, the representation of objects has come a long way since their very early depictions. Still life, one of the principal genres in Western art, is defined as a genre focusing on the depiction of inanimate objects for the sake of their qualities of form, color, texture, and composition (Encyclopaedia Britannica 2018). Still life is found across cultures that are diverse both historically and culturally. However, different periods of still life require different readings. Take for example *Xenia*, the Roman still life paintings found in the ruins of Pompeii, dated to about 500 BC. At a first glance, we see a category of still life resembling that of the later Dutch school: fruits, baskets of flowers, loaves of bread, etc. (Bryson 1990). Yet, despite the similarities in content, paintings of Xenia must be approached with different terms than paintings of the seventeenth century. Produced under different cultural conditions, the seemingly similar subject matter had different meanings for Roman viewers and for seventeenth-century Dutch audiences, and was encoded with different symbolic and semantic charges.

It was during the seventeenth and eighteenth centuries that representation of inanimate objects had become an autonomic genre. In Modern art, still life and the representation of objects were revisited and reinterpreted by the Impressionists, the Cubists, the Surrealists, and Pop artists, to give but a partial list. It is thus clear that the representation of objects has intrigued society across different eras. After all, objects are an inherent part of society, carrying with them the 'thoughts and ideas of a culture' (Bryson 1990, 20).

Only about 100 years after the 'electrification' of our objects (Moore and Simon 1999), and not even 30 years of Internet, our objects are about to join us on the web. In fact, after the World Wide Web and universal mobile accessibility, the Internet of Things represents the most potentially disruptive technological revolution in our lifetime. With 50 to 100 billion things expected to be connected to the Internet by 2020, we are now experiencing a paradigm shift in which everyday objects become smart and interconnected (Feki et al. 2013).

As we enter a new era of objects, old questions resurface and give rise to new ones; How would this new generation of objects represent the thoughts and ideas of the culture that created them? What symbolic and semantic identity will they have? What would be their new role in the society they are bound to transform? How would these objects be represented, or rather, since we are dealing with smart objects, would they represent themselves, and if so how? To answer these questions, we first need to understand this new transformation of objects, by indicating the cultural, social, and economic factors whose convergence allowed the rise of this new paradigm.

The Internet of Things: What Is It and What is it Good For?

The term Internet of Things (henceforth IoT) describes several technologies and research disciplines that enable the Internet to reach out into the real world of physical objects. Existing technologies like RFID (Radio Frequency Id), short-range wireless communications, real-time localization, and sensor networks are becoming increasingly pervasive, making the IoT a reality.

One definition of the IoT is: 'Things having identities and virtual personalities operating in smart spaces using intelligent interfaces to connect and communicate within social, environmental and user contexts' (Internet of Things 2020; RoadMap for the Future 2008). The term itself was originally coined in 1999 by Kevin Ashton, a manager at Procter & Gamble at the time, as the title of a PowerPoint presentation which was meant to attract the attention of business executives. In that presentation, Ashton proposed to integrate RFID chips into the inventory of a certain brand of lipstick which was perpetually out of stock due to holes in the data about that lipstick's supply chain (Cisco 2014). Kevin Ashton was then asked by MIT to start a group—the Auto-ID Center—which would research RFID technology. Since then, it has turned into a global organization.

The IoT, which originated as a solution to everyday problems of everyday objects, is bound to transform the way we think of our objects and relate to them. The advances in this field provide for increasingly active and autonomous artefacts. It will allow our everyday objects to attain agency and become relevant actors in the physical world, and thus influence everyday object-subject experience (Haiyan Jia et al. 2012). This technology will influence the very way we interact with one another and the way we relate to our environment, since these processes are very much determined by the technological resources that we have at hand.

The Current Economic and Social Context of the IoT

To understand some of the implications of connected objects, it is productive to look at the socio-economic conditions that contributed to the rise of the IoT. Although talk about the IoT has been around for a while, the actual materialization of this hyped technology has so far resulted in mostly gimmicky products: for example, a tweeting toaster that lets you roast twitter messages onto the toast's surface, or a connected oven that suggests a number of recipes or ingredients once you turn it on. Is this the disruptive cutting-edge technology that we have been anticipating?

The answer lies, in part, in the power of the network. A tweeting toaster, by itself, is not likely to change much. It becomes interesting only once it participates in a network of toasters that may actually be connected to a network of supermarkets, insurance companies, health providers, and so on. Imagine a toaster which is directly linked to your GP and your health insurer, alerting them every time you exceed your recommended daily intake of carbohydrates. This might result in an immediate increase in your health insurance payments, or even your GP remotely blocking the use of your toaster, or any other such seemingly absurd scenarios.

The concept of objects connecting through networks is not very new (Weiser 1991). For a few decades we have been surrounded by objects with computer chips embedded in them.[1] The evolution from merely embedded objects to objects that can also communicate with each other was only awaiting the right conditions. Moreover, prototypes for connected objects such as printers and coffee machines[2] have been around for dozens of years. Therefore, the inevitable question is why now?

A Convergence of Technological, Economic, and Social Factors

The timing of the emergence of IoT seems to stem from the convergence of several factors; The low cost of sensors and microprocessors; the growing availability of bandwidth (the abundance of communicating objects in each household requires a broad band); the availability of Wi-Fi (all these communicating objects cannot be connected online through wires); the IPv6 protocol (which assigns a specific IP address to each object individually); the cloud technology (which facilitates the online storage of data), and big data technology, to process all these data (Jankowski et al. 2014).

Still, all these technological factors need one important catalyst to make the IoT a household technology, and that is the habituated user. Most of us carry at all times the most effective extension of our central nervous system, to borrow McLuhan's terms: namely, our smartphones. The smartphone, no longer just a phone, is the hub of myriad applications, all designed to achieve two objectives: constant connectivity and a constant information flow. In other words, we have grown addicted to being always connected, and to having all the information in the world at our disposal at all times (Forbes 2012). We are the catalyst, and we are ready to take the next step.

Connected objects are no longer simply objects designed to fulfill specific functions. The IoT transforms objects into services whose real value lies elsewhere (Kaplan 2012). Take the smart mattress, for example: this is a mattress fitted with sleep monitoring sensors (Urban Tastebud 2018), and it comes with its own designated app. Thus, the mattress is transformed into a service, providing 24/7 sleep monitoring and analysis of sleep-related data. The added value of the mattress—the sleep-related data—accumulates on some distant server, to which the user has only restricted access through the app. Another excellent example is the vacuum cleaner. From its humble beginning as an electric carpet sweeper in the nineteenth century, the early twenty-first-century vacuum cleaner has an autonomous vacuum cleaning robot. Recently, with the new addition of a web cam, it has also been turned into a home surveillance device. Thus, our relationship *to* objects is being transformed into a relationship *with* them.

This point is very well illustrated in the work of designer and researcher Simone Rebaudengo. In his video *Addictive Products* (2012), he describes a hypothetical network of seemingly ordinary toasters, connected to each other across London. They use the network to exchange personal experiences as home toasters, such as the position they occupy within their 'host' households and their relationships with their 'hosts' (note that the IoT toaster is no longer 'owned,' but 'hosted' for a duration of time dependent on both the human host and the toaster itself). This mockumentary-style video describes the IoT experience from the toaster's point of view. Rebaudeno augmented the toasters with personalities, needs, fears, goals, and even developed a toaster-adapted language to facilitate interaction with their human hosts.

Still Life and the IOT

Objects and their stories have not always been the focus of attention in the history of painting. In the history of Western art, objects began to occupy a central role relatively late.[3] As noted earlier, these objects, which until the sixteenth century appeared only in the margins of most paintings, moved to the center of the canvas during the sixteenth and seventeenth centuries in the newly formed genre of still life. This shift can be explained by the increasing urbanization of European society, a society that started to turn its attention to its housing, material goods, trade, and knowledge.

Svetlana Alpers (1984) describes how Dutch art embodied a truly visual culture. Dutch visual culture served and energized a system of values in which meanings were not 'read' but 'seen' and new knowledge was visually recorded.[4] For the Dutch, images were the arena in which they represented themselves to themselves. In Holland during that period, images proliferated everywhere, beyond the limitations of what might have been considered 'art' at that time. Images were printed in books, woven into home textiles, and printed onto tiles (Alpers 1984).

Alpers establishes a link between the technological innovations of that period, mainly in the field of optics, and the interest in representing the effect and impact of these new technologies of vision in Dutch painting of that time. She describes both the scientific and artistic interest in light reflections on everyday objects such as glasses and textiles. The dishes, fabrics, and other objects represented in these paintings are taken out of their ordinary context to be placed at the center of attention, analysis, and contemplation. To prove her point, Alpers presents the influence of contemporary seventeenth-century images on the work of Constantijn Huyghens, a seventeenth-century Dutch physicist, mathematician, astronomer, and inventor who made great contributions to the field of optics. In his account of his own scientific work, Huyghens associates images with sight and seeing, focusing on the new knowledge made available through lenses—the latest technology of that period. Seventeenth-century Dutch art provides an excellent example of the way a new technology transforms our relationship with objects, and the way it is represented in the artefacts of that period. Moreover, in seventeenth-century Holland objects had transformed beyond their original function. They acquired a meaning that bound them to the socio-economical context of that period, attesting to the wealth of their owners. The value of those objects lies in the story that they tell about their owner. In other words, their value lies in the data they convey.

Just as the Dutch were fascinated by the new understanding and representation of their daily environment through the new optical technology, so are we fascinated by the new possibilities of interaction with our communicating objects. We imagine new contexts for our objects, placing them at the center of our attention and seeing them in a new light. Furthermore, the IoT shifts the focus with regard to human presence, moving it from center stage to the background. Much as still life is characterized by its 'exclusion of the human form and its seeming assault on the value and prestige of the human subject' (Bryson 1990, 60),[5] IoT makes human presence and actions redundant, as clearly illustrated by autonomous cars. Where once there was a human driver, her presence, or even remote control, are no longer necessary.

Conceiving the Art Object in a Data-Driven Culture

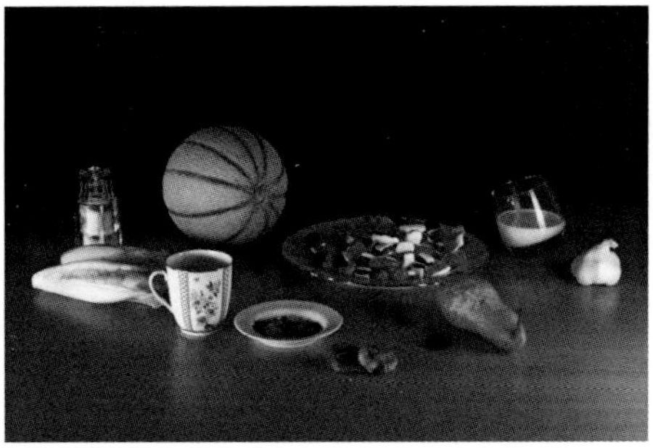

Fig. 1. Olga Kisseleva, *Mathematique* (2010). From the series *Divers Faits.* (Plate 39, p. 327)

In the previous section we described the shift of attention from the human figure to objects, in the visual representations of seventeenth-century Holland. This shift can be seen as part of the changes in the symbolic and functional identity of objects at the time. As demonstrated above, a similar change is underway today. Therefore, it would be interesting to imagine the impact of this change on the representation of objects as well as on human-object and inter-object relationships.

Yet, the changes in the symbolic meaning of the object become even more complex when considering the transformation of the art object itself. Functionality in the twenty-first century is mainly based on accumulated data and reciprocal human-object and inter-object communication (Wired 2014). This makes the IoT both a conceptual and a technological innovation, conceived and designed to accommodate and enhance these communication and data-driven functionalities. At the same time, art objects do not follow these rules, since, by definition, their functionality lies in the realm of the aesthetic. So, how will these new paradigms affect our art objects? How will this novel generation of objects and the subsequent object culture affect the aesthetic function itself? How can we represent, through art objects, the culture of 'data-fied' objects? Could there be such a thing as a 'smart' art object?

These questions are at the heart of our artistic endeavor: to create a network of art objects, based on the art objects themselves. In what follows we will describe the artistic process generated by these questions.

Natures Mortes, Live Data: From Art Objects to Art Networks

Divers Faits Series

The photographic series *Divers Faits* (Kisseleva 2010) (Fig. 1) inspired by seventeenth-century Dutch art, presents a repertoire of visual allegories. Each still life photograph in the series 'conceals' a subplot, telling a story through each item. In addition, each item has its own individual narrative, emerging not only through the visual composition in the image, but also through a dictionary of the symbolic significance of the different items, and a short story accompanying each image which uses references and keywords taken from that dictionary. The still life images of *Divers Faits* can be read through the many contemporary semiotic and emotional layers of meaning specific to the objects present in the composition.

The allegoric compositions in *Divers Faits* constitute networks of symbolic meanings. Therefore, it was only natural to rely on them for encoding the artworks into a larger network. Our purpose was to work with these connections, to translate and represent them in a way that reflects the culture and values of IoT, which are inter-object communication and exchange of data. We wanted to imagine a network of the series' photographs, a network which is an artwork in itself, whose implementation constitutes another work in the exhibition.

Together with an interdisciplinary team of students from the departments of art, design, and engineering at Shenkar College of Art and Design, we began by imagining scenarios of connectivity for these photographs. These scenarios reflect the semiotic, semantic, narrative, and artistic values of each photograph. We sought to create a network that would allow each photograph (or art object) to preserve its own identity and to have its own presence while interacting with other objects in the network. The network we envisioned would not only connect the artworks to each other, but it would also connect the artworks to their viewers. We wanted to make the human spectators an integral part of this network.

Challenges and Solutions

Our pursuit proved to be especially challenging: we were not only developing a new medium for artistic work, but we were working with technology still in its initial phase, its full potential not yet realized. Moreover, we had to imagine scenarios based on a technology which is all about efficiency and functionality, for art objects that are all but functional.

The challenge we faced was multi-faceted. Our aim was to connect a selection of art objects into a network of values inspired by the visual allegories in the photographs. We further wanted to explore the way art objects can be represented as data-generating objects with a participatory role in this particular network, as well as on the Internet.

Turning to the realm of implementation—the moment in the world of software code when abstract structures and algorithms become real—we wanted to figure out how art objects could interact with each other and with viewers and what kind of data they would exchange. This issue comprises two intertwined questions. The first is the question of function or content, namely, what would the objects communicate to each other and to the viewer? The second was the question of form, or language, namely, how would the objects relay this content? These questions were followed by the technical issues of connecting the photographs to a server, and the medium that they would use to represent themselves.

To create a network and establish a dynamic relationship between all the participating images, as well as the viewers, we realized that we had to analyze and articulate the symbolic, semantic, and artistic traits of each photograph. We had to provide the images with a 'personality,' so to speak. Together with Ran Kasmy Ilan, the curator of the exhibition, we imagined potential personalities for the selected photographs. We comprised a list of qualities such as 'timid,' 'childish,' 'arrogant,' 'team player,' and 'loner.' These qualities were based on the photographs' given narratives and dictionary of allegories, and on our readings of the images. We then graded each photograph on a 1:10 scale for each quality, reflecting the relevance of that quality to a specific photograph. This rating technique yielded photographs of different 'personalities' or values: some were more inclined to share their resources ('team players'), others were shy and tended to keep to themselves. For example, the traits we attributed to the image in figure 1 were: friendly, childish, and shy. The idea was that just as in the case of all living things, where personality traits are manifest through constant interaction with members of one's network, so would the personality traits of the photographs be revealed through their interaction with each other and with the viewers. This part of the network still remained to be established.

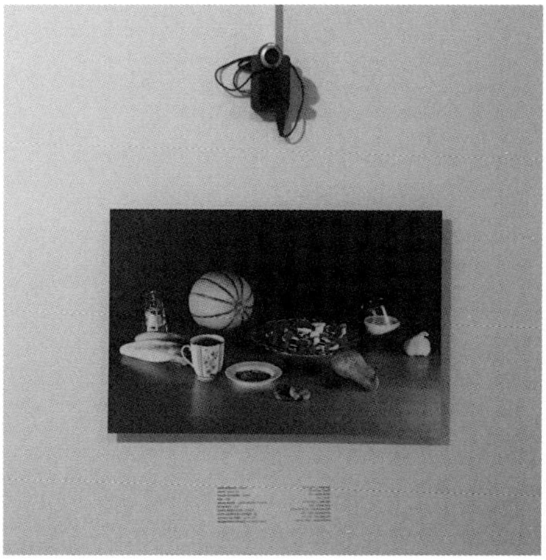

Fig. 2. Installation view: *Still Life, Live Data,* Tsila Hassine, Olga Kisseleva, Artists' Residence, Herzlliya. (Plate 40, p. 327)

The Network-As-Artwork

While our first step was to give the art objects virtual representations and to transform them into virtual entities, we now had to come full circle and finish this process by giving the virtual entity, namely our interactive network-as-artwork, a visual, physical, and concrete representation in the space of the exhibition. For although the Internet of art objects is in itself a virtual, digital entity, it still addressed a 'real,' physical audience, experiencing the installation in a 'real' physical space. Thus, the interactive art network became an in-situ installation, adapted to the special conditions and the context of the gallery.[6]

To better understand the type of interaction we wanted to establish as part of the artwork, we first had to analyze the real-life encounter between spectator and artwork, while respecting the intimate space created between them. Our strategy was to measure, as discretely as possible, the amount of time spectators spent in front of each artwork. The rationale was obvious: the time a spectator spends in front of an artwork can serve as a quantitative indicator for her engagement with the work and could be interpreted as the extent of participation in the network. We wanted to use this rationale to bring the artwork to life through the spectator's gaze. Our aim was to give a visible representation to this engagement, and to ultimately extrapolate it from the intimate space between an artwork and its audience.

To assess viewers' engagement with the artworks, we placed cameras above each photograph (Fig. 2), measuring the time viewers spent in front of it. This input provided us with an insight into the unmediated interaction between viewer and artwork. While these data allowed us to integrate the spectator into the network, we still had to figure out a way for the artworks themselves to acknowledge the spectators and react to them.

Realizing that an exhibition space has two major categories of resources, physical structure (walls, ceiling) and technical infrastructure (light, electricity), we were curious which of these resources would be best suited as a medium for physically encoding the interactive activity of the network, i.e., the reactions of the artworks themselves. We finally chose light—inherently the basis for today's communication technology[7]—as the medium through which the artworks would engage with the viewers in this exhibition. Thus, the lighting system in the gallery was designed to illuminate each artwork by a wi-fi-controlled light bulb (Philips HUE), with an intensity varying between 5 and 100%. The intensity of light illuminating each artwork could vary depending on different parameters.

Once the lighting system was set up, we started to experiment with the interactive environment. Our first experiment with basic interaction between viewer and photograph was to correlate each photograph's 'popularity'—measured by the accumulated observation times—with the intensity in which it was being illuminated. This way, the spectator and her spectatorship became technically integrated into the network installed in the gallery. Yet, this integration did not provide a two-way correspondence between viewers and artworks.

The final step was to integrate spectator and spectatorship into a dynamic relationship with the participating images (as described in 3.2). Based on the set of traits we defined for each photograph, we articulated its interactive behavior to reflect its 'personality.' The point of departure was the direct relationship between the accumulated viewership of each photograph and amount of light the photograph receives. But since different personalities may react differently to being in the limelight, we wished to incorporate these differences into the equation, and allow them to affect the illuminating pattern of the different photographs.

To illustrate how this actually worked, let's take for example a photograph that we defined as 'friendly' and 'sharing.' Once at the center of attention (i.e., once a viewer was standing in front of it), it would share its amount of light or illumination time with other photographs, which were defined as its 'friends' (in effect, upon viewing this 'friendly' photograph, not only would its lighting increase but also the lighting of a few other photographs, which were predefined as its 'friends'). Or take an 'arrogant' photograph: once in front of an audience its lighting would increase very quickly to reach full luminosity and then return to darkness as if by a switch. A timid photograph, on the other hand, upon encounter with a viewer, would be lit very slowly, never reaching maximum luminosity, only to return to darkness swiftly and discretely. Thus, a whole language of light was created for the use of this network of art objects and their audience.

Light in the Tradition of Interactive Exhibition Spaces

Our choice of light as a medium of communication is not rooted only in the history of communication technology, but also in art history. In the history of interactive spaces in particular, light has been extensively used as a medium for interaction between visitor, space, and artwork.[8] Experiments with interactive spaces, where users become part of the artwork and the exhibition space, started in the 1960s. A noted example is Robert Rauschenberg's *soundings* (1968). In this work, Rauschenberg, together with a team of engineers, installed a 36-foot-long sculpture made of three layers of Plexiglas. The front layer was partially mirrored, and behind it were two more layers of Plexiglas with images of wooden chairs on them. Lights of different intensities were placed behind the Plexiglas screens, so that, when illuminated, the images became visible through the mirror. Each light reacted to a different sound level, such that children's high-pitched voices triggered different lights than lower pitched voices of adults.

Another prominent example of interactive installation employing light as part of the interaction is Rafael Lozano Hemmer's *Pulse Room* (2006). In this installation, Lozano-

Hemmer used about 100 light bulbs (the exact number varies between installations), uniformly hung from the ceiling and completely filling the space. A hands-on interface in the room detected the pulse of those visitors who chose to hold its handles. Once the visitor released the handles, all the lights were briefly turned off and a lightbulb started flashing with the newly recorded heart rate. At any given time, the installation showed the recordings of the most recent participants (Lozano-Hemmer 2006).

Yet, *Natures Mortes, Live Data* (2017) differs from these examples in several ways. It departs from Rauschenberg's installation, where at a given time a single object responds to a single viewer, as it induces interaction between objects and spectators in a connected network. And, while in Lozano-Hemmer's *Pulse Room* the interaction is single-channeled—the visitor's pulse affecting a single lightbulb—in *Natures Mortes, Live Data* the visitor is fully integrated into the space. Her interaction with the exhibition is monitored and stored in the database in terms of the amount of time and attention that she has spent on the different photographs. This, in turn, affects the lighting scheme of the space, thus affecting the exhibition in its entirety.

Ultimately, in *Natures Mortes, Live Data* the artworks acquire a certain agency, by responding to the choices the visitors make in the space. By endowing the art object with agency, albeit limited, this installation provides a preliminary answer to the question raised earlier in this chapter, namely what would the art object *do* at the IoT era? In this work, our answer is pretty straightforward: At the era of IoT art objects will be looking at us looking at them. While the genre of still life painting is about new technologies of looking, *Natures Mortes, Live Data* is about new technologies perceiving us while we are looking at art objects.

Notes

1. Examining our daily routine, we find about 130 embedded objects we encounter on a daily basis, which range from sliding doors and ATM machines to coffee makers.
2. The first coffee machine to be connected to the internet was named 'The Trojan room coffee machine'. A video camera, connected to the internet, was placed in front of a coffee machine, and uploaded its images about three times a minute to a custom-created web page online. The image was grayscale, and so was the coffee. Thus, Cambridge computer lab researchers could track the progress of the coffee in the pot and come down for coffee only once the camera showed that the pot contained enough coffee (Cambridge Computer Lab, 1995).
3. Apart from the Roman 'Xenia' paintings, that had a rather ceremonial meaning (Bryson 1990).
4. This may be best understood in comparison with Italian renaissance painting of that period. Whereas Italian painters gave shape to some grand narratives, rife with textual

interpretation, the Dutch painted 'descriptions,' unrelated to any grand narratives. The Dutch present their pictures as describing the world seen, rather than as imitations of significant human actions (Alpers 1984).

5. In some cases, the human form, although absent from the painting, is present through hints embedded in the represented objects.
6. It would be incorrect to think that conventional art exhibitions are not interactive. Every experience consisting of an encounter between two different entities is in itself an interaction (as defined in Oxford Dictionaries Online 2018) In fact, there are three principal elements that establish interaction when looking at a painting: the spectator, the space of viewing, and the object that is viewed. Meaning is located where these three dynamic elements intersect (Helmers 2008).
7. Today's fast broadband internet infrastructures are based on optic fiber cables (FastMetrics 2018).
8. Since the 1930s artists have been exploring the different phenomenological, interactive, and performative qualities of light spaces. In 1930, László Moholy-Nagy created the *Light-Space- Modulator*, which is a kinetic sculpture that explores the phenomena of light and movement (Arozqueta 2014).

References

Alpers, Svetlana. 1983. *The Art of Describing Dutch Art in the Seventeenth Century*. London: Murray.

Arozqueta, Claudia. 2014. "Connecting Realities: Rafael Lozano-Hemmer's Pulse-based Works." *CITAR Journal of Science and Technology of the Arts* 6 (December): 21–28.

Bryson, Norman. 1990. *Looking at the Overlooked: Four Essays on Still Life Painting*. London: Reaktion.

Caple, Chris. 2006a. *Conservation skills: judgement, method, and decision making*. London: Routledge.

——. 2006b. *Objects: Reluctant Witnesses to the Past*. New York: Routledge.

——. 2012. *Preventive Conservation*. New York: Routledge.

Cisco's Technology News Site. 2014. "Kevin Ashton, Father of the Internet of Things & Network Trailblazer." https://newsroom.cisco.com/feature-content?type=webcontent&articleId=1558161.

Encyclopedia Britannica Online. n.d. "still life painting." https://www.britannica.com/art/still-life-painting.

European Union, Information Society & Media, European Platform on Smart Systems Integration (EPoSS). 2008. *Internet of Things in 2020: Roadmap for the Future*. https://docbox.etsi.

org/erm/Open/CERP%2020080609-10/Internet-of-Things_in_2020_EC-EPoSS_Workshop_Report_2008_v1-1.pdf

Feki, M.A., F. Kawsar, M. Boussard, and L. Trappeniers. 2013. "The Internet of Things: The Next Technological Revolution," *Computer* 46: 24–25. doi:10.1109/MC.2013.63.

Helmers, Marguerite. 2008. "Framing the Fine arts Through Rhetoric", in *Defining Visual Rhetorics* edited by A.C. Hill and M. Helmers. London: Lawrence Erlbaum Associates.

Lozano-Hemmer, Rafael. 2006. *Pulse Room.* http://www.lozano-hemmer.com/pulse_room.php.

Jankowski, Simona, James Covello, Heather Bellini, Joe Ritchie, and Daniela Costa. 2014. "IoT Primer: The Internet of Things: Making Sense of the Next Mega-Trend." Goldman Sachs, Equity Research. https://www.goldmansachs.com/insights/pages/internet-of-things/iot-report.pdf.

Jia, Haiyan, Mu Wu, Eunhwa Jung, Alice Shapiro, and S. Shyam Sundar. 2012. "Balancing Human Agency and Object Agency: An End-User Interview Study of the Internet of Things." *Proceedings of the 2012 ACM Conference on Ubiquitous Computing*:1185–88. DOI=http://dx.doi.org/10.1145/2370216.2370470.

Kaplan, Frédéric. 2012. *La métamorphose des objets.* Nouvelle édition. Limoges: Fyp Editions.

Kisseleva, Olga. 2010. *Divers faits.* https://www.digitalartarchive.at/database/general/work/divers-faits.htm.

McLuhan, Marshall. [1964] 1967. *Understanding Media: The Extensions of Man.* London: Sphere Books.

Moore, Stephen and Julian L. Simon. 1999. *The Greatest Century That Ever Was 25 Miraculous Trends of the Past 100 Years, Executive Summary.* Washington DC: Cato Institute. http://bib-purl.oclc.org/web/43117/pa364.pdf.

Oxford Dictionaries Online, n.d. "Interaction." https://en.oxforddictionaries.com/definition/interaction.

Rifkin, Jeremy. 2015. *The Zero Marginal Cost Society: The Internet of Things, the Collaborative Commons, and the Eclipse of Capitalism.* New York: Palgrave Macmillan.

Rebaudengo, Simone. 2012. "Addictive Products: The Story of Brad the Toaster." Vimeo video. https://vimeo.com/41363473.

Schwab, Klaus and Davis, Nicholas. 2017. *Shaping the Fourth Industrial Revolution,* with a foreword by Satya Nadella. New York: Penguin Random House.

"10 Key Benefits of Fiber Optic Internet Service." 2018. *FastMetrics.* https://www.fastmetrics.com/blog/tech/10-benefits-fiber-optic-internet/.

Toonders, Joris. 2014. "Data Is the New Oil of the Digital Economy." *Wired.* https://www.wired.com/insights/2014/07/data-new-oil-digital-economy/ 2014.

Urban Tastebud. 2018. "5 Best Smart Mattresses – Buyer's Guide and Reviews." https://urban-tastebud.com/best-smart-mattresses/.

Weiser, Mark. 1991. "The Computer for the 21st Century." *Scientific American* 265: 94–95, 98–102, 104.

Unmanned Imaging of the Anthropocene

Aaron Burton

> 'My path leads to the creation of a fresh perception of the world. I decipher in a new way a world unknown to you.'
> Dziga Vertov *Kino-Eye* (1923)

It is dizzying to reflect that only five decades ago NASA released the first photographic image of the entire Earth. The Applications Technology Satellite ATS-3 was tasked with 11 scientific experiments, one of which, led by the 'father' of satellite meteorology Dr Vernor Suomi, was to successfully photograph, in near real-time daylight, pictures of the Earth's atmospheric system (NASA SSDCA 2018). The series of images produced by the satellite's camera system on November 10, 1967 marked the first complete color photographs of this planet. One year later, on board the Apollo 8 mission in 1968, astronaut William Anders, with his fellow crewmen Jim Lovell and Frank Borman, took the famous *Earthrise* image (NASA 2017). These inaugural images of the Earth have both been credited with highlighting the vulnerability of the planet as well as alluding to the responsibility of humanity to maintain its habitability.

Placed side by side, the ATS-3 satellite image and Anders' photograph represent markedly different perspectives: one by a machine, the other by a human. This difference arguably offers an important ontological and epistemological juncture—The Earth and Our Earth—empirical instrumentation versus an overtly human-centric narrative of Western civilization: progress, God, the individual, and (in this instance the United States of America's) cultural superiority. Put more simply, the ATS-3 satellite was recording data whilst the astronaut was recording personal experience. Information on one hand, storytelling on the other.

The role of photography and film to reflect and intervene in our world continues to offer a fascinating paradigm of truth, beauty, and persuasion. Now more than ever these alignments are paramount to effectively mediating the global environmental crisis and animating social change. The divided international response to the Anthropocene is perhaps indicative of the clouded status of contemporary scientific knowledge, further evidenced

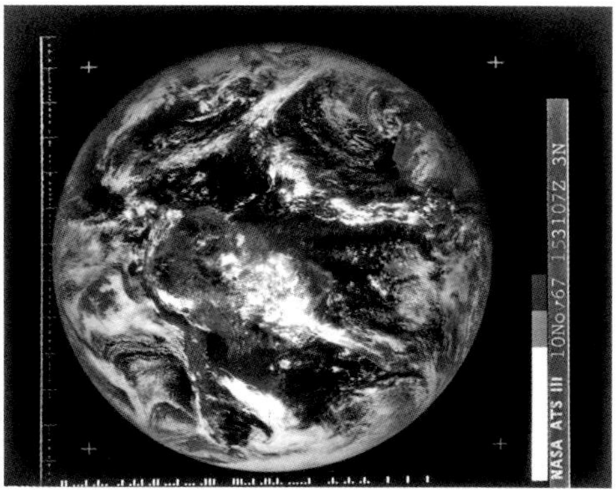

Fig. 1. One of the first color photographs of the whole Earth, shot from the ATS-3 satellite on November 10, 1967.
Image credit: NASA. (Plate 41, p. 328)

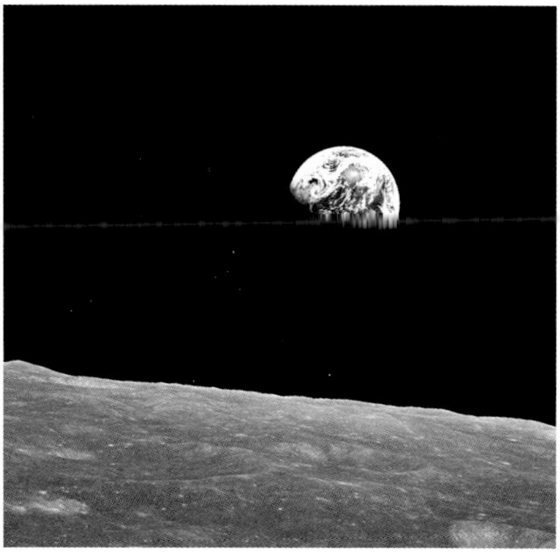

Fig. 2. *Earthrise* was taken on December 24, 1968 by astronaut William Anders during the Apollo 8 mission.
Image credit: NASA. (Plate 42, p. 329)

in acute cases of Lysenkoism and 'fake news' promulgated by political leaders and power brokers around world. But it may, in what Timothy Morton (2013) has aptly termed the 'hyperobject' proportions of immanent ecological crisis, reflect the deterministic impotence to the trajectory of modern capitalism or analogous forms of apathy such as 'catastrophe fatigue' (Truscello 2018, 259). As modernist convictions in truth fall by the wayside, so too do assumptions around representation.

This chapter examines the challenge of representing environmental issues, specifically how the human-centric perspective which has dominated the history of photography and film, is now being supplemented by nonhuman perspectives. In the face of a New Climate Regime, to borrow Bruno Latour's term (2017), exploring the porousness between human and nonhuman points of view reflects a shift in understanding existence on Earth. With satellite imagery at our fingertips, remote sensors, action cameras, and drones in our tool kits, consumer imaging technologies are encouraging the capture of nonhuman experiences that are effectively rearranging our paradigms of representation and transforming our 'scopic regimes'. Martin Jay's (1988) elaboration of scopic regimes considers visual perception as historically located and necessarily malleable, as informed by but not limited to cultural, technological, and phenomenological influences. While unmanned technologies are celebrated and even fetishized for their data-gathering capabilities, I argue it is their distinctly non-human provenance, proxemic arrangements, and experiential operation that present an opportunity for new ways of seeing. I draw on contemporary films and media art practices that demonstrate innovative directions in storytelling by way of such encounters, namely: Werner Herzog's *Encounters at the End of the World* (2007), Simon Faithful's *Escape Vehicle No. 6* (2004), Lucien Castaing-Taylor and Verena Paravel's *Leviathan* (2013), and Korakrit Arunanondchai's ongoing series, *Painting With History In A Room Filled With People With Funny Names* (2013–17).

The analysis draws on a critical framework for interpreting and producing contemporary media by way of an expanded understanding of the art-historical concept of *provenance*. Provenance is a necessary framework for considering the textual, contextual, and fluxional qualities of media circulation. As the aforementioned images of Earth reveal, removing the human from behind the camera significantly alters the provenance and meaning of the image; nonhuman actors are foregrounded—such as the weather, material objects, plants, and other animals—and there is a leveling of existence. Provenance becomes a critical framework for evaluating the success and authenticity of media and reveals shortcomings across the history of representation on environmental issues, in particular the challenge to humanist traditions of documentary to render an analogous level of *intimacy* towards nonhuman subjects.

The five decades since 1967 are an important historical period to examine for a number of reasons: firstly, it is arguably since then that humans have been sufficiently aware of our environmental impact and have nevertheless failed to curtail the Great Acceleration of atmospheric carbon dioxide (Lewis and Maslin 2015, 177; Crutzen et al. 2007, 618); secondly, the late 1960s manifest a similarly dramatic upheaval of politics, culture, and visual culture to that which we are witnessing now. It is at this point in history that *intimacy* becomes a defining quality of visual storytelling, taking the lead from Walter Benjamin, shaped by lived experience as distinct from information.

Unveiling distinctions between information and storytelling has been a crucial thematic response by the humanities to questions posed by the Anthropocene (Ghosh 2017; Le Guin 2017). Imaging technology treads a particularly unique path through this dichotomy. Unmanned imagery presents a new arrangement of phenomenological, spatial, and experiential exchange. I argue that, in addition to the camera's indexical and empirical capabilities of relevance to the natural sciences and cultural histories, it is its physical distance from the subject that mimics human intimacy and garners affect, interpreted here as *proxemics*. In contrast, unmanned cameras open up the exploration of nonhuman distance, by way of satellites in space, remote action cameras, and taking flight with drones. This new scopic regime, in contrast to the psychological and ocular centricity of modern imagery, is participatory and experimental. The act of representation is of unseen possibilities and speculative ways of being, potentially opening up new ways to experience and understand the fragility of life on Earth.

I employ the conceptual framework of *provenance* as a critical method for establishing authenticity across modes of storytelling tied to both aesthetics and social intervention (Burton 2014). An expanded understanding of the art-historical term questions how the artist or storyteller incorporates the passage of production and 'creative treatment' into the meaning of the work itself. In a similar vein to John Dewey's (1934) advocacy for phenomenology in fine art, *provenance* connects notions of authenticity to idiosyncratic associations across disparate stages of the creative process. Within a documentary paradigm, dependent on the indexical nature of photography, I argue that the individual provenance of the storyteller is inherent, and symptomatic of interpreting film and photography, as anticipated by auteur theory. Current technology offers a discernibly distinct provenance and creative and ethical pathways through the production and storytelling process. In other words, provenance defines the creative process. The uniqueness of the process, the encounters, contingencies, and instigations establish the authenticity and ethical integrity of the production. As ethnographers and social scientists are keenly aware, the act of research itself animates social and cultural influence, and all of our actions in producing, participating in, and distributing media contribute to meaning and lived realities. *Provenance* is an effort to acknowledge the journey as integral to the meaning of a work and open pathways for creative intervention in knowledge creation.

Storytelling with Satellites

A consideration of the provenance of the two aforementioned photographs of Earth reveal the unique attributes of unmanned imagery. The provenance of both images is tied to NASA's space programs, which are undoubtedly reflective of ideological agendas, but the images themselves have emerged from contrasting purposes. The seemingly straightforward meteorological aims of Dr Suomi's 1967 experiments on board the weather satellite ATS-3, namely the successful operation of a Multicolor Spin Scan Cloud Camera (MSSCC), are in stark contrast to the deeply politicized agenda of the Apollo 8 mission. Earthrise was the trophy shot of winning the Space Race and conquering the New Frontier, as the astronaut photographer William Anders recently remarked, 'We didn't go the moon for Teflon pans or lunar rocks (...) We went to the moon to beat the "dirty Commies"' (in Silder 2008). The politics extends to the religious overtone of the mission; when on Christmas Eve 1968 the Apollo 8 crew entered lunar orbit they held a live broadcast to what was reportedly the largest audience to ever simultaneously listen to a human voice, and the crew's message was a reading of the first ten verses of the book of Genesis: 'In the beginning God created the heaven and the earth (...).' Evidently, this was not a purely scientific or technological voyage, nor was it an exercise in worshipping Gaia. Historian Neil Maher remarks on the political significance of Earthrise,

> By figuratively depicting Kennedy's New Frontier in its sloping lunar surface, the Apollo 8 photograph helped extend America's Manifest Destiny into the ultimate wilderness—outer space' (2004, 528).

Even the composition of Earthrise can be interpreted as triumphant colonization—the moon's surface largely positioned in the foreground against the distant Earth underscores the incredible feat of the mission. Interestingly, Maher presents a similar comparison between Earthrise and the famous 1972 image taken on Apollo 17 known as the Blue Marble. Maher contrasts the Cold War motivations of Earthrise with the influence of the growing environmental movement and NASA's subsequent shift in focus from outer space to back on Earth (2017, 135). The Blue Marble, like the ATS-3 image, presents a straight composition of the circle of the Earth. Removing the lunar surface, Maher argues, effectively erases the American frontier narrative and instead presents identifiable features of Africa, the Middle East, and Antarctica (2004, 529). The composition and perspective of Earthrise on the other hand is ocular-centric; it is as though we are seeing what the astronaut sees. Placing ourselves in the photographer's shoes instantly conveys the incredible prospect of being one of the first humans to exit Earth's orbit, but also alludes to the unremarked significance of having a highly educated white thirty-something American male point-of-view.

Both Earthrise and Blue Marble were photographs taken by astronauts with specialized 70mm Hasselblad equipment. In an audio recording when Earthrise was taken, astronaut Frank Borman can be heard giving a light meter reading of 1/250th at *f*11. The fraction of a second as 'decisive moment,' as Henri Cartier-Bresson called it, is an experience of time we easily relate to, like the blink of an eye. In contrast to both of these photographs *taken* by astronauts, the flat, descriptive, even flawed aesthetic of the ATS-3 image is devoid of human attributes. The ATS-3 used an experimental 'spin-cloud' camera entailing a 24-minute color scan using the rotation of the satellite to synchronize with the Earth. This single image, produced over 24 minutes, is arguably removed from blinking human experience, calling to mind the ghastly clamps and brace technology used in nineteenth-century studio portraits to hold sitters still for long exposures. Approximately three months after the ATS-3 produced its initial images in November 1967 the red and blue channels failed and the satellite continued to produce black and white images until its decommissioning in 1978. Unmanned technologies are vulnerable, their life is finite, and our control over them is strained, offering appropriate parallels to how humans experience the natural world.

Such origin stories highlight the significance of production but fail to consider the dynamic passage of the artefacts and their surrounding discourses. The late 1960s were culturally momentous. A snapshot of political events of the period is important to contextualize the environmental movement, a cursory list might include: the Vietnam War; tensions in the Middle East resulting in the Six-Day War; genocidal-scale killings under the Indonesian Communist Purge; the South African Anti-Apartheid movement; civil rights under the banners of the National Organization for Women (NOW) and Black Panther Party; in France the May 1968 unrest; and in Australia the Aboriginal Land Rights movement was underway. Photography and film were helping to re-frame our societies and our relationships to each other. It is thought-provoking to reflect which of these historical events shaped the contemporary documentary genre and spurred critical theories on media, and which events have left little or no visual record.

The early Earth photographs were circulated widely across television, newspapers, and magazines, as both news stories and commercial pictures (Maher 2017, 125). In 1966, environmental activist and psychedelic artist Stewart Brand could be found selling pinback buttons reading 'Why Haven't We Seen a Photograph of the Whole Earth Yet?'. Brand was on an LSD-inspired campaign for the release of an image he believed would galvanize the public into caring for the environment (Kirk 2007, 40–41). The late 1960s can be interpreted as the beginning of mainstream awareness of a looming environmental crisis, thanks largely to influential publications from the period, such as: Rachel Carson's *Silent Spring* (1962), Buckminster Fuller's *Operating Manual for Spaceship Earth* (1967), and Paul Ehrlich's *The Population Bomb* (1968). The counter-cultural movement was gaining momentum and Stewart Brand embodied the energy- and spirit-provoking mainstream culture. In 1967, when he finally got hold of the ATS-3 Earth image, he used it for the cover of the first edition of the Whole Earth Catalog.

The legacy of the Whole Earth Catalog has recently been celebrated in numerous exhibitions, publications, and cultural references, such as: MoMA library's 2011 exhibition, *Access to Tools: Publications from the Whole Earth Catalog, 1968–1974*; HKW's 2013 exhibition *The Whole Earth: California and the Disappearance of the Outside*; and the late Steve Jobs' reference to it as a personal inspiration and as a kind of prototype for the Internet in a speech he gave to commencing students at Stanford in 2005. According to historian Andrew Kirk, Brand's use and advocacy of images of Earth from space 'were a harbinger of changing views about the role of technology and the environment' (2007, 41). The tools and literature the catalog promoted envisioned technological advancement as crucial to sustainable living into the future; however, in a clear distinction to contemporary consumerism (Cubitt 2017, Huesemann and Huesmann 2011), there was an emphasis on 'appropriate technology,' designed to be both functional and environmentally sustainable. Maher outlines how Brand's campaigning was successful in influencing NASA, and by the early 1970s there was sufficient internal and external pressure for the agency to place a greater emphasis on monitoring the environment at home:

> Beginning in the 1970s NASA began remedying such environmental anxieties not only by both redirecting its satellites back down toward Earth to collect global environmental data, but also by superimposing this newly acquired global data onto scientific images that hinted at older, more culturally familiar photographs of Earth from space (2017, 135).

For NASA, communicating their findings highlighted a fundamental deficiency in satellite and unmanned imagery, which on their own simply conveyed data and lacked the story to provide meaning. Stewart Brand and The Whole Earth Catalog provided a vibrant narrative for the Earth as seen by the ATS-3 to encapsulate environmental concern.

Information and Narrative

The distinction between information and story is perhaps one of the greatest challenges to representing the Anthropocene. Walter Benjamin raised similar concerns in his 1936 short essay on Nikolai Leskov. Benjamin lamented the demise of storytelling as a symptom of modernity, and if we reflect on contemporary habits in which millions of us spend hours a day scrolling through social media feeds, his concerns about information-driven distraction were not misplaced. The success of social media is an important consideration for gauging a contemporary scopic regime. The phenomenon is arguably the result of the shallow experiential qualities it affords. Fifty years since those first glimpses of Earth and we have now come to expect hyper-realistic high definition images from NASA's daily Instagram feed (@nasagoddard), such as the 2015 Earthrise tribute composite image taken by the Lunar Reconnaissance Orbiter (LRO). But these images lack a meaningful or purposeful narrative and are readily swallowed up by the information stream.

A more ubiquitous engagement with satellite imagery is our everyday engagement with Google Earth. In a notable effort to narrativize our view from space, Google Earth Engine recently compiled 45 years of NASA's Earth-monitoring Landsat satellite images. The project sifted through over 900 terabytes of data, identifying the best pixel from each image and stitching them together to provide a cloudless, aberration-free visual timeline of changes to the Earth's surface. Recognizing the need for stories around these timelines, Google partnered with Time magazine to produce an ongoing series of time-lapse animations featuring particular locations with profound landscape alteration since the first Landsat satellite images of 1972 (Time 2016).

Planet Labs, a team of ex-NASA garage start-ups, led by CEO Will Marshall, now manage the largest fleet of Earth imaging satellites in history, with 200 satellites in orbit. Planet Labs contextualize their work as part of a Global Sensing Revolution, stating 'We are basically starting to extend a nervous system all around the planet that will help us to understand how it's changing' (Marshall 2018). The ring of orbiting satellites is imaging every spot of earth, every day, with applications ranging from monitoring environmental change, agriculture, industry outputs, water levels, disaster relief, to predicting stock market investment. This is undoubtedly an incredibly powerful and useful tool; however it feels like a cruel joke to provide a digital 'nervous system' for the planet as it suffers irreparable damage.

The imagery we receive from satellites is data-driven, determined by empirical instrumentation, in the same way as other scientific imaging tools and remote sensors, such as near-infrared and 3D Lidar models. While they often radiate a techno-aesthetic allure, they are ultimately devoid of relatable human experience. For Benjamin stories are experientially informed, their testimonial quality is in contrast to information, which is unrelatable and lacks intent or purpose. Satellite imagery has become commonplace, yet the disheartening timeline of the Anthropocene is in stark contrast to the zesty technophilia of The Whole Earth Catalog. Living in this era of immense quantities of raw data and information, the question becomes: how can visual storytellers meaningfully integrate these instruments into their narratives?

To highlight the scope and ambition of what Benjamin was advocating, Werner Herzog would surely be one of the most appropriate auteur filmmakers to liken to a contemporary master storyteller. As Herzog outlines in the opening narration of *Encounters at the End of the World* (2007), he was invited to Antarctica by the National Science Foundation but promised he wouldn't make another film about penguins. His film explores the eccentric 'professional dreamers' who inhabit McMurdo Station. In one revealing sequence we are introduced to glaciologist Douglas MacAyeal who begins by relating a recurring dream of walking on top of the massive B-15 iceberg he studies, adrift like 'a vagabond floating in the ocean.' MacAyeal explains how, unlike Shackleton and the early explorers, we now understand the ice as a 'dynamic living entity' and he presents the camera with a three-tone greyscale time-lapse animation of satellite imagery of sea ice around Antarctica. The graphic fills our screen, we hear a rising cello, the camera zooms in on the phantasmagoric time-lapse of icebergs dancing around the monitor, and the scientist expresses his fears, 'Certainly on a gut level, it's going to be frightening to watch what happens to these babies once they get north.' The image cross-dissolves to black and white archival film of Sir Ernest Shackleton smiling and stumbling around a tacky recreation studio set. Herzog chimes in,

> What environment would the men of Shackleton's expedition encounter if they returned in the next life?
> Would there be any ice left?
> Would our only modern recourse be to create ice with machines?

Cut to a young cook who then explains the crises in McMurdo when the Frosty Boy ice-cream machine runs out. The sequence concludes with a wide shot of the settlement and Herzog's hallmark delivery:

> From the very first day we just wanted to get out of this place. McMurdo has climate-controlled housing facilities; its own radio station; a bowling alley; and abominations such as an aerobics studio and yoga classes. It even has an ATM machine. For all these reasons I wanted to get out into the field as soon as possible.

Herzog adeptly choreographs cinematic tropes to relay his deeply personal and profound experiences. Rather than looking at 'nature,' as a documentary about penguins might, *Encounters at the End of the World* presents the charming absurdities of humans forging an unnatural life in order to understand nature better. Satellite imagery in this context, like the archival footage, interviews, and musical score, is drawn upon as another element, even contingency, in a wider montage of human media and histories of knowing.

Proxemics and Action Cameras

Celebrated photojournalist Robert Capa's mantra, 'If your pictures aren't good enough, you're not close enough,' is often called upon as a hot tip for photography enthusiasts to take a few steps closer before clicking the shutter. Reduced subject distance generally produces a cleaner composition. However, as cultural anthropologist Edward T. Hall theorised (1963), human spatial arrangements are far more significant than compositional elements. Hall proposed the concept of *proxemics* for his investigation into the impact of spatial relationships on interpersonal communication. In photography, spatial relationships reveal the contract between the photographer and the subject. It is the reason why long lens paparazzi images of celebrities are voyeuristically fetishized and, conversely, why the up-close war photographs by James Nachtwey are so affecting. Charlie Chaplin knew this when he insisted on shooting jokes in long shot and emotional scenes in close up (Giannetti 1990, 64). The other notable attribute of proxemics is that it applies to most recording media; audio microphone distance for instance, such as the palpable wildlife recordings by Chris Watson, is produced with closely positioned or attached lapel microphones.

Subject-distance is arguably the most revealing indicator of provenance. Consider Diane Arbus' oeuvre of 'fringe-dwelling' subjects and the way her waist-high camera, often too close or too far away for conventional portraiture, as Max Kozloff observed (2017, 24), reveals a nonjudgmental, dignified, and sincere relationship. It is only through the emergence of *intimacy* that individuals can reveal a fidelity to the viewer, effectively breaking down 'otherness.'

The announcement of the Anthropocene has confirmed a problematic distance between humans and nature. The 'othering' of nature in Western civilization, as Latour points out (2017, 17), is demonstrated in the historical oddity of singular perspectivism that grew out of the Renaissance, specifically, depicting reality as a God-like arrangement, and compositions of nature as objects within the frame (Berger 1972, 16). John Berger, however, credits the introduction of photography as liberating representation from the singular perspective and timelessness of painting, offering multiplicity across time and space: "What you saw depended upon where you were when. What you saw was relative to your position in time and space" (1972, 18). Yet the history of photographic representation of the environment is largely oriented around ideas of the sublime, devoid of human meddling. Historian Andrew Kirk has suggested that, at least in the USA, the early lobbying successes of the wilderness preservation movement have been at the expense of more environmentally sustainable urban development (2007, 19). I sense this might also be the case in the history of representation, in which the celebration of pristine wilderness conceals our impact on global ecosystems. This distinction is reflected in the unspoilt animal kingdoms of nature photography and wildlife documentary. This is evidenced in examples such as Ansel Adams' stunning black and white landscapes of the Yosemite Valley and their role in Sierra Club's campaigning for state preservation, or David Attenborough's enthralling narratives of animal survival.

Social documentary is not spared in this struggle to portray an intimacy with nature. One of the most revealing attempts to honestly portray our relationship to the environment is the Documerica project spanning five years from 1972 to 1977. The Environmental Protection Agency's ambitious project featured over 70 photographers who produced an archive of over 22,000 images. Documerica director Gifford Hampshire envisioned a photo-documentary project similar to the influential Farm Security Administration (FSA), with the distinction it would be 'photographer-led' and left quite open to subject matter provided they were illustrative of environmental concerns. Unfortunately, as Bustard (2013) outlines, numerous shortcomings of the project led to an earlier than planned closure and lacklustre public influence. In addition to mundane subject-matter, Bustard points to the hundreds of aerial photographs in the project's files and a distinct lack of human focus:

> While aerial photography can be a useful way to depict air pollution and suburban sprawl, many photographers fell back on this easy solution—a solution that distanced the photographer from the 'human connection' (…) Another of the project's shortcomings was that, with only a few exceptions, photographers did not take viewers inside homes. This might not seem important to a project on the environment, but such failures avoided an exploration of how individuals contributed to the environmental crisis through their lifestyle and economic decisions (Bustard 2013, 6).

These shortcomings allude to what Maher describes as issue-oriented environmentalism which dominated the 1970s (2017, 135). However, Bustard also observes shortcomings in both storytelling qualities and proxemics. The aerial images were overly descriptive and lacked experiential qualities, and images of people were similarly taken at a removed distance, outside the home interiors and intimate spaces which distinguished the success of the FSA.

A fundamental aesthetic and conceptual opportunity in unmanned imaging is to turn our attention away from the ocular and psychoanalytic 'gaze,' of looking *at* the subject, and instead consider the proxemics of a more participatory scopic regime. Where satellite imagery reveals global ecosystem dynamics, action cameras attend to the Earth's objects and inhabitants. I use the term 'action camera,' commonly referred to by the brand name GoPro, to denote consumer-level cameras fixed on things other than our hands and eyes. Attachment as a proxemic arrangement offers an unprecedented intimacy with objects and their experiences.

Escape Vehicle No 6, Simon Faithfull's 2004 performance video commissioned by Arts Catalyst, follows the journey of an unmanned chair attached to a weather balloon as it floats to near space (around 30,000 meters) before spinning out of control and losing contact. The empty chair offers the daunting prospect of taking a seat on this journey. By projecting ourselves onto the chair we become complicit in the performance; we are collaborators rather than spectators. As the glitchy camera image is tossed around we realize the weather has taken over the production, *that* worldly constant we used to take for granted is foregrounded, and the human construct à la Joseph Kosuth is dismantled before plummeting to the ground. At the upper reaches of the atmosphere, having been freed from the burden of worldly furniture, the camera takes a final look around, scanning the spherical horizon of the planet, a final goodbye before losing transmission. We can only speculate what happens from there.

Placed in the context of representing the environment, the legacy and re-enactments of *Escape Vehicle No 6* are particularly revealing. Inspired by Faithfull's performance, in 2009 electronic goods company Toshiba controversially produced their own version for an advertisement with a working budget of 3 million pounds and the ultimate goal to sell High Definition television monitors. The economic and consumer-driven imperative of Toshiba's advertisement, in stark contrast to Faithfull's performance, underscores the significance of provenance to inform meaning. Furthermore, video recordings of the weather balloon experiment have since been reproduced by 'content creators' around the world and shared on YouTube. Many of these are high school and university students exploring the physics and material attributes of the environment, such as a Hello Kitty doll launched at Cornerstone Christian School for a 7th grade science project investigating the effects of altitude on air pressure and temperature, or Operation Skyfall by a group of Harvard University students sending the first hamburger into space. A Toy Robot, Thomas the Tank Engine, a donut, and a can of beer, are among the wide range of objects starring in similar expeditions on YouTube. Collectively, these objects sent to near-space at first seem to represent a profound lack of respect for the natural environment, sending plastic rubbish to the furthest reaches of our atmosphere, but in the context of experiential storytelling I wonder if these kinds of experiments prompt a much deeper consideration of wastefulness, materiality, and accountability than the equivalent unseen discharge of the same objects into landfill and waterways.

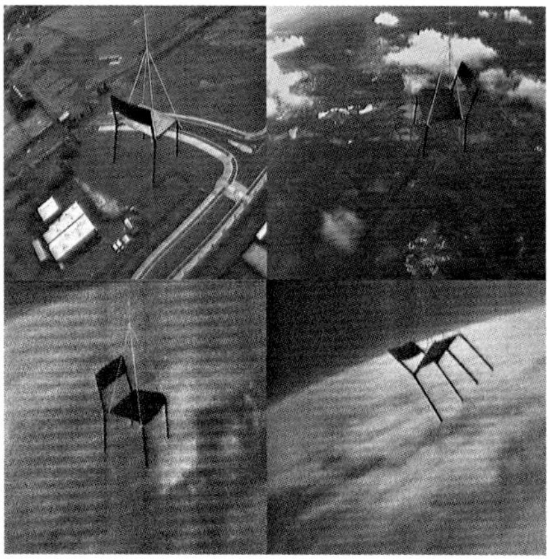

Fig. 3. Simon Faithfull, *Escape Vehicle no. 6*, 2004, video stills.
Courtesy the artist (Plate 45, p. 330)

The affordability and flexibility of action cameras have made them a standard feature of all levels of screen production, yet for the most part their aesthetic application mimics traditional ocular-centric film language. There is, however, a growing trend in experimental productions embracing the spirit of Dziga Vertov's *kino-eye*, freed from 'human immobility' (1984, 17). The Sensory Ethnography Lab (SEL) at Harvard University, established in 2006 as a collaboration between anthropology and visual arts departments, has championed such sensory explorations of the world. Lucien Castaing-Taylor and Véréna Paravel's *Leviathan* (2012) is perhaps the most well-known and critically reviewed production to emerge from the lab. The feature-length documentary takes a fishing trawler as its subject, or vehicle, and presents the audience with a richly layered series of audio-visual perspec-

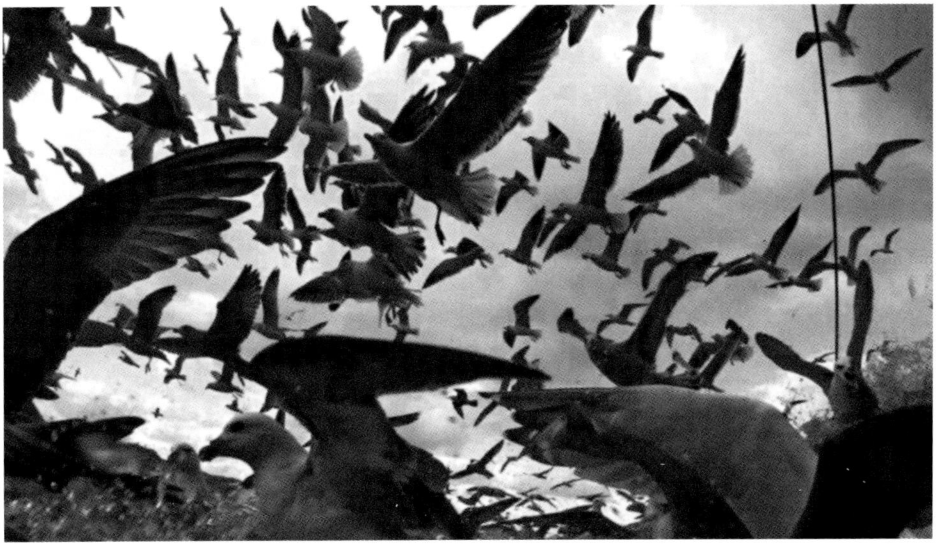

Fig. 4. Lucien Castaing-Taylor and Véréna Paravel, *Leviathan*, 2012, video still. Courtesy Cinema Guild.

tives of what we assume to be its usual working activity. In the tradition of observational ethnographic films, such as those by David MacDougall, *Leviathan* features no voice-over narration or intertitles, no 'talking heads,' no musical score, nor clear narrative. Instead the audience is immersed into the *experience* through often lengthy but highly considered visual and audio compositions. Visually, what makes Castaing-Taylor and Paravel's film so revelatory and innovative is their deployment of numerous action cameras to fixtures, objects, and bodies around the trawler. As Catherine Russell observes,

> I would describe this as a new kind of experience for the spectator in which the GoPro camera is given a role, and a value, in the documentation of labor. It enables us to go where no camera has gone before (2015, 31).

The documentation of labour in *Leviathan* is not limited to human toil but also the vessel, the stresses on its chains and nets as it forges a destructive path through the ocean, the fish and their final breaths, and the competing cries from swarming seagulls. The intimate proxemic distance provided by the action cameras presents an unprivileged distribution of labour on this brutal voyage.

The recurring theme of immersion in *Leviathan* extends to the filmmakers as ethnographic field-researchers. Provenance, or related ideas around authenticity and pedigree, has for a long time been fundamental ethnographic research; in other words, what *right* you have to tell the story. For instance, Jean Rouch's mentors Marcel Griaule and Germaine Dieterlen both advised him that a researcher needs to spend at least 20 years experiencing a culture before one can achieve a 'deep knowledge' of a society's 'system of thought' (Henley 2009, 10). Leaving the methodologies and ideological hurdles of ethnographic immersion to play out within the discipline, worth noting here is the inversion of conventional scripted filmmaking. Castaing-Taylor and Paravel 'did their time' over six trips on the trawler of between nine days and three weeks (MacDonald 2015, 407), and the resulting composition emerges from their unique experiences. As with Faithfull's performance, *Leviathan* resists overt political sentiment through encouraging the audience to interpret their own experiential understanding from the media. Nevertheless, the traces of the storyteller cling to the story; in Benjamin's analogy, 'the way the handprints of the potter cling to the clay vessel' (1968, 92).

Tracing the material provenance of *Leviathan* leads to the immersive cinematic experience. Featuring richly textured Dolby 5.1 surround sound by Ernst Karel and Jocob Ribicoff, and similarly refined color grading, the film is crafted for cinema theaters and the corresponding 'haptic visualities' they afford (Marks 2000, xii). Our current scopic regime, however, no longer assumes the cinema as the ideal 'output' for filmmakers interested in challenging audience experiences. Other projects by Castaing-Taylor and Paravel, and across the SEL catalog, are oriented towards art gallery installations and respond to different sensory alignments (MacDonald 2015, 398).

Drones and the Supernatural

Drones are in many ways the offspring of satellites and action cameras. They depend on satellite communication to determine their spatiality and we experience their agency and materiality through the attached camera. Drones are now frequently used in ecological research, offering closer details than satellites, such as tree foliage analysis, and seeing through lenses and locations usually out of human reach (Weule 2016). In more allegorical settings, the proxemic distance drones hover at is similarly caught between the heavens and the earth, even the naming of the iconic 'Phantom' by DJI alludes to this supernatural domain.

Korakrit Arunanondchai, in the series *Painting with History in a Room Filled with People with Funny Names*, taps into the spirit-like perspective of drones. The films feature conversations between the artist and a drone-spirit Chantri about life-cycles and rebirth, on which drones are expert, with regular flat batteries, lost connections, software updates, and aviation accidents. Arunanondchai creates a fictional cosmos out of his distinct cultural

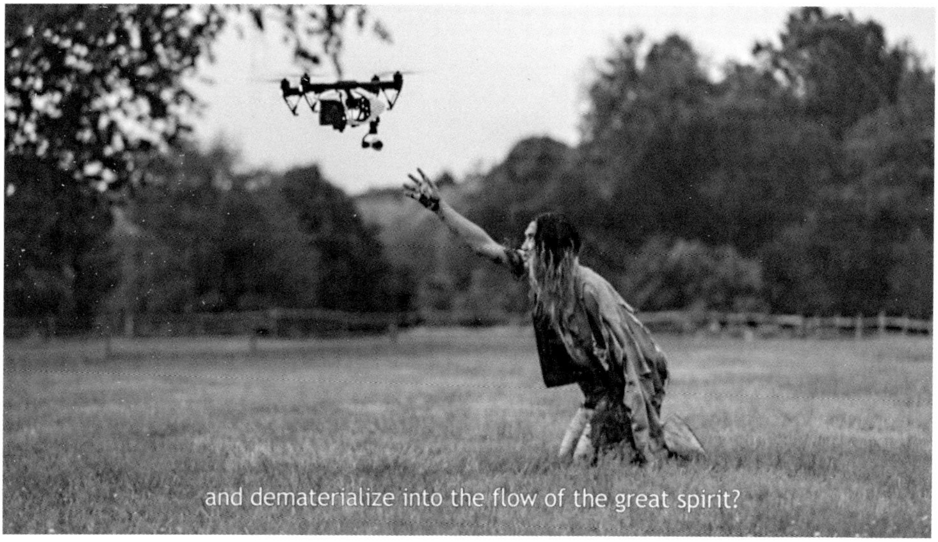

Fig. 5. Korakrit Arunanondchai, *with history in a room filled with people with funny names 4*, 2017, video still. Courtesy Carlos/Ishikawa.

background, loaded with references to Thai animism, pop culture, consumerism, globalization, art history, and personal trials. Hip denim clad youth wander through rice paddies, handle crocodiles, dance to urban beats, among other happenings, and the artist performs his trademark splashing of brightly colored paint across his body. The mesmerizing fusion of music video editing, bass-heavy electronic music, and scenes from Bangkok city's urban jungle reminiscent of *Bladerunner*, provide the unlikely texture for an exploration of a contemporary spirituality, one in which the natural world is deeply intertwined with urban existence. The drone-spirit Chantri is our filmmaking guide, interweaving humans and nature, but not as data gatherer, rather as a sage-like being with its own limitations and life force. Arunanondchai's films are usually featured within elaborate installations and curated spaces, extending the material emphasis of the screen narratives into immersive physical experiences. *With history in a room filled with people with funny names* 4 (2017) features a sculptural 'garden' composed of soil, dead vegetation, seafood waste, with LED lights and fibre optics intricately wired like a vascular system. Indigo dye fills a pond like blood, the material synonymous with Bangkok denim adorned by the artist and his co-stars, but here the desirable blue resin is a politically charged referent to colonial exploitation, globalization, and environmental abuse (Larmon 2017). Like sending plastic toys to space, the urban materials of Arunanondchai's films and installations can be unsettling; his immersive universe confronts the distinction of 'nature' and human hierarchies within it.

Unmanning the Anthropocene

The progress of humanity, the refinement and efficiency of our tools and knowledges, is revealing a dead end. The Anthropocene has profound consequences for life on our planet. It represents the significant failure of humans as responsible stewards of Earth. It also presents the imminent challenge of doing something about it. All of our faculties are being called on to comprehend both the current implications of the Anthropocene and alternative directions. If there is any hope to extract from the geological periodization of our influence, it is in the realization that the last five decades have witnessed rapid changes in our behavior, understanding, connectivity, production, and cultural values.

Social documentary is often credited for its role in cultural upheavals and social awareness, yet its successes in adequately addressing environmental issues are arguably limited to examples of wilderness and wildlife preservation and romanticized primitivism. Pursuing new 'ways of seeing' might seem like falling for the modernist trap, but what I hope this discussion has revealed is the need to compose ourselves, our actions, and communication beyond particular technologies, ideologies, or even historical precedence. Provenance reveals multiple layers and dimensions of potential intervention and activation.

Contemporary scopic regimes are complicit in the problem of modern consumption, but visual storytelling plays an important role in altering perceptions. Removing the human from behind the camera significantly alters the provenance and meaning of the image. As we have seen, nonhuman actors, such as the weather, objects, and animals, are foregrounded, and there is a non-hierarchical relationship between the object and the viewer. Unmanned imaging technologies offer an unprecedented range of spatial encounters, from impossibly close to impossibly far, opening up new ways to experience the fragility of our ecosystems. Drones, action cameras, and satellites are similarly caught up in cycles of life and death and the mysteries of in-between. Storytelling may always rely on human experience and interpersonal proxemics, but unmanned imaging is offering the potential for a new degree of ecological intimacy and the potential for new forms of storytelling and meaning-making.

References

Benjamin, Walter. 1968. "The Storyteller: Reflections on the works of Nikolai Leskov." In *Illuminations*, edited by Hannah Arendt. 83–110. New York: Harcourt, Brace & World.

Berger, John. 1972. *Ways of Seeing*. London: British Broadcasting Corporation.

Brand, Stewart (ed.). 1968. *The Whole Earth Catalog: Access to Tools*. San Francisco CA: Portola Institute.

Burton, Aaron. 2014. "Provenance in Personal Documentary: My Mother's Village." PhD diss., University of New South Wales.

Bustard, Bruce I. 2013. *Searching for the Seventies: The Documerica Photography Project*. London: Giles.

Crutzen, Paul J., John R. McNeill, and Will Steffen. 2007. "The Anthropocene: Are Humans Now Overwhelming the Great Forces of Nature?" In *Ambio* 36(8): 614–21.

Cubitt, Sean. 2017. *Finite Media: Environmental Implications of Digital Technologies*. Durham NC: Duke University Press.

Dewey, John. 1934. *Art as experience*. New York: Minton.

Dierderichsen, Diedrich, and Anselm Franke (eds.). 2013. *The Whole Earth. California and the Disappearance of the Outside*. Berlin: Sternberg Press.

Encounters at the End of the World. 2007. Motion picture. Directed by Werner Herzog. Silver Spring: Discovery Films.

Escape Vehicle no. 6. 2004. DVD. Directed by Simon Faithfull. London: Arts Catalyst.

Fox, Robert J., Terri Gregory, Russell Hall, Jean Phillips, and Tony Wendricks (eds.). 1998. *Verner E. Suomi 1916–1995: A Man For All Seasons*. Madison WI: University of Wisconsin, Space Science and Engineering Center.

Giannetti, Louis. 1990. *Understanding Movies, 5th edition*. Englewood Cliffs CA: Prentice Hall.

Ghosh, Amitav. 2017. *The Great Derangement*. Chicago IL: University of Chicago Press.

Google Earth Engine. 2016. https://earthengine.google.com.

Hall, Edward T. 1963. "A System for the Notation of Proxemic Behaviour." In *American Anthropologist*, no. 65. 1003–26.

Henley, Paul. 2009. *The Adventure of the Real: Jean Rouch and the Craft of Ethnographic Cinema*. Chicago IL: The University of Chicago.

Huesemann, Michael, and Joyce Huesemann. 2011. *Techno-Fix: Why Technology Won't Save Us or the Environment*. Gabriola Island: New Society Publishers.

Jay, Martin. 1988, "Scopic Regimes of Modernity." *Vision and Visuality*, edited by Hal Foster. Seattle WA: Bay Press. 3–27.

Jobs, Steve. 2005. "2005 Stanford Commencement Address." Stanford CA. https://youtu.be/Hd_ptbiPoXM.

Kirk, Andrew. 2007. *Countercultural Green: The Whole Earth Catalog and American Environmentalism* Kansas City KA: University Press of Kansas.

Kozlof, Max. 2017 "New Documents Revisited." In *Arbus Friedlander Winogrand: New Documents, 1967*, edited by Sarah Hermanson Meister, 21–27. New York: Museum of Modern Art,

Larmon, Annie Godfrey. 2017. "Korakrit Arunanondchai at C L E A R I N G." *Art Viewer*, April 4, 2017. http://artviewer.org/korakrit-arunanondchai-at-c-l-e-a-r-i-n-g/.

Latour, Bruno. 2017. *Facing Gaia: Eight Lectures on the New Climatic Regime*. Cambridge: Polity Press.

Leviathan. 2012. Motion picture. Directed by Véréna Paravel and Lucien Castaing-Taylor. Brooklyn NY: The Cinema Guild.

Le Guin, Ursula K. 2017. "Deep in Admiration". In *Arts of Living on a Damaged Planet*, edited by Nils Bubandt, Elaine Gan, Heather Swanson, and Ann Tsing. Minneapolis MN: University of Minnesota Press.

Lewis, Simon L., and Mark A. Maslin. 2015. "Defining the Anthropocene." In *Nature*. 519 (March 12): 171–80.

MacDonald, Scott. 2015. *Avant-Doc: Intersections of Documentary & Avant-garde Cinema*. New York: Oxford University Press.

Maher, Neil. 2004. "Neil Maher on Shooting the Moon." In *Environmental History*, no. 9.3. 526–31.

Maher, Neil. 2017. *Apollo in the Age of Aquarius*. Cambridge MA: Harvard University Press.

Marks, Laura. 2000. *The Skin of the Film: intercultural cinema, embodiment, and the senses*. Durham NC: Duke University Press.

Marshall, Will. n.d. "Reimagine Change From Space: Using Satellites To Image The Entire Planet." Presentation on *Salesforce*. https://www.salesforce.com/video/183624/.

McHale, Brian, *The Cambridge Introduction to Postmodernism*. New York: Cambridge University Press.

Morton, Timothy. 2013. *Hyperobjects: Philosophy and Ecology after the End of the World*. Minneapolis MN: University of Minnesota Press.

The Museum of Modern Art. n.d. "Access to Tools: Publications from the Whole Earth Catalog, 1968–1974." https://www.moma.org/interactives/exhibitions/2011/AccesstoTools/.

NASA. 2017. "Apollo 8: Christmas at the Moon". https://www.nasa.gov/topics/history/features/apollo_8.html.

NASA. 2017. "Earthrise". https://www.nasa.gov/multimedia/imagegallery/image_feature_1249.html.

NASA Goddard Space Flight Center (@nasagoddard). 2015. "Nasa's Lunar Reconnaissance Orbiter (LRO) recently captured a unique view of Earth from the spacecraft's vantage point in orbit around the moon." Instagram photo, December 24, 2015. https://www.instagram.com/p/_pO0zet_Bg/.

NASA Space Science Data Coordinated Archive. 2018. https://nssdc.gsfc.nasa.gov/nmc/experimentDisplay.do?id=1967-111A-01.

Planet Labs. 2018. https://www.planet.com.

Russell, Catherine. 2015. "Leviathan and the Discourse of Sensory Ethnography: Spleen et ideal." In *Visual Anthropology Review*. 31(1): 27–34.

Schwerdtfeger Library Space Science and Engineering Center. 2009. "ATS-III Image Collection". https://library.ssec.wisc.edu/spinscan/about.php.

Silber, Kenneth. 2008. "Apollo 8: Flying into Space History during a Moment of Public Enthusiasm." In *Scientific American*. December 22, 2008. https://www.scientificamerican.com/article/apollo-8-space-history/.

TIME. 2016. "Timelapse." http://world.time.com/timelapse/.

Truscello, Michael. 2018. "Catastrophism and Its Critics: On the New Genre of Environmentalist Documentary Film." In *Interrogating the Anthropocene. Palgrave Studies in Educational Futures*, edited by Jan Jagodzinski, 257–75. Cham: Palgrave Macmillan.

Vertov, Dziga. 1984. *Kino-eye: The Writings of Dziga Vertov*, edited by Annette Michelson, trans. Kevin O'Brien. Berkeley CA: University of California Press.

Weule, Genelle. 2016. "Drones in science: Rising beyond pretty pictures." *ABC News*, August 18, 2016. https://www.abc.net.au/news/science/2016-08-18/drones-in-science-rising-beyond-pretty-pictures/7320738.

with History in a Room Filled with People with Funny Names 4. 2017. Motion picture and mixed media installation. Directed by Korakrit Arunanondchai. London: Carlos/Ishikawa.

List of Contributors

Hava Aldouby is senior lecturer at the Department of Literature, Language, and the Arts, The Open University of Israel. Her main field of research is moving-image art, looking at experimental cinema, video, and media arts. She is the author of *Federico Fellini: Painting in Film, Painting on Film* (Toronto: University of Toronto Press, 2013), and has contributed to *Ori Gersht: History Reflecting* (Boston MA: MFA Publications, 2014), as well as to *Borderlines: Essays on Mapping and the Logic of Place* (Sciendo, De Gruyter Poland and Da'at Hamakom, 2019), and to the *Wiley Blackwell Companion to Fellini* (forthcoming 2020), in addition to academic papers on experimental cinema and video art. Her current projects include behavioral and physiological experiments investigating haptic engagement with digital artworks, carried out in Parma, Italy, and in Tel Aviv, and a monograph on old master painting as it is reflected and refracted in contemporary video art.

Grant Bollmer is Assistant Professor of Media Studies at North Carolina State University, and an Honorary Associate of the Department of Media and Communication at the University of Sydney. He is the author of the books *Inhuman Networks: Social Media and the Archaeology of Connection* (2016) and *Theorizing Digital Cultures* (forthcoming). He is currently working on two projects, one on the varieties of materiality in media theory, and another on the history of empathy in psychology, aesthetics, and digital media.

Adam W. Brown is an internationally recognized conceptual artist whose work incorporates art and science hybrids including living and biological systems, robotics, molecular chemistry, and emerging technologies that take the form of installation, interactive objects, video, performance, and photography. Brown's creative research is informed by a background in Intermedia, a philosophy that provides a framework for breaking down and combining different models of thought and bringing together disparate disciplines, leading to the establishment of new forms of research and creative activity. To this end, most of his creative and research endeavors are collaborative in nature. Brown is currently an Associate Professor at Michigan State University where he created a new area of study called Electronic Art & Intermedia. He is also a Research Fellow at the Institute for Digital Intermedia Arts at Ball State University, and serves as an Artist in Residence for the Michigan State University BEACON (Bio/Computational Evolution in Action Consortium) project, funded by the NSF. Previously he was an Associate Professor at the University of Oklahoma, where he developed an electronic art program called Symbiotic Media. He received his BA, MA, and MFA degrees from the University of Iowa. His most recent project, *The Great Work of the Metal Lover* (with Dr. Kazem Kashefi), is an artwork that sits at the intersection of art, science, and alchemy. The piece received an Honorary Mention and was exhibited at Ars Electronica 2012 and received an Award of Distinction from Vida 14. *Origins of Life: Experiment #1*.x, is a working scientific experiment using simulated lightning, heat, and primordial gases that has been repositioned as an art installation (with Dr. Robert Root Bernstein). The 'open source' project, which invites contributions and participation from other scientists, builds on Miller's 1953 iconic experiment. In 2011 the piece was selected as part of Ars Electronica and the Synth-ethic exhibition in Vienna. In 2012, Brown and Root-Bernstein received a grant from the National Science Foundation to continue this project. His earlier work *Bion* (with Dr. Andrew Fagg) makes reference to an individual element of primordial biological energy identified as 'orgone' by the scientist Wilhelm Reich (1897–1957). The interactive installation is a sensor network composed of more than 1,000 mass-produced, three-dimensional glowing and chirping autonomous robots called bions. In 2006 Brown was selected as an emerging artist to exhibit this piece in Archival to the Contemporary, Six Decades of the Sculptors Guild, and that same year it was also selected for SigGraph in Boston. In 2010 the work appeared in the Brazilian Biennial Emoção Art.ficial 5.0 in Sao Paulo. Brown's work has been written about widely in publications such as the New York Times, Wired, Nature, Sculpture Magazine, Washington Post, Forbes, The Proceedings of the National Academy of Sciences (PNAS), Discover, and the Huffington Post.

Aaron Burton is a lecturer in Media Arts at the University of Wollongong, Australia. Burton's doctoral thesis 'Provenance in Personal Documentary: *My Mother's Village*' (2014) draws on the disciplines of photography, film, ethnography, and contemporary art. The feature-length documentary film produced as the creative component to this PhD has screened internationally and is distributed through Ronin Films. Burton has wide-ranging industry experience in film-making and photography. His creative media-arts practice includes numerous prizes and exhibitions, such as; 2012 Colombo Art Biennale, *Hatched 07* National Graduate Show (PICA), Brisbane Airport Fresh Cut 2009 (IMA), and the inaugural Jeremy Hynes Award 2009 (IMA). Research interests include the role of provenance in contemporary art, recasting science communication, and merging experimental ethnography with personal documentary—manifesting in the film *Sunset Ethnography* (2014) with professors Michael Taussig and Stephen Muecke.

Wendy Jo Coones directs the Exhibition Strategies Division at the Department for Image Science at the Danube University in Krems, Austria. She received a Fine Arts degree in New Genres at the San Francisco Art Institute. After receiving an M.Ed. in Educational Research & Philosophy, she began working in museums as an exhibition developer and visitor researcher of international cultural and scientific exhibits. Institutions where she has worked range from a Science & History Museum in Texas, a Space Science Center in California to the Max Plank Institute for the History of Science in Berlin. Since 2005 she has been on the academic staff at the Department for Image Science responsible for curricula development, teaching working professionals, and support of research initiatives. She is responsible for Master of Arts programs related to digital cultural life, its histories and futures, and is currently primary coordinating staff for an EU-funded Erasmus Mundus European Master of Excellence program in Media Arts Cultures. An upcoming publication with DeGruyter Press as co-editor *Museum and Archive on the Move* includes her text Museum on Mars: re-define, re-auratize, re-territorialize. She has worked on the Archive of Digital Art (formerly Database of Virtual Art)) since 1999.

Derek Curry is Assistant Professor at Northeastern University in Boston and is an artist-researcher whose work addresses spaces for automated decision-making systems. His recent work has addressed automated decision-making processes used by automated stock trading systems and Open Source Intelligence gathering (OSINT). His artworks have replicated aspects of social media surveillance systems and communicated with algorithmic trading bots. Derek earned his MFA in New Genres from UCLA's Department of Art in 2010 and his PhD from the State University of New York at Buffalo. He is currently an Assistant Professor at Northeastern University in Boston. http://derekcrry.com

Jennifer Gradecki is a PhD candidate in Visual Studies at SUNY Buffalo and a visiting professor at Northeastern University in Boston. She is an artist-theorist who aims to facilitate a practice-based understanding of socio-technical systems that typically evade public scrutiny. Using methods from institutional critique, tactical media, and information activism, she investigates information as a source of power and resistance. Her investigations have focused on Institutional Review Boards, financial instruments, and, most recently, technologies of mass surveillance. She received her MFA in New Genres from UCLA in 2010. www.jennifergradecki.com

Tsila Hassine is an artist and researcher, a PhD candidate at the Art&Media laboratory, Paris 1 Pantheon Sorbonne. Her work examines the shifts in contemporary culture brought about by accelerated digitization processes. Her works question the price we pay for constant connectivity. She holds a BSc in Pure Mathematics from Tel Aviv University, an MFA in Media Design from the Piet Zwart Institute in Rotterdam, and she was a research fellow at the Jan van Eyck Academie in Maastricht. Her work has been exhibited at the Contemporary Art Center, Tel Aviv; The Israeli Center for Digital Art; the Jerusalem Science Museum; Transmediale, Berlin; Van Abbemuseum, Eindhoven; Smart Project Space, Amsterdam; Gasworks, London; the Venice Biennale, and more.

Jens Hauser is a postdoctoral fellow in art & bio- and natural sciences/technology, at IKK & Medical Museion, Copenhagen University. He is a Paris- and Copenhagen-based art curator and media studies scholar focusing on the interactions between art and technology. He holds a dual research position at both the Department of Arts and Cultural Studies and the Medical Museion at the University of Copenhagen, and is a distinguished affiliated faculty member of the Department of Art, Art History and Design at Michigan State University. His curated exhibitions include *L'Art Biotech* (Nantes, 2003), *Still, Living*(Perth, 2007), *sk-interfaces* (Liverpool, 2008/Luxembourg, 2009), *the Article Biennale* (Stavanger, 2008), *Transbiotics* (Riga 2010), *Fingerprints…* (Berlin, 2011/Munich/2012), and *Synthethic* (Vienna, 2011).

Jonas Jørgensen is Assistant Professor at the University of Southern Denmark (SDU). He is formally trained as a physicist (BSc) and an art historian (BA, MA) at Copenhagen University and Columbia University (New York). He did his PhD on soft robot aesthetics at the IT University of Copenhagen as a member of the Robotics Evolution and Art Lab (REAL). Jørgensen is currently involved in establishing a newly founded Research Center for Soft Robotics at the University of Southern Denmark.

Olga Kisseleva is an international artist, and the founder of the Art & Science laboratory at the University of Paris 1 Panthéon Sorbonne. She conducts research on creative processes in conjunction with contemporary science and technology. This work was presented at the MoMA (New York, USA), the Pompidou Center (Paris, France), the National Center for Contemporary Art (Moscow, Russia), the ARC (Paris, France), KIASMA Helsinki, Finland), the Consortium (Dijon, France), the Musee Nacional Centro de Arte Reina Sofia (Madrid, Spain), the Venice Biennials, Istanbul, Dakar, Tirana, Rennes, Moscow, and Berlin.

Ryszard Kluszczyński is Full Professor and Head of the Department of Electronic Media, University of Łódź. Kluszczyński (PhD, media art scholar, writer and curator) investigates issues of new media arts and cyberculture, contemporary art theory, avant-garde, and recent art practices with a focus on interactions between art, science, and technology. His recent book publications are *Human Traits. Patrick Tresset and the Art of Creative Machines* (2016); *Guy Ben-Ary: Nervoplastica. Bio-robotic Art and its Cultural Contexts* (2015); *Ken Feingold: Figures of Speech* (2014); *Meat, Metal & Code/Contestable Chimeras: Stelarc* (2014); *Robotic Art and Culture. Bill Vorn and His Hysterical Machines* (2014); *Wonderful Life: Laurent Mignonneau + Christa Sommerer* (2012); *Crude Life. The Tissue Culture & Art Project: Oron Catts + Ionat Zurr*(2012); *Towards the Third Culture. The Co-Existence of Art, Science and Technology* (2011); and *Interactive Art. From Artwork-Instrument to Interactive*

Spectacle, 2010. In 1990–2001 he was a Chief Curator of Film, Video, and Multimedia Arts in the Center for Contemporary Art – Ujazdowski Castle in Warsaw. In 2010 he co-curated with Tsutomu Mizusawa 'Beyond Mediations,' the main exhibition of the Second International Biennale of Contemporary Art in Poznan, Poland. Since 2011 he has held the position of the Artistic Director of Art + Science Meeting Project in the Centre for Contemporary Art in Gdansk. His most recent exhibition there was Masaki Fujihata's *Augmenting the World* (2017).

Daniel Landau is a media artist, researcher, and lecturer. He completed his MA degree in music-technology at the Royal Conservatory in The Netherlands, where he lived and worked for close to a decade. He is currently a PhD candidate at the Department of Media, School of Arts, Design and Architecture, Aalto University, Helsinki, Finland. Landau's work resides in the intersection of Art, Technology, and Science—exploring the complex relationship between body and technology. Core to his work is the attempt to trace techno-political processes and their impact on social and private spaces. His work has been presented in major venues, museums, and festivals worldwide including Jerusalem, New York, London, Amsterdam, Berlin, Tokyo, Paris, Hungary, and Mexico City. Between 2012 and 2016 Daniel led the Media Studies Department at the Midrasha Faculty of the Arts, Beit Berl Academic College, and is currently a Senior Research Fellow at the Interdisciplinary Centre, Herzliya. Daniel is the Co-founder and Director of Oh-Man, Oh-Machine, an art, science, and technology platform that includes an international conference, workshops, and a research lab.

Manuela Naveau is an artist and curator at Ars Electronica Linz. Together with the artistic and managing director Gerfried Stocker she developed Ars Electronica Export. She teaches at the University of Art and Design Linz, as well as at the Paris Lodron University in Salzburg, and Danube University Krems. Her research investigates networks and knowledge in the context of computer-based artistic practice. Her book *Crowd and Art – Kunst und Partizipation im Internet* (Crowd and Art – Art and Participation on the Internet) was published in 2017 by Transcript Verlag, Germany. The book is based on Naveau's PhD dissertation, which received the Award of Excellence from the Austrian Ministry of Science, Research and Economy in 2016. www.manuelanaveau.at; www.crowdandart.at.

Ziv Neeman holds a BFA in Fine Arts from Bezalel Academy of Art & Design, and a PhD in literature and cultural studies from Columbia University, New York. His postdoctoral period was carried out at the Michigan Society of Fellows. His current interdisciplinary interests include: critical theory, digital studies, STS, and contemporary art.

Andrea Pinotti is Professor of Aesthetics at the University of Milan. His research focuses on image theories and visual culture studies; collective and social memory in connection with memorialization and monumentality; psychological and phenomenological aesthetics; empathy theories. He has been fellow of various international institutions (among which are the Italian Academy at Columbia University, the Warburg Institute in London, the Zentrum für Literatur- und Kulturforschung in Berlin, and the Institut d'études avancées in Paris). Among his recent publications are the volumes *Empathie. Histoire d'une idée de Platon au post-humain* (Paris: Vrin 2016) and *Cultura visuale. Immagini sguardi media dispositivi* (with Antonio Somaini, Torino: Einaudi 2016).

Paul Sermon is Professor of Visual Communication at the University of Brighton, School of Art, Grand Parade, Brighton, UK. Since 2014 he has produced five new public installation artworks, participated in two international AHRC workshops, and led an AHRC follow-on research project, as well as presenting five new peer reviewed conference papers. Born in Oxford, England, 1966, Sermon studied a BA(Hons) Fine Art degree under Professor Roy Ascott at Newport School of Fine Art from September 1985 to June 1988 and studied for an MFA Fine Arts degree at the University of Reading, England, from October 1989 to June 1991. He was awarded the Prix Ars Electronica Golden Nica, in the category of interactive art, for the hyper-media installation 'Think about the People now' in Linz, Austria, in September 1991. He produced the ISDN videoconference installation 'Telematic Vision' as an Artist in Residence at the ZKM Center for Art and Media in Karlsruhe, Germany, from February to November 1993. He received the Sparkey Award from the Interactive Media Festival in Los Angeles for the telepresent installation 'Telematic Dreaming' in June 1994. From 1993 to 1999 he worked as Dozent for Media Art at the HGB Academy of Fine Arts, Leipzig, Germany. During this time he continued to produce further interactive telematic installations including 'Telamatic Encounter' in 1996 and 'The Tables Turned' in 1997 for the Ars Electronica Centre in Linz, and the ZKM Media Museum in Karlsruhe. From 2000 to 2013 he was Professor of Creative Technology at the University of Salford, School of Arts & Media. From 1997 to 2001 he was Guest Professor for Performance and Environment at the University of Art and Design Linz, Austria.

Yvonne Volkart is a lecturer in art theory and researcher at the Academy of Art and Design, FHNW Basel. Her current projects include *Ecodata – Ecomedia – Ecoaesthetics. The Role and Significance of Technologies and Technoscientific Methods in the Arts for the Perception and Awareness of the Ecological* (with Marcus Maeder, Rasa Smite, and Aline Veillat); *Times of Waste*. Her finished projects are *RhyCycling. Aesthetics of Sustainability in the Basel Border Area*; *Cyborg Bodies*, *Mediaartnet* (ZHdK/ZKM Karlsruhe); She was curator at Shedhalle Zurich (2009–12). Living in Zurich, she regularly writes for *Springerin*. Her texts and curatorial projects are *Eco_Visionaries*, a touring exhibition and book, with Sabine Himmelsbach and Karin Ohlenschläger; "From Trash to Waste. On Art's Media Geology", *Texte zur Kunst*, December 2017; "Art and Ecology in the Technosphere" (2017), in M. Maeder (ed.), *Kunst Wissenschaft Natur*; *Ecologies of Existence. Art and Media beyond the Anthropocene* (conference with Christoph Brunner, 2016), *Subverting Disambiguities* (with A. Hoffmann, 2012); *Ecomedia. Ecological Strategies in Today's Art* (with S. Himmelsbach and K. Ohlenschläger, 2007). https://www.fhnw.ch/de/personen/yvonne-volkart.

Index

Color Plates

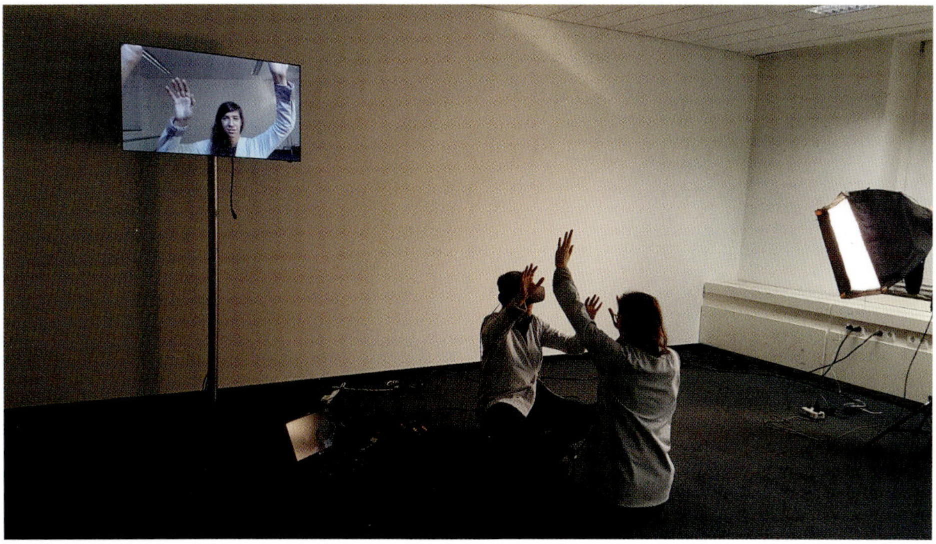

Plate 1. Recording session B3 Festival – November 2017, Frankfurt (p. 49)

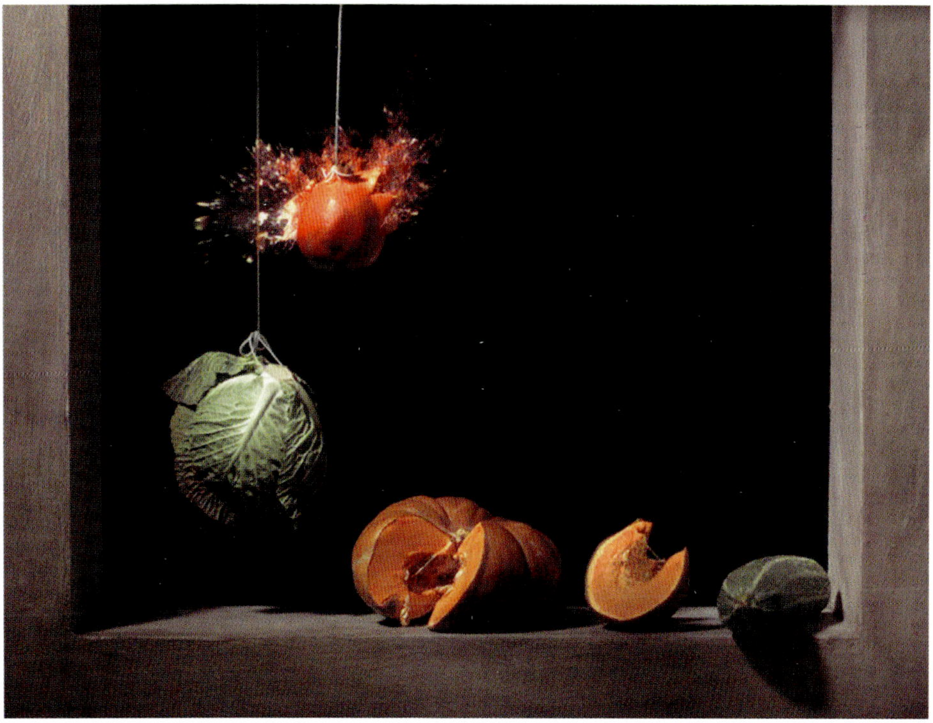

Plate 2. Ori Gersht, *Pomegranate*, *Off Balance*. 2006. Archival ink print, 30 × 40 cm.
Image courtesy of the artist (Fig. 1, p. 63).

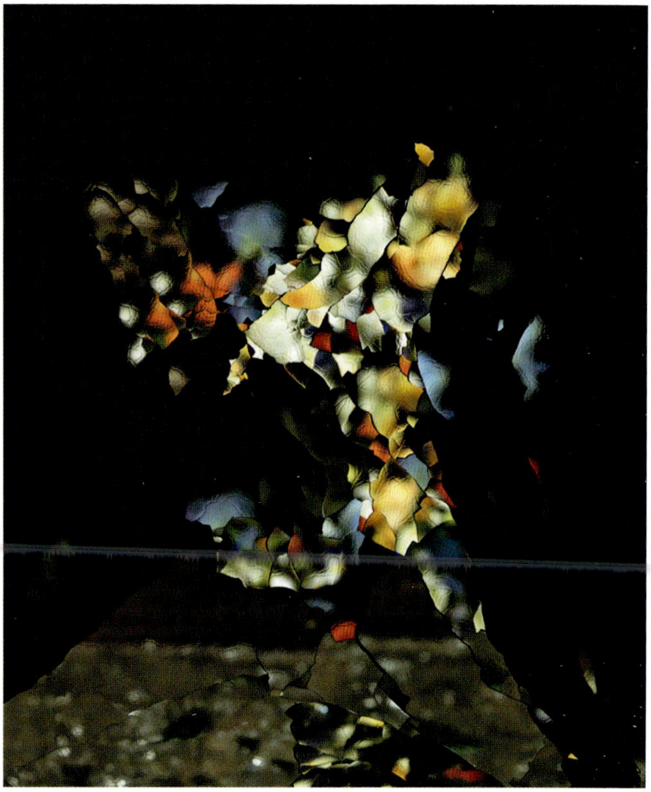

Plate 3. Ori Gersht, *On Reflection, Material 11*, 2014. Still image from HD video.
Image courtesy of the artist (Fig. 2, p. 64).

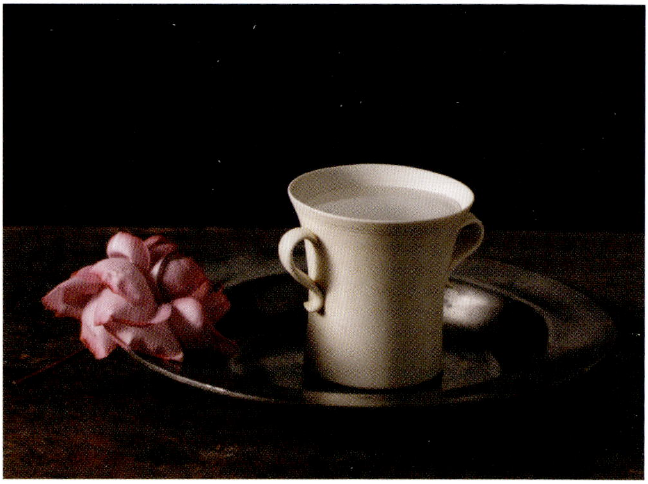

Plate 4. Ori Gersht, from the series *New Orders* (2018). Image courtesy of the artist. (Fig. 3, p. 65).

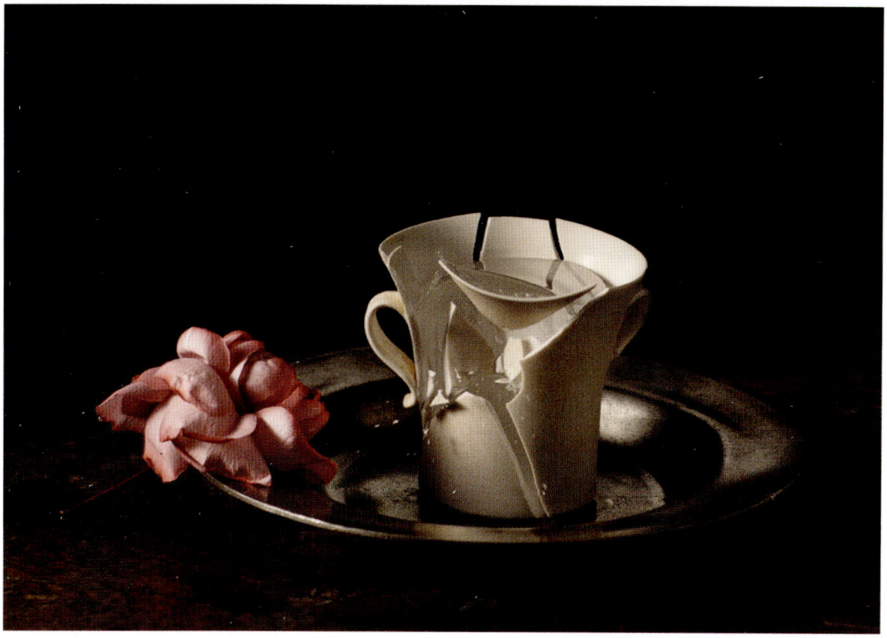

Plate 5. Ori Gersht, from the series *New Orders* (2018). Image courtesy of the artist. (Fig. 4, p. 65).

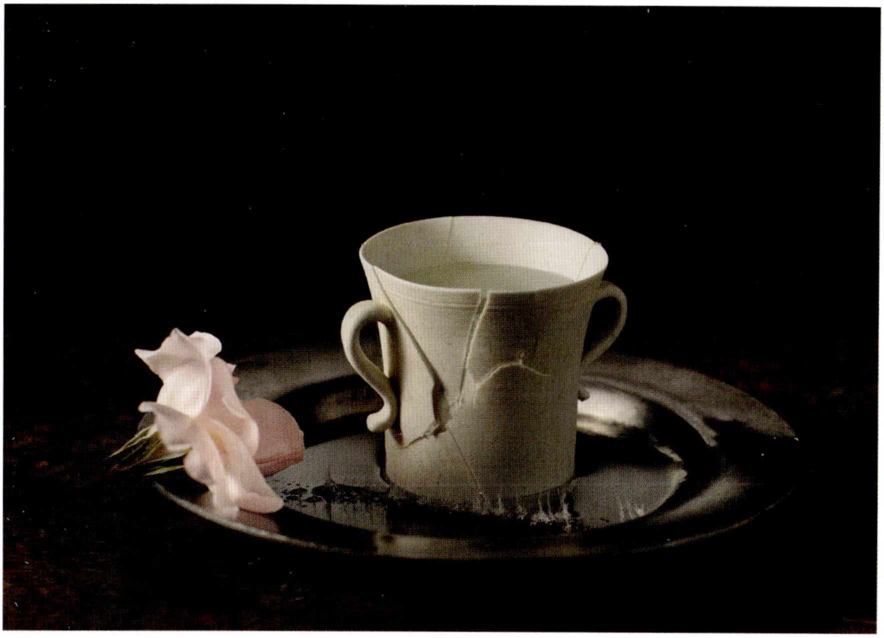

Plate 6. Ori Gersht, from the series *New Orders* (2018). Image courtesy of the artist. (Fig. 5, p. 65).

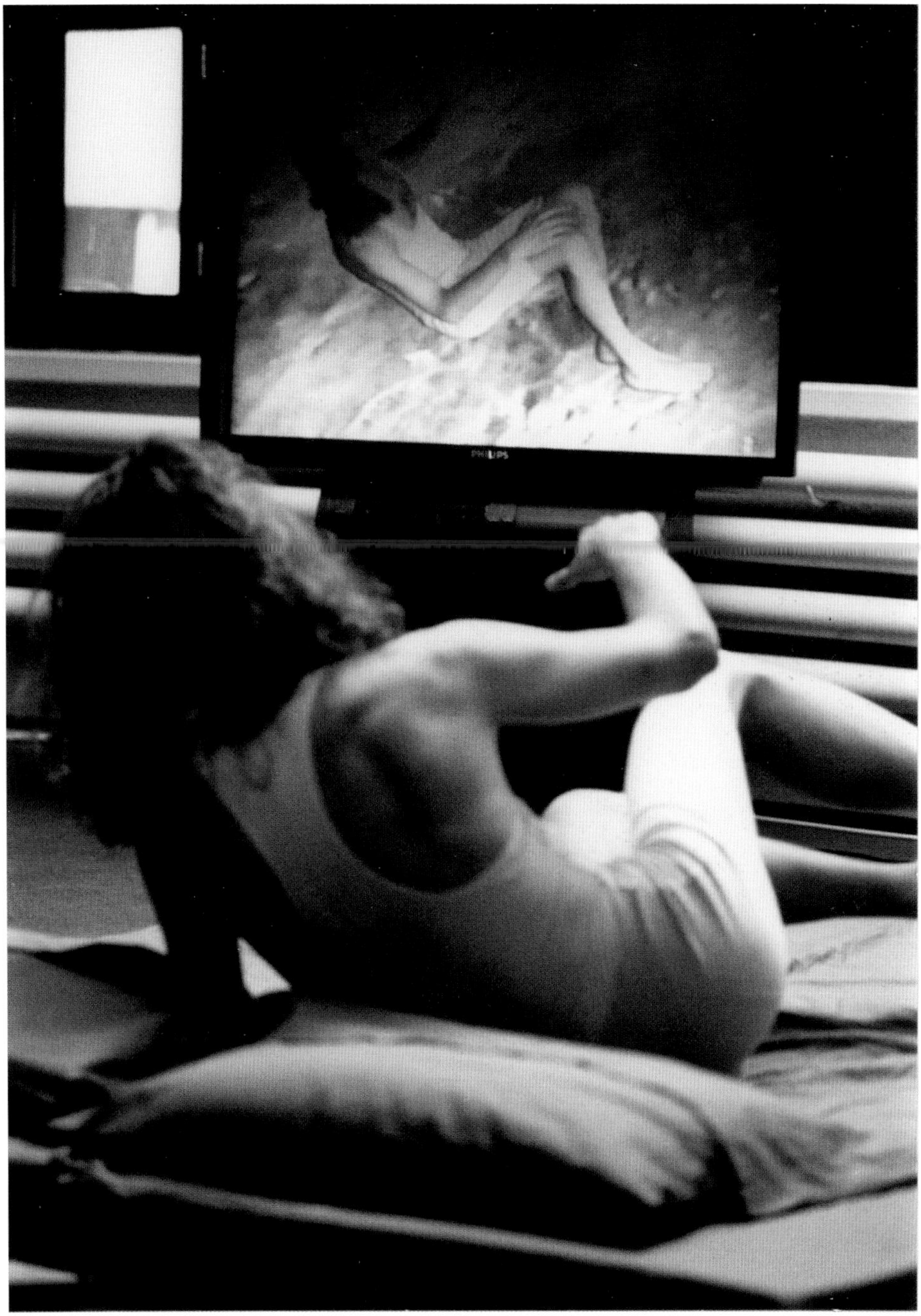

Plate 7. Paul Sermon participating as a performer in 'Telemtic Dreaming' at the Telegalleria.
(Fig. 1, p. 77)

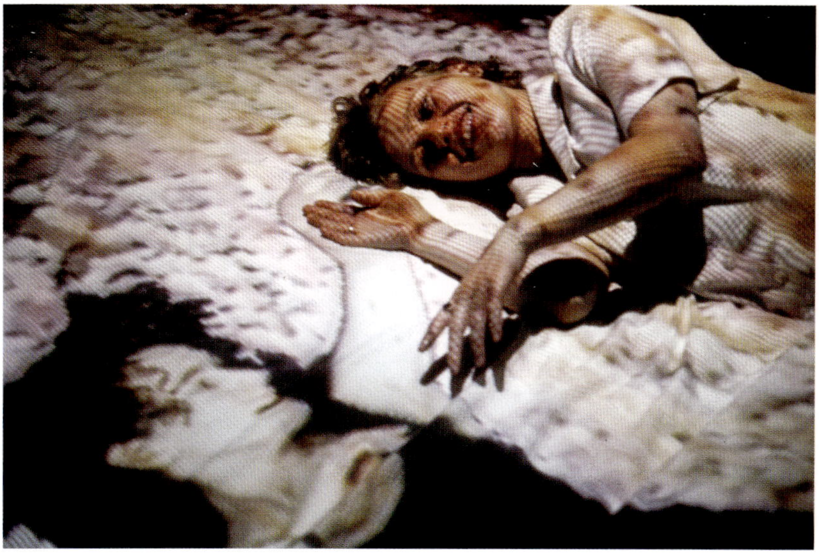

Plate 8. Projected image of Susan Kozel with a gallery participant in 'Telematic Dreaming.' (Fig. 2, p. 79)

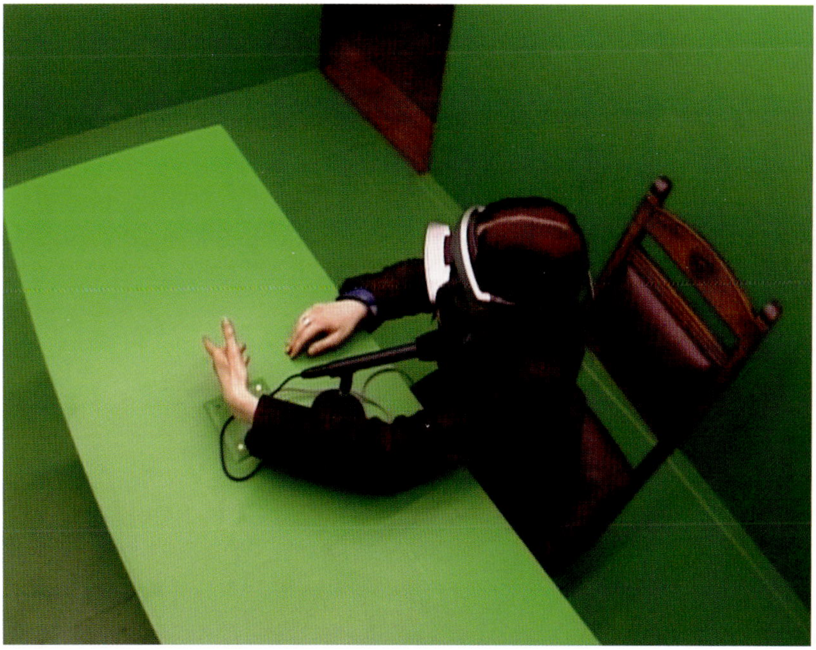

Plate 9. Participant seated at the green table interacting and conversing in 'Peace Talks.' (Fig. 3, p. 81)

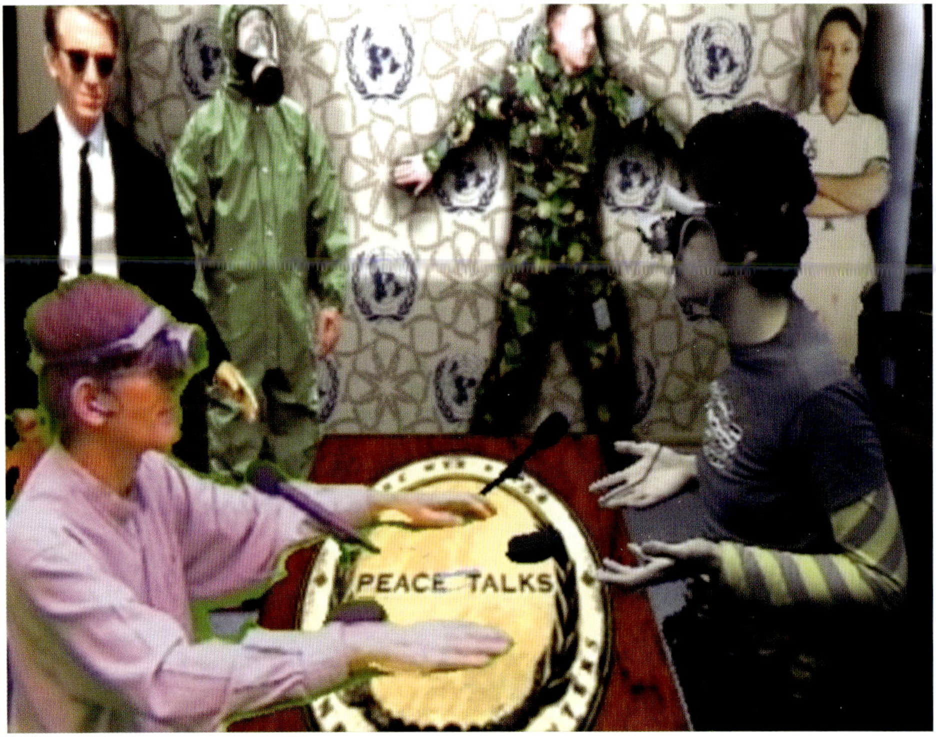

Plate 10. Line-out video still of participants conversing in the virtual 'Peace Talks' environment.
(Fig. 4, p. 82)

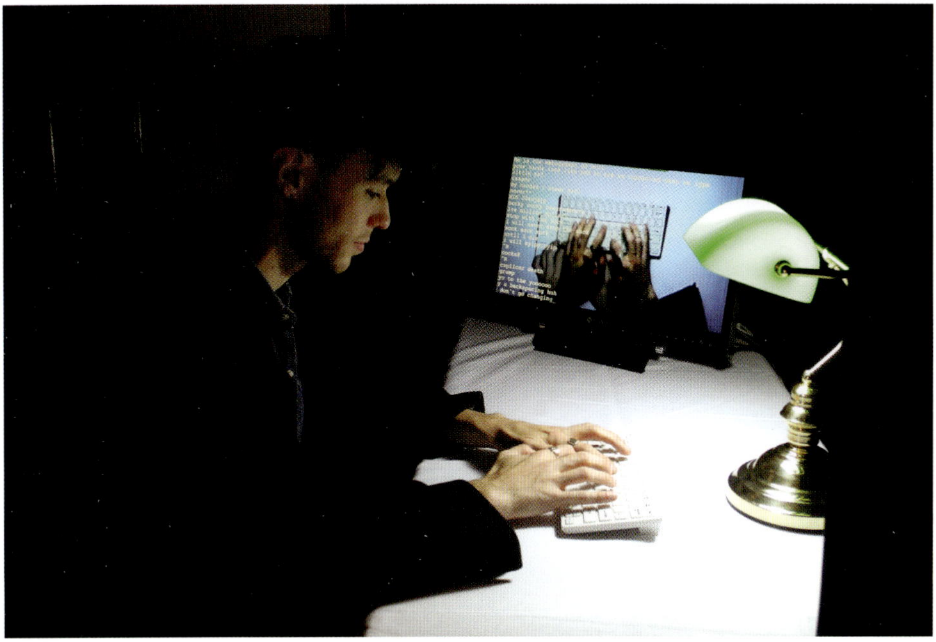

Plate 11. Camera capturing image of participant's hands on the keyboard below in 'Touched.
(Fig. 5, p. 84).

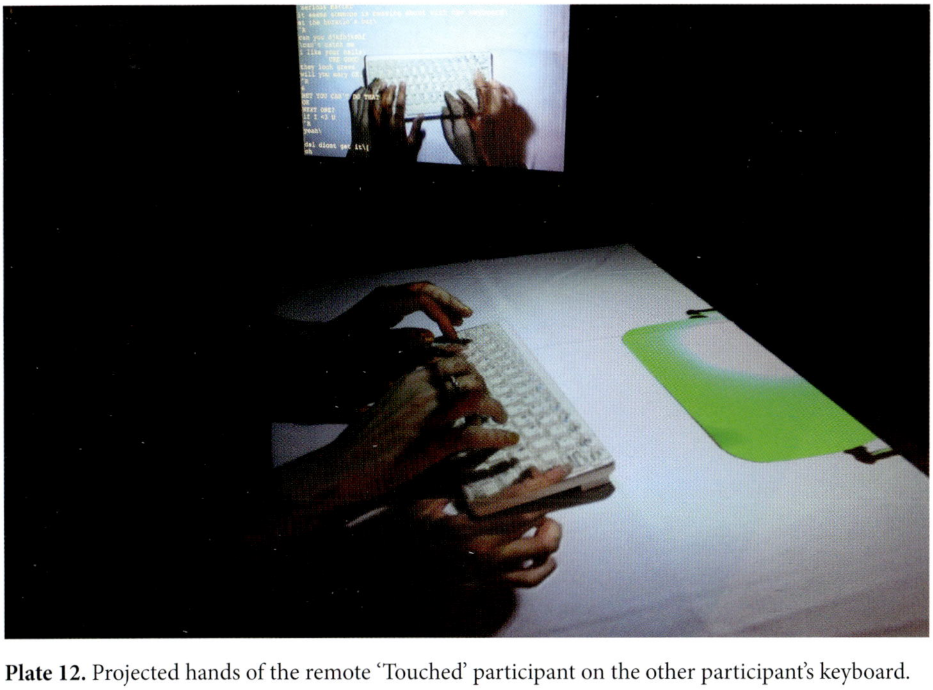

Plate 12. Projected hands of the remote 'Touched' participant on the other participant's keyboard.
(Fig. 6, p. 85)

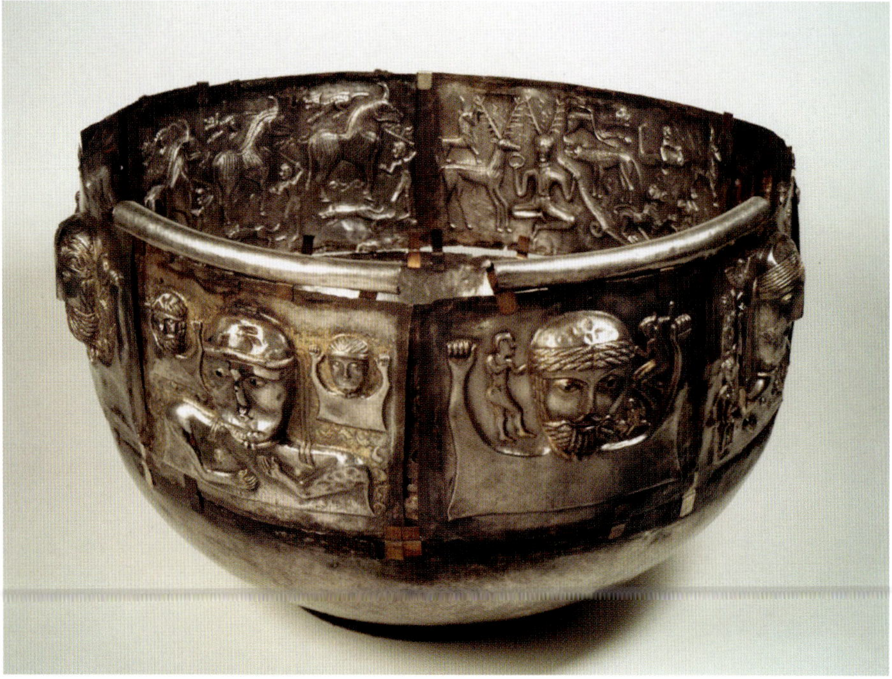

Plate 13. Gundestrup Caldron. European Iron Age silver vessel assembled between 150 BC and 1 BC, found 1891 in Gundestrup, northern central Denmark.

Photo: Nationalmuseet (Wikimedia CC-BY-3.0).

(Fig. 3, p. 102)

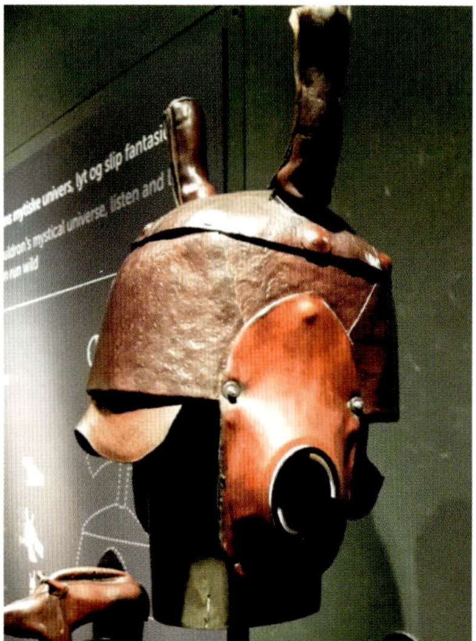

Plate 14. Gundestrup Helmet. Leather "Viking style" helmet, which covers the eyes of the wearer and integrates headphones that play the exhibition's audio description of the relief wall.

Photo: Mie Ellekilde, with kind permission.

(Fig. 4, p. 102)

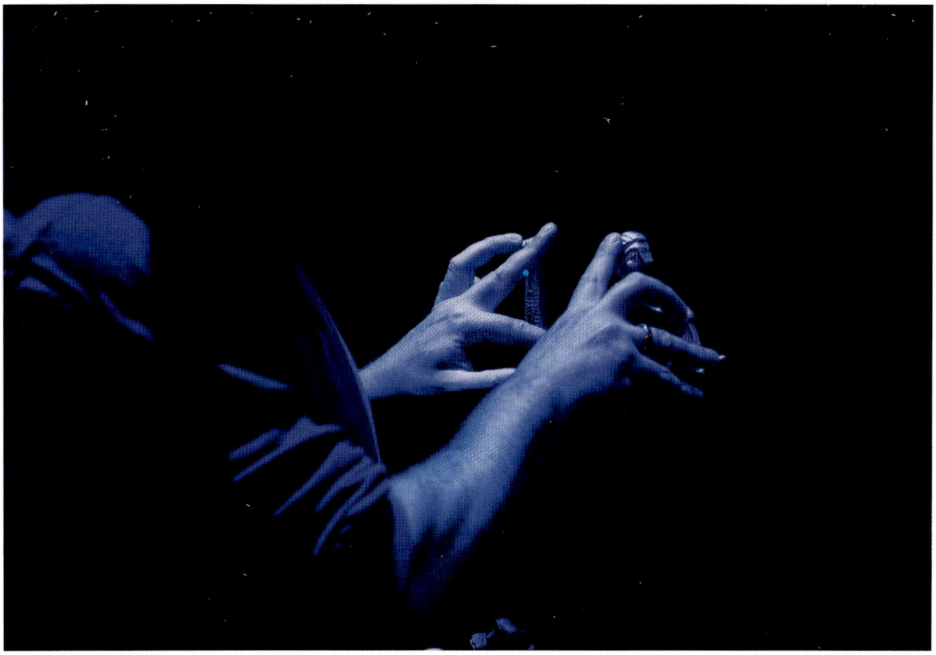

Plate 15. Detail view of sensing relief figures of Gundestrup Caldron installation at the Moesgaard Museum. Photo: Moesgaard Museum, with kind permission. (Fig. 6, p. 103)

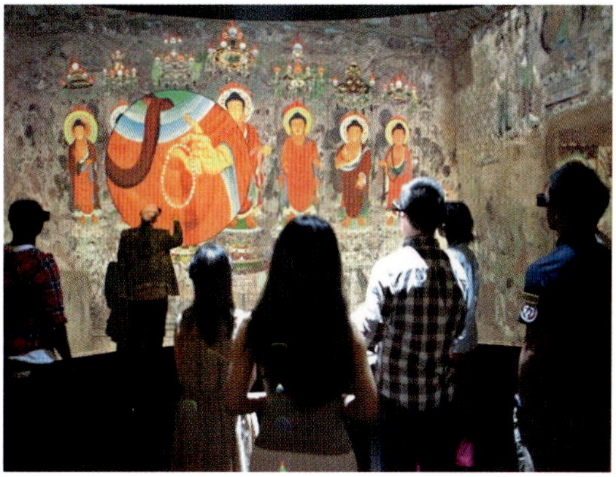

Plate 16. *Pure Land 360*. Close-up view highlighting 3-D interface with detailed information. Jeffrey Shaw, Sarah Kenderdine, Cédric Maridet (sound). With kind permission of the artists. (Fig. 8, p. 106)

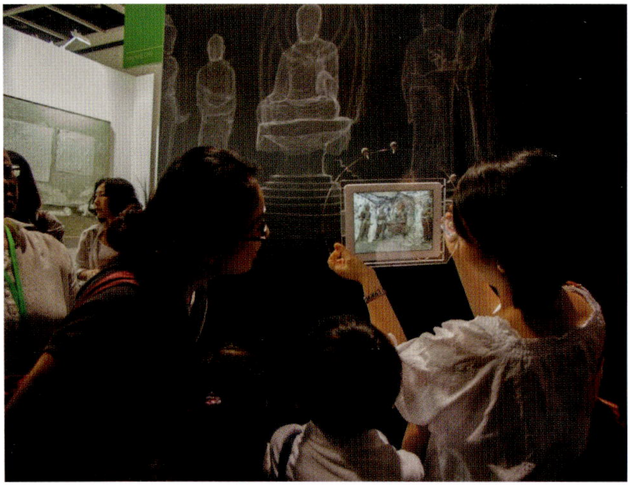

Plate 17. *Pure Land AR.* Close-up and view of images and content as seen by the viewer via tracked ipad. Jeffrey Shaw, Sarah Kenderdine, Leith Chan, (with kind permission of the artists.) (Fig. 10, p. 107)

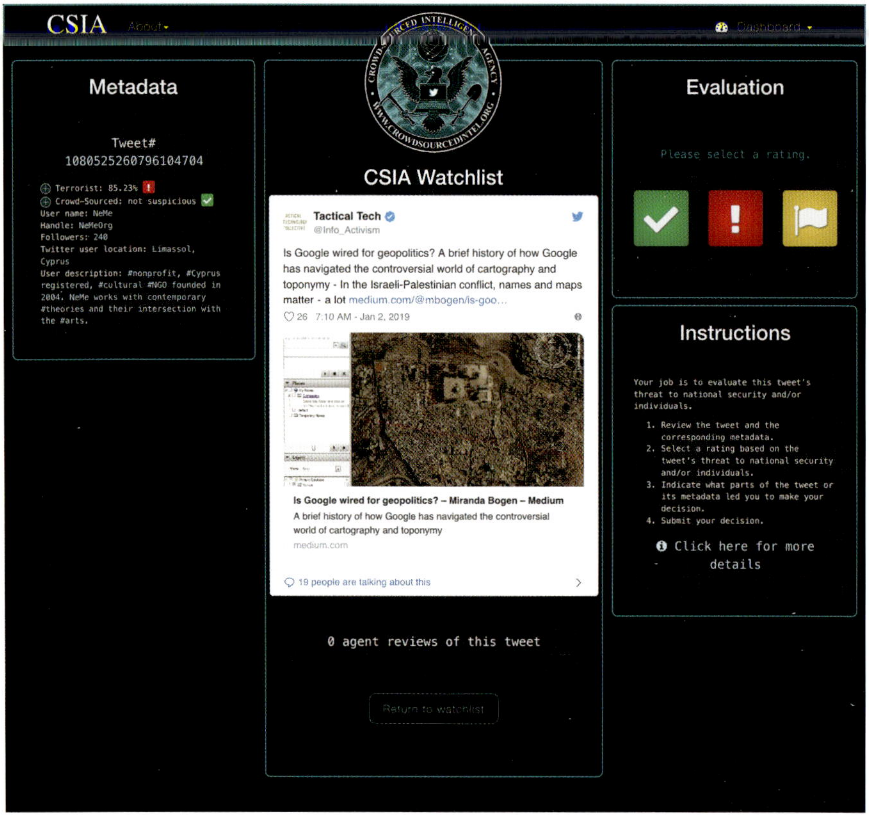

Plate 18. The CSIA Watchlist. (Fig. 2, p. 123)

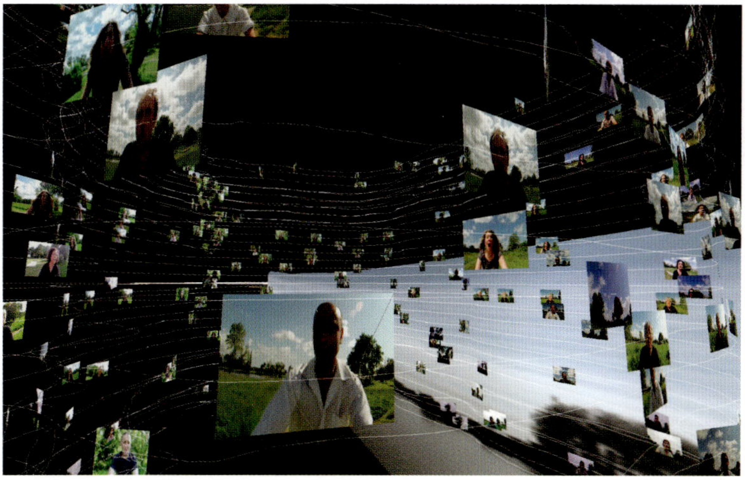

Plate 19. Masaki Fujihata, *Voices of Aliveness.*

(Fig. 1, p. 147)

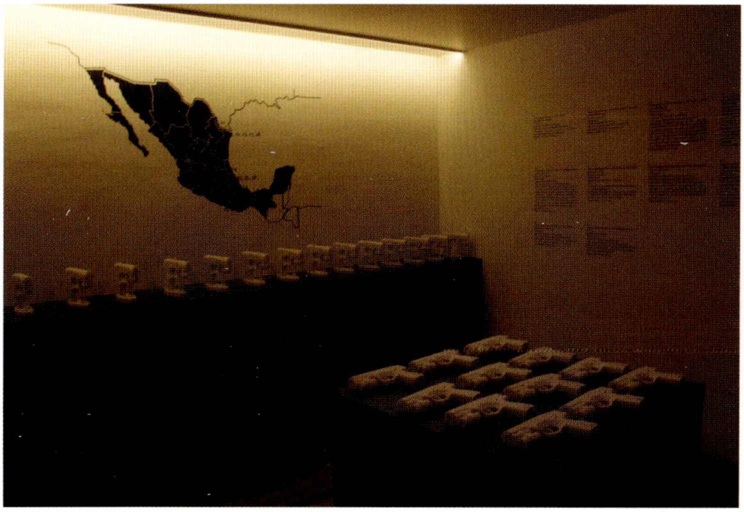

Plate 20. Luz María Sánchez, V.U(n)f_1.01 (2017). Multi-channel asynchronous sound installation: 40 portable digital speakers* in the shape of a Caracal 9mm pistol [3D prints], 40 Micro-SD cards, 40 mp3 sounds, modular structure. Dimensions: variable. Duration: undetermined. Installation view: Modos de oír. Prácticas de arte y sonido en México. 28 November 2018 – 31 March 2019. Laboratorio de Arte Alameda, Mexico City. Mexico.
Photo credit: Marcos Ysair Pérez Botello. (Fig. 4, p. 154)

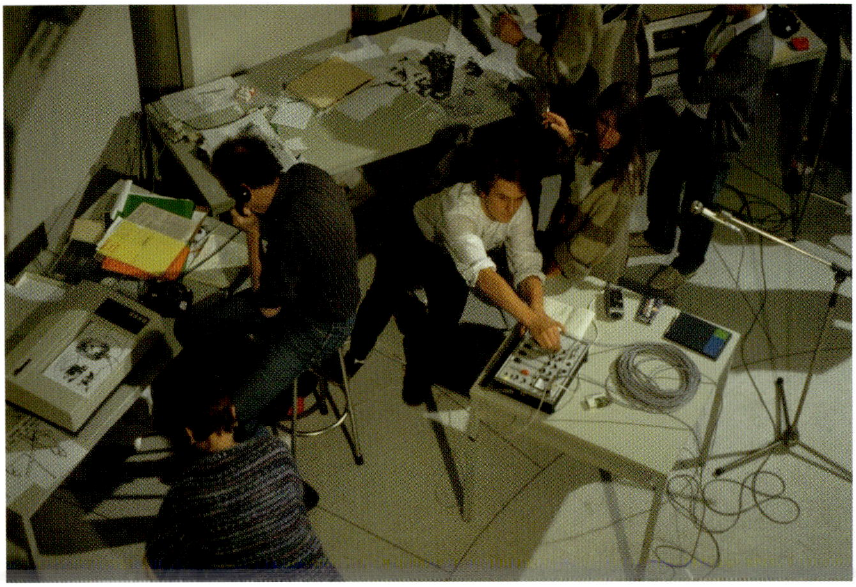

Plate 21. Robert Adrian X, *Die Welt in 24 Stunden*, Ars Electronica Festival, Linz, 1982. Photo: Sepp Schaffler. (Fig. 2, p. 164)

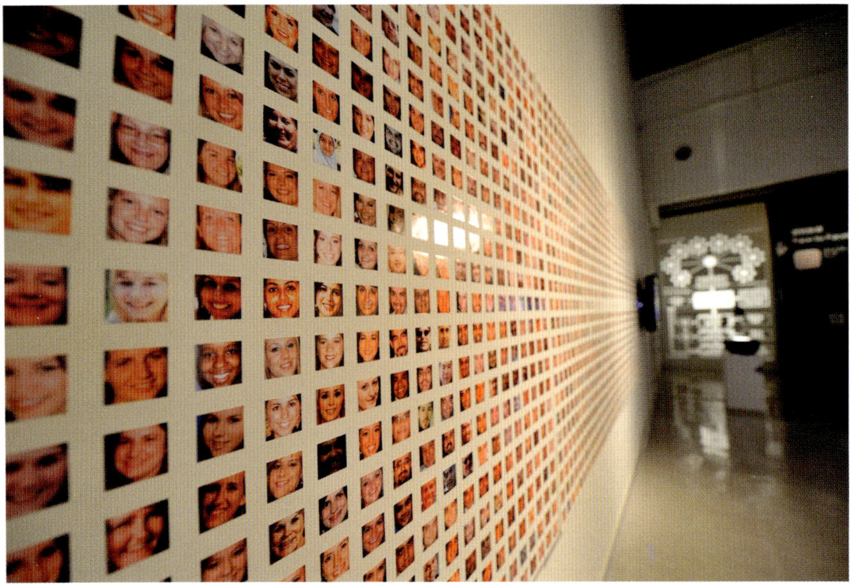

Plate 22. Paolo Cirio and Alessandro Ludovico, *Face to Facebook,* Ars Electronica Export in Taichung 2012. Photo: David Sun / Chi River Art International. (Fig. 5, p. 169)

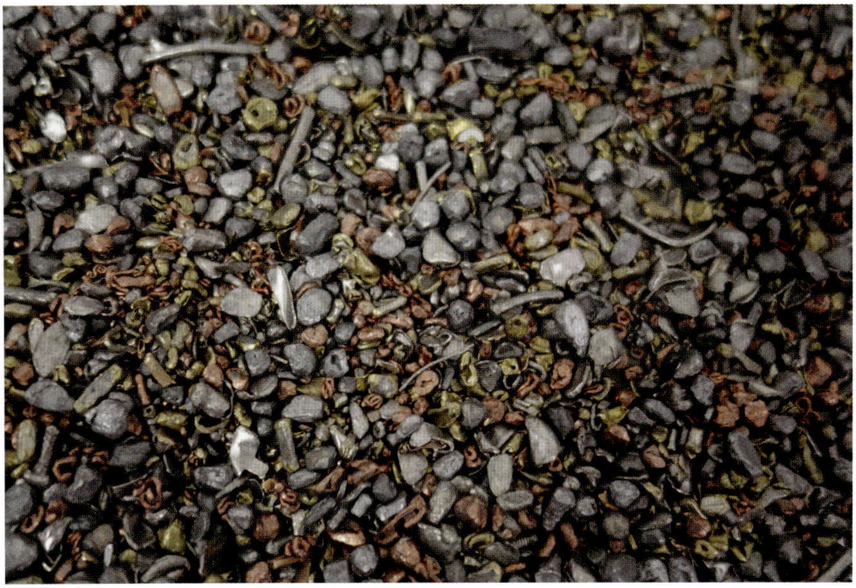

Plate 23. Granules resulting from the recycling process of e-waste at Immark AG Regensdorf/ Switzerland, 2015. Courtesy of Team *Times of Waste.* (Fig. 2, p. 180)

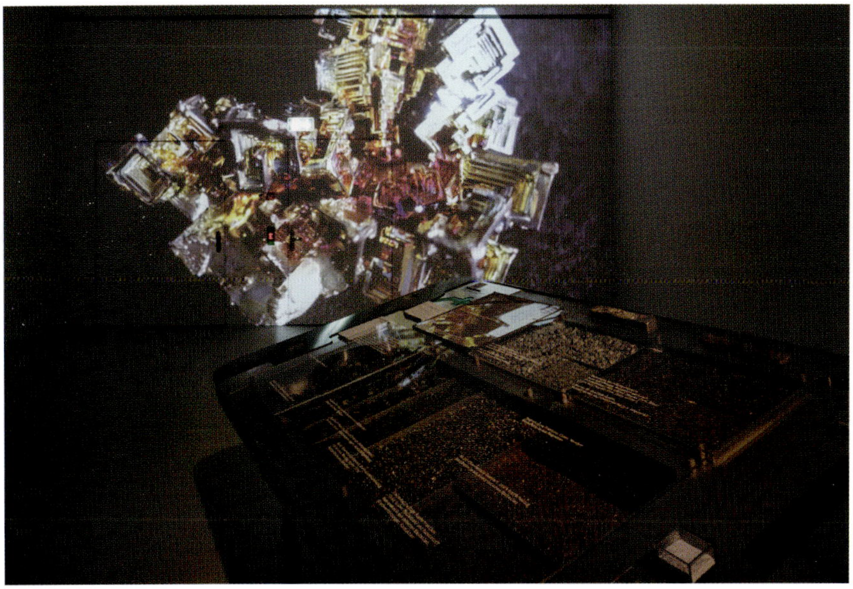

Plate 24. Research team 'Times of Waste': Exhibition Times of Waste—The Leftover, Museum der Kulturen, Basel 2017. Courtesy of Team Times of Waste, © Omar Lemke. (Fig. 3, p. 181)

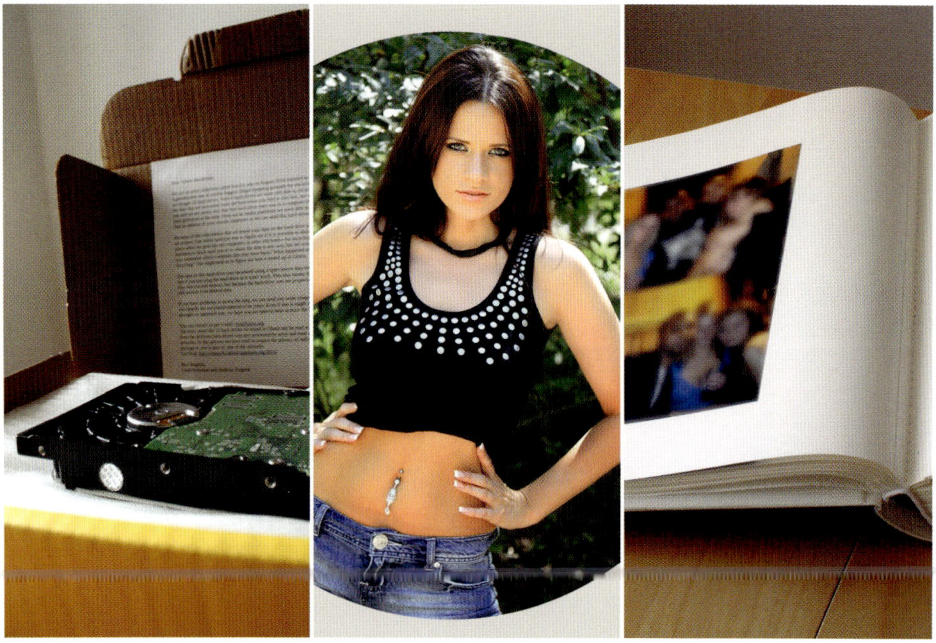

Plate 25. KairUs: *Forensic Fantasies*, 2016. Courtesy of KairUs. (Fig. 4, p. 184)

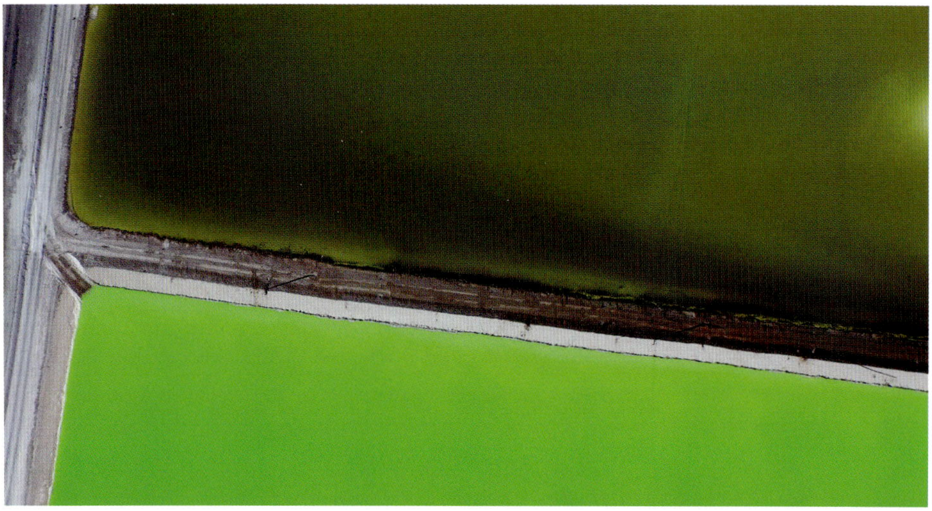

Plate 26. Unknown Fields Division: *We Power Our Future with the Breastmilk of Volcanoes / Lithium Dreams*, 2015/18. Courtesy of Unknown Fields Division. (Fig. 5, p. 186)

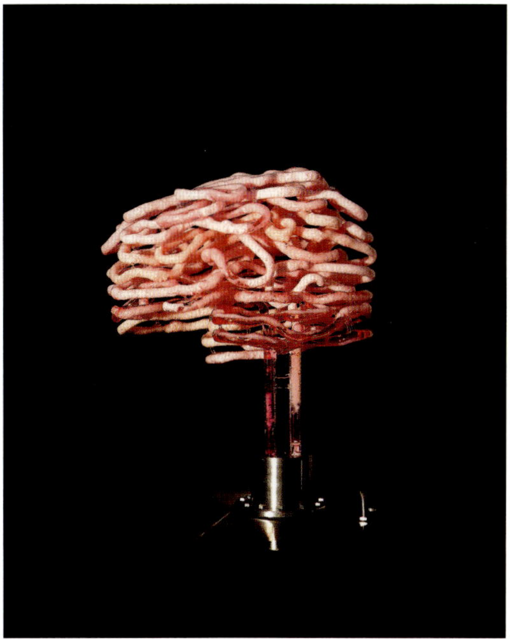

Plate 27. Thomas Feuerstein, *Pancreas* (2009), photo by the artist. (Fig. 1, p. 205)

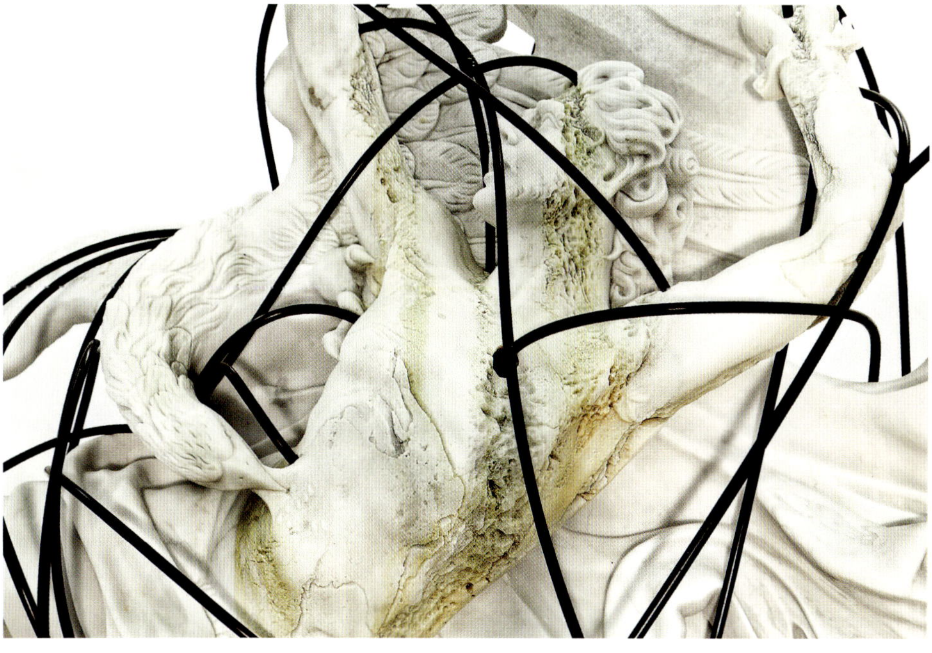

Plate 28. Thomas Feuerstein, *Prometheus Delivered* (2017), photo by the artist. (Fig. 2, p. 205)

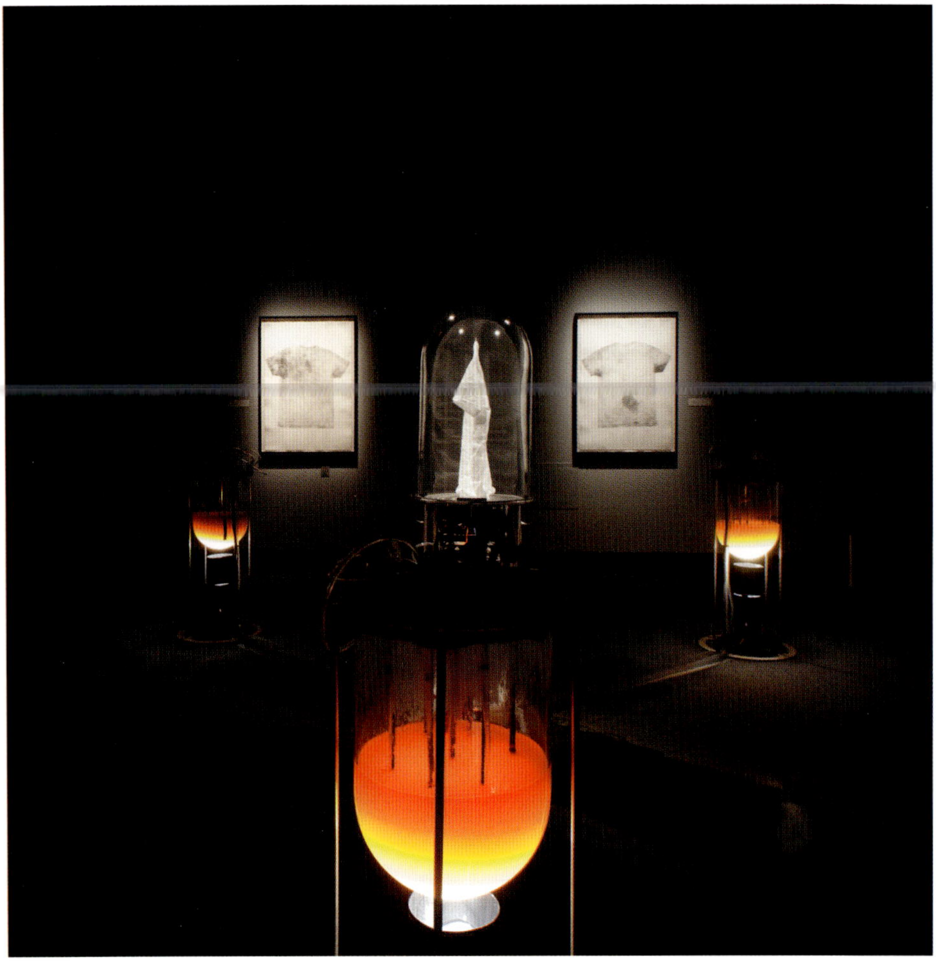

Plate 29. Paul Vanouse, *Labor* (2019), Burchfield Penny Art Center, photo by Tullis Johnson.
(Fig. 3, p. 207)

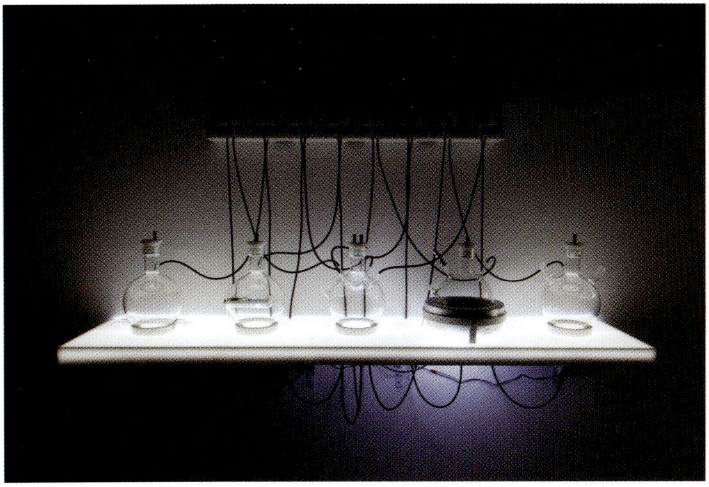

Plate 30. *ReBioGeneSys*, 2015. *ReBioGeneSys* is a fully functioning scientific experiment capable of being reconfigured into any real or imaginary world, be it Venus, Titan, our prebiotic Earth, or Middle Earth. By combining all the scientific research on the origins of life in one set of integrated processes, *ReBioGeneSys* creates 'mashed-up' extreme minimal ecosystems theoretically capable of forming the self-organizing chemistries necessary to produce semi-living molecules and perhaps even protocells. (Fig. p. 216)

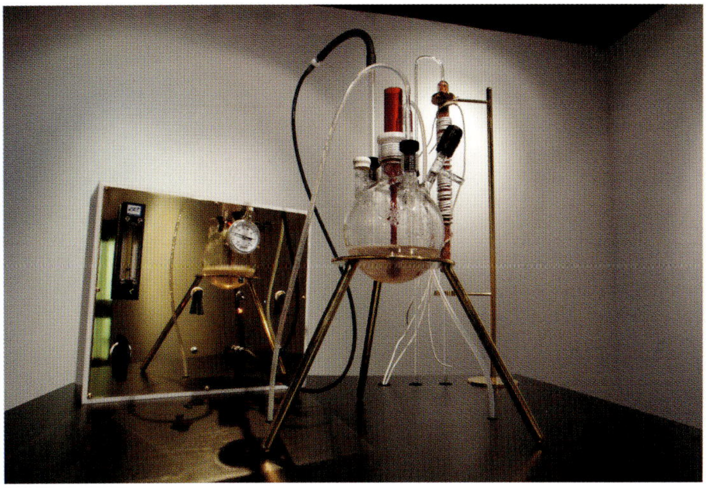

Plate 31. *The Great Work of the Metal Lover*, 2012. Gold production is accomplished by the pairing of a highly specialized metallotolerant extremophilic bacterium (*Cupriavidus metallidurans*) and an engineered atmosphere contained within a customized alchemical bioreactor. The extreme, minimal ecosystem within the bioreactor forces the bacteria to metabolize high concentrations of toxic $AuCl^3$ (gold chloride), turning soluble gold into usable 24K gold. (Fig. p. 217)

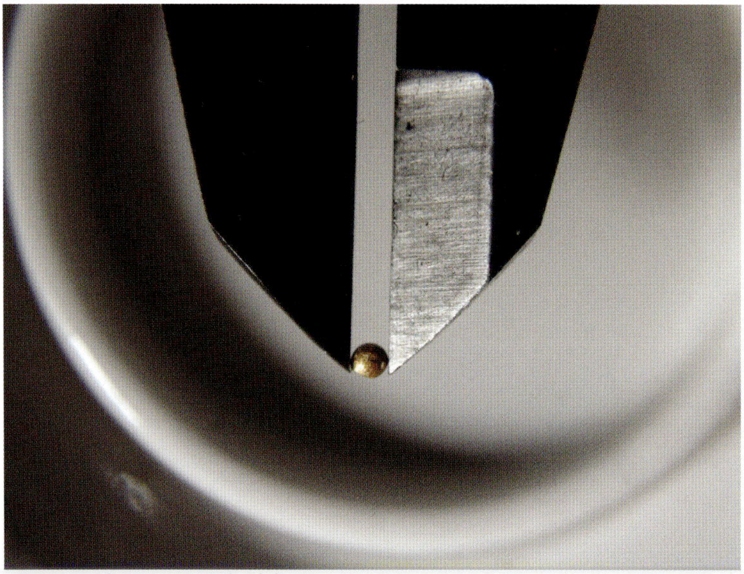

Plate 32. *The Great Work of the Metal Lover*, 2012. 150 mg of 24K gold(Au) produced after 2 weeks of running the bioreactor. (Fig. p. 217)

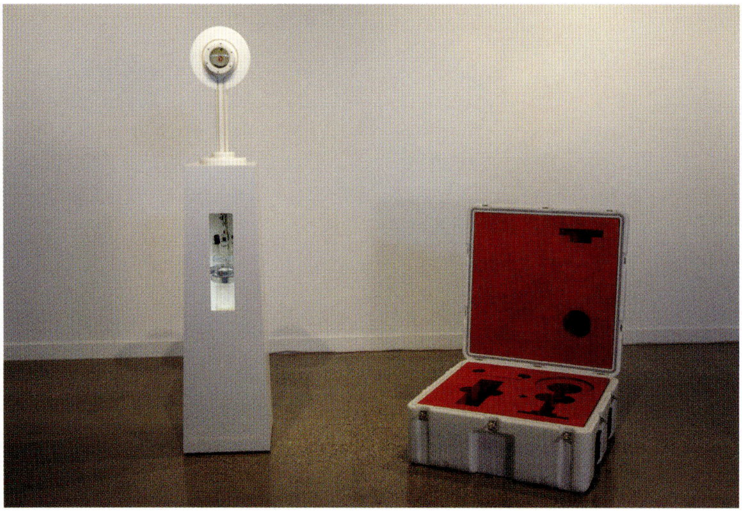

Plate 33. *[ir]reverent: Miracles on Demand*, 2019. Full Installation view. 'Bleeding' Eucharist Host, *Serratia marcescens*; Custom PTFE (Teflon), glass, and stainless-steel Monstrance/incubator, bioreactor, stainless-steel Luna and control system; White acrylic polymer pedestal with glass window; White ABS shell case with red velvet interior. (Fig. p. 221)

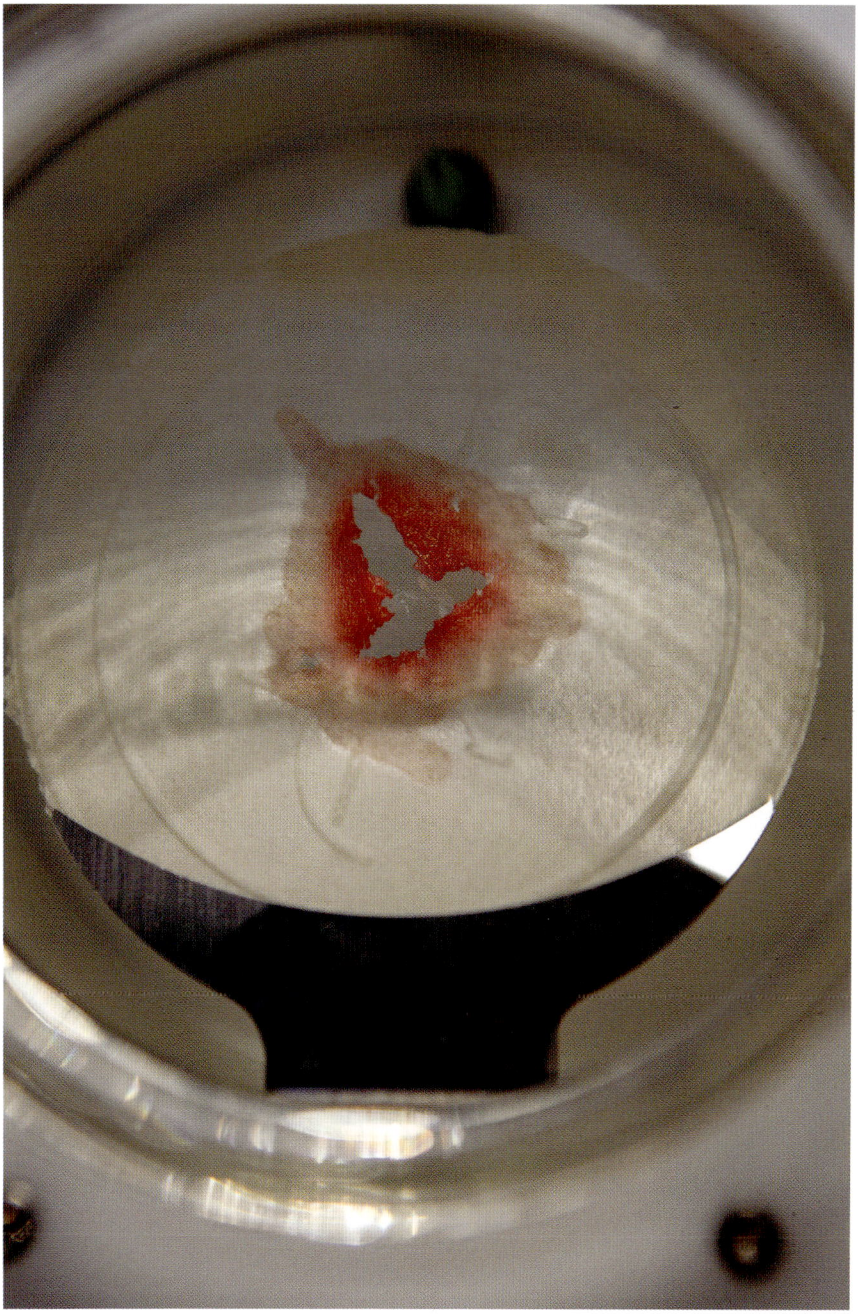

Plate 34. *[ir]reverent: Miracles on Demand*, 2019. Detail. 'Bleeding' Eucharist Host, *Serratia marcescens* and stainless-steel crescent-shaped Luna. (Fig. p. 221)

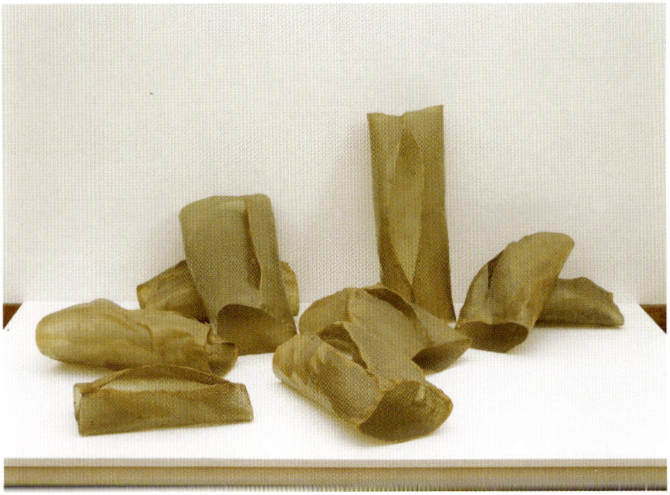

Plate 35. Eva Hesse, *Tori* (1969). Philadelphia Museum of Art. Polyester, resin, and fiberglass on wire mesh. Largest of nine units 119.4 × 43.2 × 38.1 cm. Purchased with funds contributed by Mr. and Mrs. Leonard Korman, Mr. and Mrs. Keith Sachs, Marion Boulton Stroud, Mr. and Mrs. Bayard T. Storey, and with other various funds, 1990, 1990, 121-1-9, © The Estate of Eva Hesse. Courtesy Hauser & Wirth. (Fig. 4, p. 228)

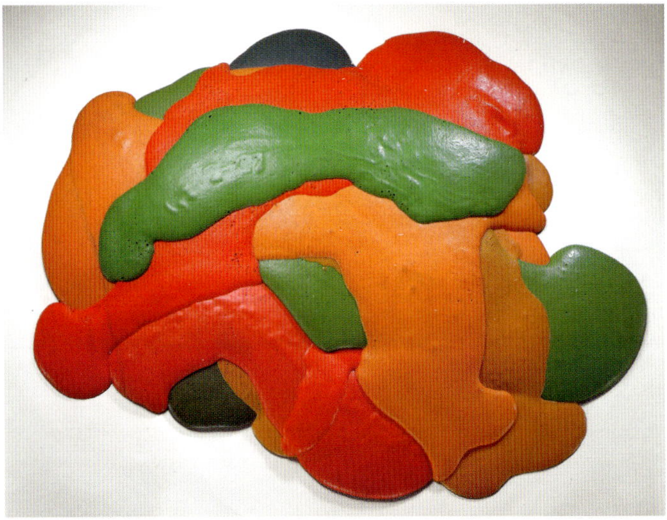

Plate 36. Lynda Benglis, *NIGHT SHERBET A* (1968). Dayglo pigment, phosphorescence and poured polyurethane foam. 5 × 48 × 60 1/2 inches (12.7 × 121.9 × 153.7 centimeters). © Lynda Benglis / DACS, London / VAGA, New York / VISDA, Copenhagen. Courtesy Pace Gallery and Thomas Dane Gallery. (Fig. 5, p. 229)

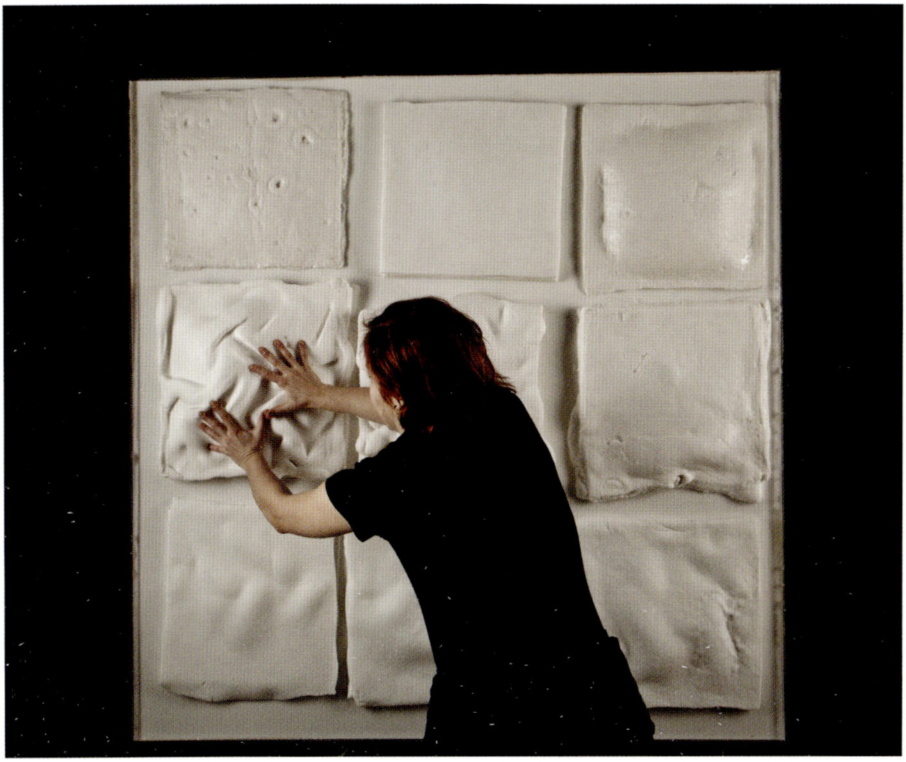

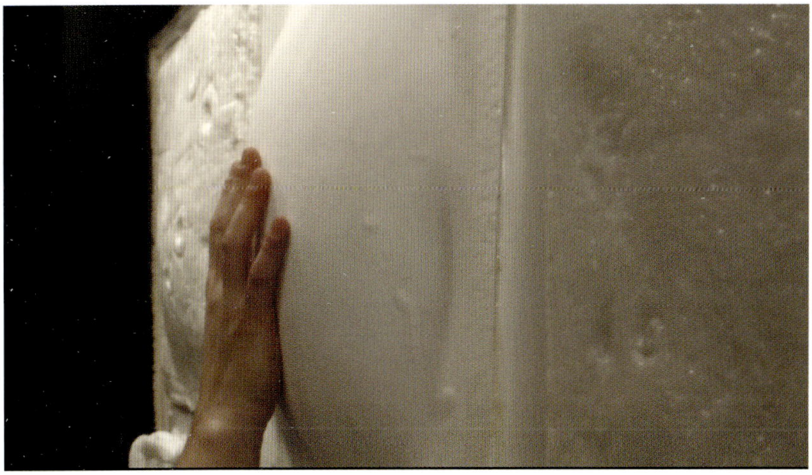

Plate 37. Ece Budak and Ozge Akbulut, *BRALL* (2015). Silicone on polycarbonate panel, 145 cm × 145 cm. © Ece Budak and Ozge Akbulut. Photo and video still © Baris Dervent, Murat Ugurlu. Courtesy of the artists. (Fig. 6, p. 231)

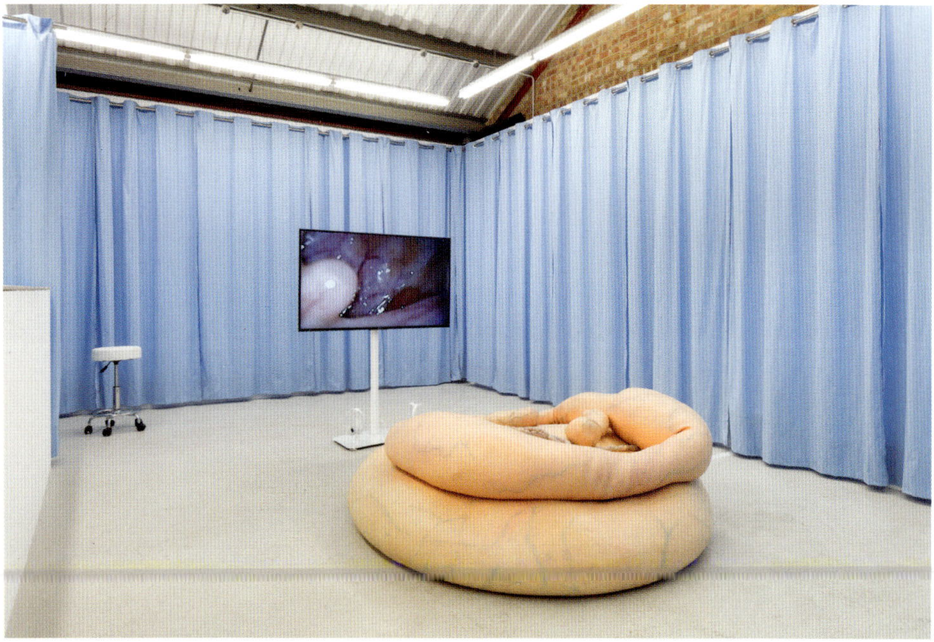

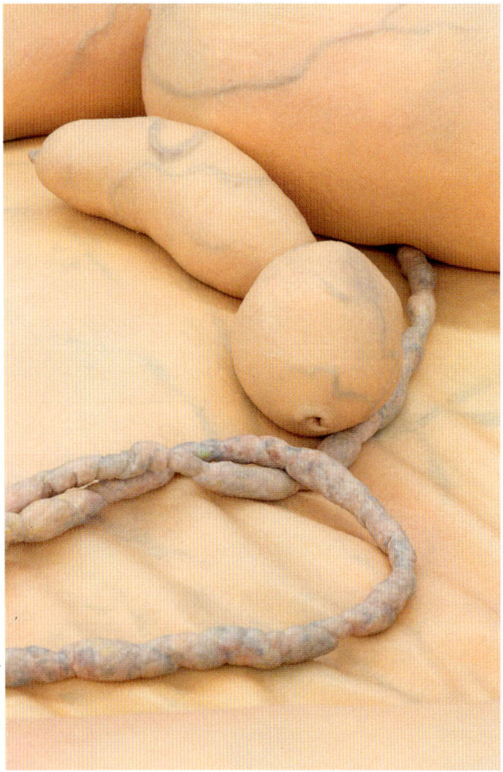

Plate 38. Stine Deja and Marie Munk, *Synthetic Seduction* (2018). Installation view, Annka Kulty's Gallery, London.
© The artists. Courtesy of the artists and Annka Kulty's Gallery. (Fig. 7, p. 233)

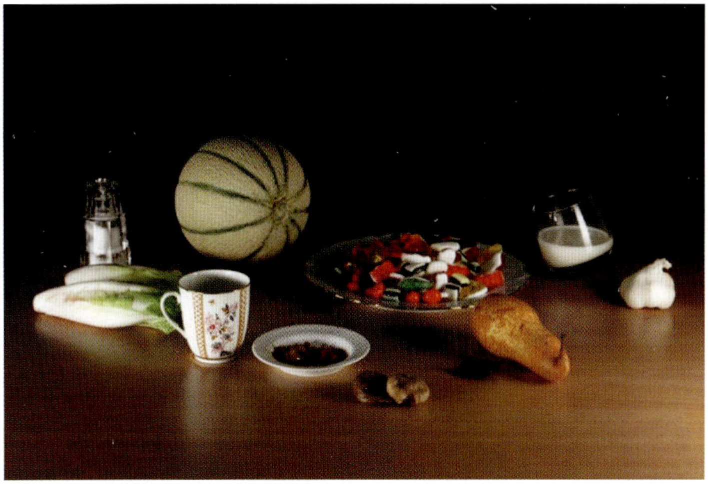

Plate 39. Olga Kisseleva, *Mathematique* (2010). From the series *Divers Faits.* (Fig. 1, p. 249)

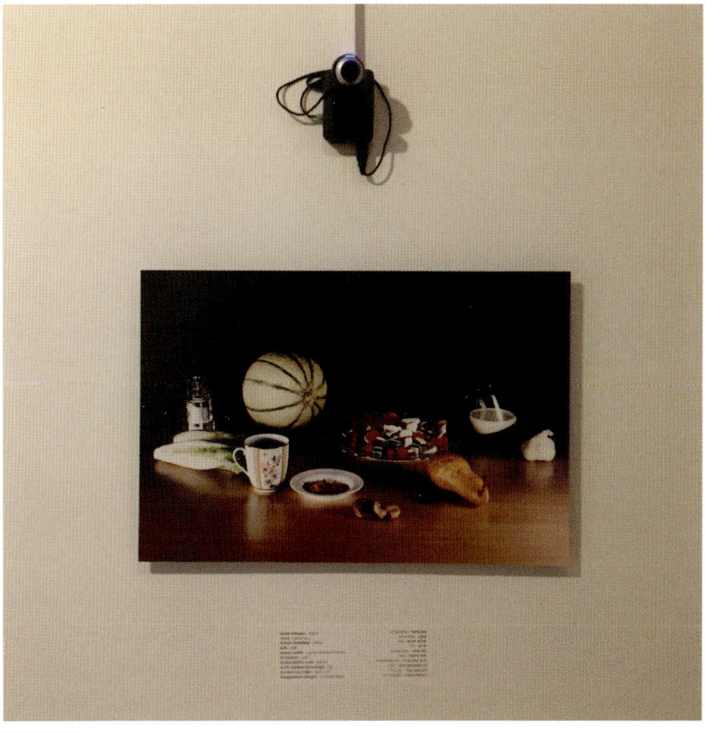

Plate 40. Installation view: *Still Life, Live Data,* Tsila Hassine, Olga Kisseleva, Artists' Residence, Herzlliya. (Fig. 2, p. 251)

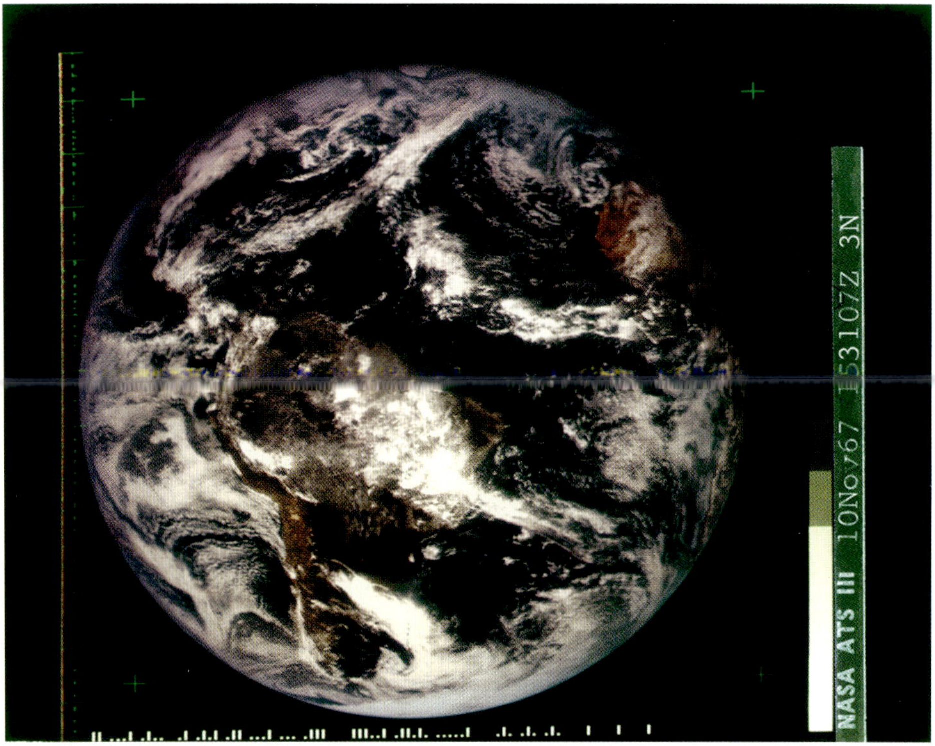

Plate 41. One of the first color photographs of the whole Earth, shot from the ATS-3 satellite on November 10, 1967. Image credit: NASA. (Fig. 1, p. 258)

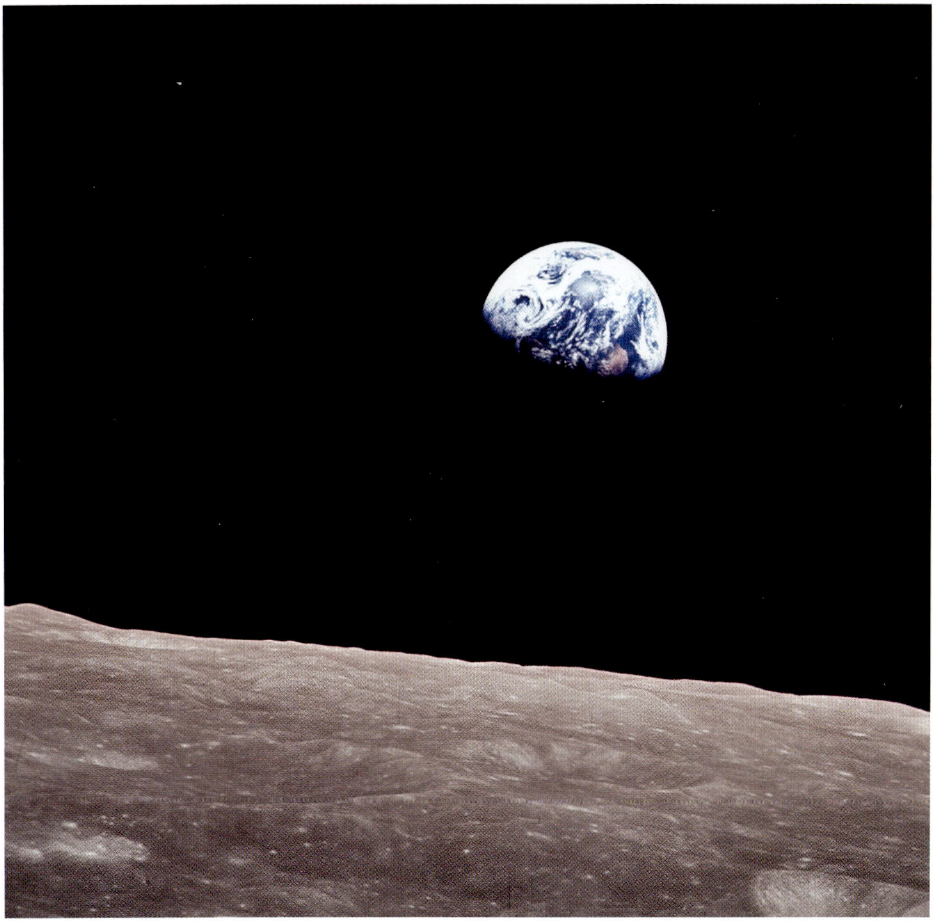

Plate 42. *Earthrise* was taken on December 24, 1968 by astronaut William Anders during the Apollo 8 mission. Image credit: NASA. (Fig. 2, p. 258)

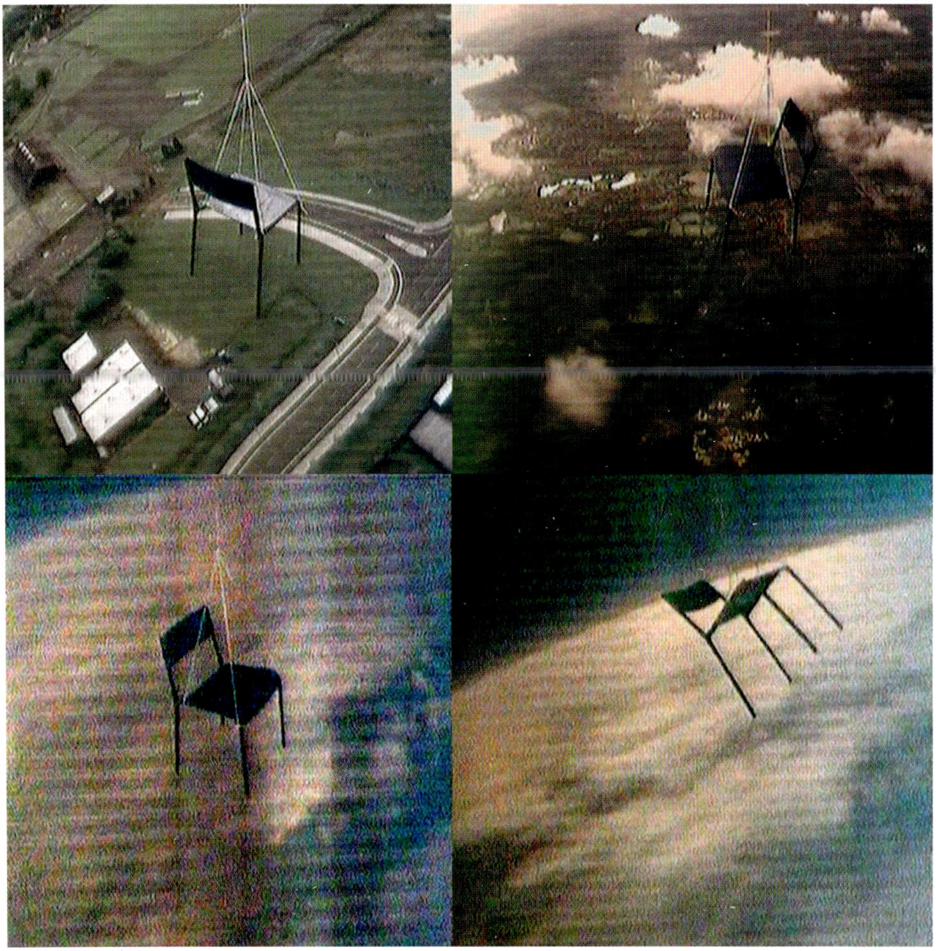

Plate 43. Simon Faithfull, *Escape Vehicle no. 6*, 2004, video stills. Courtesy the artist. (Fig. 3, p. 268)